Quantum Mechanics of Fundamental Systems: The Quest for Beauty and Simplicity

Marc Henneaux • Jorge Zanelli
Editors

Quantum Mechanics of Fundamental Systems: The Quest for Beauty and Simplicity

Claudio Bunster Festschrift

Springer

Marc Henneaux
Université Libre de Bruxelles
Service Physique théorique
et mathématique
1050 Bruxelles
Campus de la Plaine
Belgium
henneaux@cecs.cl
henneaux@ulb.ac.be

Jorge Zanelli
Centro de Estudios Cientificos
Valdivia, Chile
z@cecs.cl

ISBN: 978-0-387-87498-2 e-ISBN: 978-0-387-87499-9
DOI: 10.1007/978-0-387-87499-9

Library of Congress Control Number: 2008942059

Printed on acid-free paper

springer.com

Contents

Part 1 OPENING REMARKS

Greetings 3

Opening Lecture 11
Marc Henneaux

Part 2 CONTRIBUTED PAPERS

On the Symmetries of Classical String Theory 17
Constantin P. Bachas

Eddington–Born–Infeld Action and the Dark Side of General Relativity 27
Máximo Bañados

Light-Cone Field Theory, Maximal Supersymmetric Theories and $E_{7(7)}$ in Light-Cone Superspace 33
Lars Brink

Strongly Hyperbolic Extensions of the ADM Hamiltonian 71
J. David Brown

Black Hole Entropy and the Problem of Universality 91
S. Carlip

Sources for Chern–Simons Theories 107
José D. Edelstein and Jorge Zanelli

The Emergence of Fermions and the E_{11} Content 125
François Englert and Laurent Houart

Why Does the Universe Inflate? 147
S.W. Hawking

Kac–Moody Algebras and the Structure of Cosmological Singularities: A New Light on the Belinskii–Khalatnikov–Lifshitz Analysis 155
Marc Henneaux

Black Holes with a Conformally Coupled Scalar Field 167
Cristián Martínez

Quantum Mechanics on Some Supermanifolds 181
Luca Mezincescu

John Wheeler's Quest for Beauty and Simplicity 193
Charles W. Misner

Magnetic Monopoles in Electromagnetism and Gravity 197
Rubén Portugués

The Census Taker's Hat 213
Leonard Susskind

Static Wormholes in Vacuum and Gravity in Diverse Dimensions 267
Ricardo Troncoso

Part 3 CLOSING

Claudio Bunster: A Personal Recollection 289
Jorge Zanelli

Profile of Claudio Bunster 295

Index 303

Part 1
OPENING REMARKS

Greetings

Recorded in Santa Barbara

David Gross[1]

Hi Claudio,

Good to see you. I am sorry I am not able to be there for your birthday. Jackie and I really loved our visit last year, and I am sure we would have enormously enjoyed what I expect will be a great party, but at the moment we have a previous engagement somewhere in Thailand.

I have known Claudio since he was a mere lad of 22, a young graduate student of Johnnie Wheeler when I first came to Princeton. He then stayed on as an Assistant Professor, so I have known him for more than half of his life. After Princeton he went to Texas and a few years later, much to my surprise, returned to Chile, first on a part time basis, and then full time. I must say that at the time I was surprised and amazed that he did this. I really admired him for his courage and dedication in going back to Chile to help build science at such a very difficult and dangerous time.

I was delighted over the years to visit Claudio at the institute that he established in Santiago and later in Valdivia. I tried to do the little I could do to help him in his remarkable leadership in developing science in Chile and in healing the wounds of previous hard times.

Claudio Teitelboim is truly a great scientist and a great statesman of science. He has helped to transform Chilean science. I have enormous respect for him and all that he has achieved and I just wish that I was there in person to offer him my congratulations. I cannot be in Chile, so from afar, congratulations Claudio!

To paraphrase a well known Hebrew saying that is often said on such occasions "until 120". You are already half way there, so enjoy the rest, the second half.

Best of wishes,

Bye

David Gross

[1] Kavli Institute of Theoretical Physics, University of California, Santa Barbara, CA 93106, USA, e-mail: gross@kitp.ucsb.edu

M. Henneaux, J. Zanelli (eds.), *Quantum Mechanics of Fundamental Systems: The Quest for Beauty and Simplicity*, DOI 10.1007/978-0-387-87499-9_1,

Recorded in Cambridge, MA

Frank Wilczek[2]

Greetings Claudio, greetings friends in Chile.

Claudio, as you will know, time is an illusion. The Hamiltonian that evolves systems in time is just a constraint, and it is zero, so your 60th birthday should not be a cause for alarm and there are other branches of the wave function in which you are celebrating your 20th birthday or 30th, or 40th or whichever one your prefer.

Even on this branch of the wave function the 60th birthday is really a cause for celebration. It is a chance to celebrate what has been achieved by this time that is quite an impressive thing to contemplate.

Your contributions to fundamental physics are in retrospect even more remarkable than they seemed at the time. Thinking about abstract problems on how you quantize constraint systems or how you deal with extended objects and generalize the Dirac quantization conditions or how you understand black holes as quantum mechanical objects. These things to which you contributed so much and focused interest on, have proved to be some of the most unlikely, yet rich and fertile fields of theoretical physics in recent decades.

You have also of course founded the institute in Chile, which has been an extraordinary place for intellectual adventure, not only in theoretical physics but in topics that have proved to be, again, amazing choices of things to focus on. Glaciology now is at the forefront of interest in the world's problems of climate change and what we are going to do about it and understanding it and of course, understanding how the mind works is going to be the great occupation, I'm sure, of science in the later parts of the twenty-first century.

Besides intellectual achievements and setting up other people's intellectual achievements, you have had direct effect on people's lives, family, friends, coworkers, and in future years you'll have the joy not only of extending your own adventures, but of watching their adventures as the solution of time marches on.

So you can look back with satisfaction and look ahead with anticipation.

Happy Birthday!

Frank Wilczek

[2] Department of Physics, Center for Theoretical Physics, Massachusetts Institute of Technology, Cambridge, MA 02139, USA, e-mail: wilczek@mit.edu

Greeting from Utrecht

Gerard 't Hooft[3]

Dear Marc, Jorge and other Organisers of the Claudio Fest,

Although, regretfully, I will not be able to be physically present at this gathering, I do wish to send Claudio my very warmest wishes and congratulations for his 60th birthday. As the title of the meeting shows, 60 is the respectable age when one is beginning to be more reflective, pondering about the real essentials of our research topics: what is quantum mechanics, what is beauty and simplicity? In what directions should future searches go, and what is it that we can expect? I hope the meeting, by its informal nature, will be thought provoking. My warmest greetings also to all my friends and colleagues who did manage to be present and I hope they will bring Claudio the hommage he deserves.

In case it is still opportune: Merry Christmas and a Happy 2008,

Gerard 't Hooft

[3] Institute for Theoretical Physics, Utrecht University and Spinoza Institute, Postbus 8000, 3508 TA Utrecht, the Netherlands, e-mail: G.tHooft@phys.uu.nl

For Claudio: A 60th Birthday Greeting

Stanley Deser[4]

Dear Claudio,

Let me preface this, at times indiscreet, 60th birthday message with the consolation that, were it not for the impossible 18 h (each way) plane journeys, you would have had to suffer the insult of hearing it live!

The much appreciated invitation to your celebration made me both count and remember: we have known each other for over half your life so far, and it led me back to our first encounter, of early 1975, in Princeton. You were then already famous – the successor in the lineage of Feynman and Misner – as the latest of Wheeler's "discoveries," having reached New Jersey by the route closest to John's heart, crazy electrodynamics. Since we're also traveling backward in time, that takes care of defining Dick's way, while Charlie's (more recent) exploit was to destroy the beauty of Maxwell–Einstein theory by (re-) discovering its – horribly complicated, "already unified" – purely geometrical version. Your road was the radiation reaction problem, which you clarified so much that it forced Sidney Coleman to (sort of) do it right later, also using exotic Feynman propagators, come to think of it. What most impressed me however, in that initial meeting, was of a non-physics nature: you looked like a Jeune Premier (someone will translate) and you (claimed you) were equipping an ancient and highly unlikely-looking wreck of a yellow Land Rover for the journey home (merely to Santiago, rather than Valdivia, but still) down the (then highly unfinished) Pan-American highway. In drab, conventional, Princeton, one can imagine that these characteristics stood out most vividly! I had been invited for an informal visit by the Physics Department. I had a lot of fun – except when the whole Physics Department – from Wigner down, would close for grading Freshman physics exams (a Princeton custom, I was told!). Despite this quaint custom, you and I were able start what would become a long-standing research collaboration, whose first result was our Phys Rev electromagnetic duality paper (surviving those unsettling local mores). It was to become famous, long before duality became fashionable, but not before going through the usual "it's wrong and trivial" scoffing. It still gets quoted, and as you know, served as a basis of three of our further collaborations, with Henneaux and Gomberoff, at your Institute some two decades later, as well as of more recent papers by you and Marc, that in turn generated ones by Domenico Seminara and me. Duality has indeed evolved from breakthrough to truism, as (legend has it) good physics ideas always do!

The second big thing for us came less than 2 years after, early in 1977, when we met on a freezing day in Harvard Yard (they all are) and realized that we had both been dreaming that the then brand-new Supergravity might be the key to one of the

[4] Department of Physics, Brandeis University, Waltham, MA, USA and Caltech, Pasadena CA, USA, e-mail: deser@brandeis.edu

really big problems in ordinary General Relativity: positive energy. For background (in case there are younger members of the audience, unaware that there ever might have been such a problem), it had long been suspected – I had wasted countless years on it myself – that the GR Hamiltonian was nonnegative, and only vanished for vacuum = flat space. A proof had actually quite recently appeared, by Schoen and Yau, but it was very pure mathematics of the sort that no one whom I understood understood, if you see what I mean. For you, the flash came via the notion that SUGRA was some sort of Dirac square root of gravity, for me it was the same fact, but stated as the SUGRA algebra's relation that the Hamiltonian was the (hermitian) square of the supercharge. We compared notes, calculated some more-and it held up! I was eager to publish this final validation of SUGRA as also the savior of GR stability and to finally shake the problem, while you wisely hesitated because our proof was seemingly for the purely QUANTUM (because of the fermions) SUGRA, rather than its classical GR counterpart, and you hoped we could soon overcome that hurdle. I "won," so we had to later cede a bit of the glory to Witten, who extracted the classical content via Killing spinors, and to Grisaru who simply noted that classical GR was the $h = 0$ limit, restricted to the no external fermion sector, of SUGRA, at least formally. Nevertheless, ours represented (if I may say so) a most rewarding accomplishment of SUGRA (and of its devoted servants), providing a clear physical basis for a deep necessity.

I cannot speak as authoritatively about your many other non-research feats, such as the enormous service rendered to Chile and to Science by the Center you have so tirelessly served, starting from very lean and difficult times; many of your Northern Hemisphere colleagues have witnessed this first-hand. Certainly, the now-legendary South Pole theoretical physics conference of a few years back will eternally resonate in the hearts of all its survivors, at many levels, including its superb organization-even unto mobilizing the entire Air Force as well as commanding the Winter South Pole's weather to obey! On a personal note, Gary Gibbons and I are indebted to this meeting for having gotten us started on perhaps the world's lowest paper, if only (we hope) latitude-wise! I can, however, say a bit more about the remarkable evolution of some of the physics ideas that you have produced; this requires making a severe selection, but the four surviving my triage should give the overall flavor: In sheer SLAC citation density, of course it is BTZ that leads the list at 10^3 (and counting). That a non-dynamical theory like $2+1$ GR could be tortured into exhibiting a black hole solution is already amazing; that it then became the first entropy-explanation laboratory (triggered by Brown and Henneaux's work), and still keeps on giving, testifies to its depth. It should also be a special source of all-Chilean pride. Next, there is Regge–Teitelboim, as it is simply known, a clear and detailed exposition of the ins and outs of the dynamics of GR, that has weaned whole generations of relativists, despite its mysterious samizdat-like Italian publication. Then there is another, if less famous, Regge–Teitelboim, one that is especially close to my heart, though for impure reasons. Your work was a most original attempt to give a description of GR in terms of higher-dimensional embeddings, and it prompted a followup by Pirani, Robinson and me. We submitted ours to the Physical Review, in the days when one did not toy with that Journal's majesty; our title, "Embedding the G-String," was not only technically appropriate but also (knowingly) salacious, as we were immediately and thunderously informed: change title or you'll never publish in Phys Rev again! Though we (of course) gave in, it has always been a source of regret to all three of us to have lost such a unique chance at a swinging reputation. Luckily(?), the original title has been preserved for posterity in the Arxiv listing. Finally, Claudio has always had a venturesome weakness (before it became de rigueur) for dimensions outside the usual 4, in both directions; I think here not only of the 3 of BTZ, but of his pioneering and very fertile work in 2D GR, as well as his forays into $D > 4$.

I could go on and on, but let me rest my case here and wish one of the giants in our field a prolonged analytic continuation!

Stanley Deser

Greeting from Rome

Volodia Belinski[5]

Dear Claudio,

First of all Trayasca Chaushescu! Then I congratulate you with your 60th anniversary, with Christmas and with coming New Year. I wish you good health, good money and satisfaction in your private life. These three ingredients are basic.

Plus to this I wish you to continue your successful scientific activity, in spite of the fact that the works you already did are more than enough to leave a significant trace in Theoretical Physics.

Unfortunately I cannot come to Valdivia at January for "Claudio fest" (I have some health problems and recovering need few months from now, please remember that I am much older than you, I am 66!). Hope to see you somewhere in the next future.

Yours,

Volodia Belinski

[5] INFN, Rome University "La Sapienza", 00185 Rome, Italy, e-mail: belinski@ICRA.it

Greeting from Göteborg

Bengt Nilsson[6]

Dear organizers,

It would be great to go to Chile and see Claudio again after so many years. I last met him in Santiago, where we all met as you probably remember. Unfortunately, I will not be able to make it. So maybe I can ask you to give him my very best regards and wishes for the future.

Although I did not interact very much with him in Texas (we met just a couple of times; he was in Europe, I think, for most of the time) I do owe him a lot for creating (together with Weinberg) the postdoc position which was my first one (I had to turn down an offer from Peccei in München which was a bit tricky).

I find Claudio an extremely nice and warm person and I wish him all the best. He once took my wife and me out for dinner in Austin and he made a very special impression on both of us.

Yours,

Bengt Nilsson

[6] Fundamental Physics, Chalmers University of Technology, SE 412 96 Göteborg, Sweden, e-mail: tfebn@fy.chalmers.se

Opening Lecture

Marc Henneaux

In the name of the organizing committee, it is a great pleasure to welcome all of you to Claudio's Fest, a meeting in which we shall celebrate the scientific work accomplished so far by Claudio. It has become customary to choose the 60th birthday for such celebrations and we have followed the tradition, although we all know that a scientific career does not stop at 60 and that we shall therefore miss all the important scientific contributions that are still to come.

Your presence with us today is a clear homage that the international scientific community is paying to a great scientist. The meeting is organized by CECS, with support from the International Solvay Institutes, of which Claudio is an honorary member.

"Quantum Mechanics of Fundamental Systems: The Quest for Beauty and Simplicity" is the title that we have chosen for the conference. This is a wink to the early days of CECS, since the physics meetings organized by Claudio in Santiago in the 1980s, when the center was just created, were precisely entitled "Quantum Mechanics of Fundamental Systems." We hope that the present conference will carry the same pioneering spirit, the same freshness, the same driving enthusiasm as those heroic meetings and that we shall all remember it as one of these unusual conferences where something magic occurred.

To stimulate the discussions, we have invited more participants than speakers. This is also in perfect line with the philosophy of the CECS's early meetings where discussions were central. I would like to thank all the participants, speakers and non-speakers, for having accepted our invitation and for being with us to make this meeting a very special event indeed.

"The Quest for Beauty and Simplicity" are words that we added to the title of the original conferences in order to reflect Claudio's central inspiration in his research. Black holes, and in particular the black hole in three dimensions, as well as magnetic

M. Henneaux
Université Libre de Bruxelles and International Solvay Institutes, ULB-Campus Plaine CP231, B-1050 Bruxelles, Belgium
e-mail: henneaux@ulb.ac.be

M. Henneaux, J. Zanelli (eds.), *Quantum Mechanics of Fundamental Systems: The Quest for Beauty and Simplicity*, DOI 10.1007/978-0-387-87499-9_2,

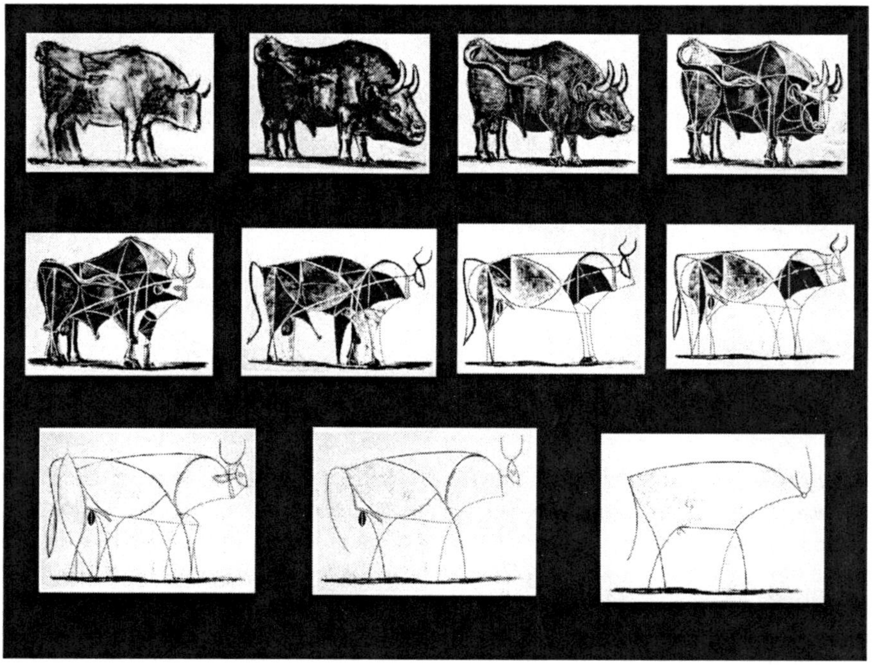

Fig. 1 The Quest for Beauty and Simplicity: Picasso's series of 11 drawings of a bull

monopoles, are for instance enigmatically beautiful, and presumably ultimately simple objects which have been recurrent themes in Claudio's investigations. The best illustration of this quest for beauty and simplicity is due to Claudio himself, who likes to show the famous series of 11 drawings of a bull due to Picasso to start some of his lectures (Fig. 1).

This is a physics conference. I believe, however, that it would be impossible to celebrate Claudio's 60th birthday without evoking his exceptional contribution to the development of science in Chile. When he decided more than 20 years ago to give up prestigious positions in the United States to come back to Chile and to build from scratch, without local support, a private research institute, I remember that many colleagues all over the world said that he was completely mad and that this enterprize was bound to fail. Time has shown that these pessimistic colleagues were wrong and that Claudio was right. CECS is in the world league of top research institutions and remarkably contributes to the international scientific visibility of Chile. If we are all here today in Chile, it is thanks to this most significant achievement.

I have known Claudio for almost 30 years now. What a long way since I arrived at Princeton in 1978 as a student! What started as a 1-year visit turned into a long-term collaboration. I think I am therefore in a good position to describe, in the name of Claudio's students and collaborators, his unique style of work that has deeply influenced us as physicists. This is not an easy exercise – to speak about a friend in his presence is never an easy exercise – but I'll try to do it. Perhaps the first lesson that he taught us is that life is too short to lose one's time on marginal problems. One should develop a good scientific taste for what is relevant and important, even if this means not following fashion.

Another lesson is that doing research is – and should be – enjoyable. It is especially enjoyable when one tries to foresee and anticipate the implications of a plausible physics result before even attempting to prove it, trying to understand whether these implications make sense. Again, life is too short and educated guessing is the fastest – and funniest – way to go ahead.

Another unique aspect of Claudio's style of work is his ability to take advantage of any situation for doing physics. I am sure that all of us have on many occasions discussed physics with him in unthinkable circumstances, be it during a motorbike trip to a lost place in Texas to fetch an angora rabbit, or on a jeep ride on a bumpy dust road to Zapallar, or in a dentist waiting room, or at an airport counter waiting for the airline to accept sending with minimum extra charge oversized and overweight luggage, or even on a risky boat trip in the middle of the night in which we almost sank. Working with Claudio is indeed fun, but requires some capacity of adaptability from his collaborators, which is not part of the standard academic training.

I will not elaborate more now on Claudio's style and on the characteristics of his work – we are all here because we know them!

Before letting the fest begin, I would just like to add a few words in Spanish – this is a première.

Claudio, vengo a Chile desde hace más de veinte años y nunca he hablado español en público. ¡Tengo que empezar a hacerlo! No hay mejor oportunidad que hoy, en tu fiesta científica. Quiero añadir a lo que dije en inglés que hay otra cosa importante que aprendí de ti : es que debemos ser capaces de tomar riesgos, que pueden parecer a veces locos, no sólo riesgos en nuestras investigaciones sino también riesgos en la orientación de nuestra carrera, de nuestra profesión, quizás de nuestra vida. No es una cosa que se aprende en círculos académicos. Se puede aprender de exploradores, de poetas. Entonces voy a concluir con dos citas, la primera del explorador francés Paul-Emile Victor que organizó expediciones al Ártico y a la Antártica, y la segunda del poeta chileno Vincente Huidobro. Comienzo con Paul-Emile Victor : "La única cosa que estamos seguros de no lograr es la que no intentamos." Y Huidobro : "Si yo no hiciera al menos una locura por ao, me volvería loco." Eres un ferviente adepto de estos principios y has convertido con tu ejemplo a muchos de tus amigos.

And now, let the "fest" begin !

Part 2
CONTRIBUTED PAPERS

On the Symmetries of Classical String Theory

Constantin P. Bachas

Abstract I discuss some aspects of conformal defects and conformal interfaces in two spacetime dimensions. Special emphasis is placed on their role as spectrum-generating symmetries of classical string theory.

1 Loop Operators in 2d CFT

Wilson loops [47] are important tools for the study of gauge theory. They describe worldlines of external probes, such as the heavy quarks of QCD, which transform in some representation of the gauge group and couple to the gauge fields minimally. More general couplings, possibly involving other fields (e.g., scalars and fermions), are in principle also allowed. They are, however, severely limited by the requirement of infrared relevance or, equivalently, of renormalizability. In four dimensions this only allows couplings to operators of dimension at most one, i.e., linear in the gauge and the scalar fields. An example in which the scalar coupling plays a role is the supersymmetric Wilson loop of $N = 4$ super-Yang Mills theory [38, 43].

The story is much richer in two space–time dimensions. Power-counting renormalizable defects in a two-dimensional non-linear sigma model, for example, are described by the following loop operators

$$\mathrm{tr}_V W(C) = \mathrm{tr}_V \, \mathrm{P} e^{i \oint_C \mathrm{H_{def}}}, \tag{1}$$

where V is the n-dimensional space of quantum states of the external probe, whose Hamiltonian is of the general form

$$\oint_C \mathrm{H_{def}} = \int ds \Big[\Big(\mathbf{B}_M(\Phi) \partial_\alpha \Phi^M + \varepsilon_{\alpha\beta} \widetilde{\mathbf{B}}_M(\Phi) \partial^\beta \Phi^M \Big) \frac{d\hat{\xi}^\alpha}{ds} + \mathbf{T}(\Phi) \Big]. \tag{2}$$

C.P. Bachas
Laboratoire de Physique Théorique, École Normale Supérieure, 24 rue Lhomond, 75231 Paris, France
e-mail: bachas@lpt.ens.fr

M. Henneaux, J. Zanelli (eds.), *Quantum Mechanics of Fundamental Systems: The Quest for Beauty and Simplicity*, DOI 10.1007/978-0-387-87499-9_3,

Here s is the length along the defect worldline C, and the Hamiltonian is a hermitean $n \times n$ matrix which depends on the sigma-model fields $\Phi^M(\zeta^\alpha)$ and on their first derivatives evaluated at the position of the defect $\hat{\zeta}^\alpha(s)$. The loop operator is thus specified by two matrix-valued one-forms, $\mathbf{B}_M d\Phi^M$ and $\widetilde{\mathbf{B}}_M d\Phi^M$, and by a matrix-valued function, $\mathbf{T}$, all defined on the sigma-model target space $\mathscr{M}$. Because $\mathrm{H}_{\mathrm{def}}$ is a matrix, the path-ordering in (1) is non-trivial even if the bulk fields are treated as classical, and hence commuting c-numbers.

The non-linear sigma model is classically scale-invariant. The function $\mathbf{T}$, on the other hand, has naive scaling dimension of mass, so (classical) scale-invariance requires that we set it to zero. The reader can easily check that, in this case, the operator (1) is invariant under all conformal transformations that preserve C. This symmetry is further enhanced if, as a result of the field equations, the induced one-form

$$\widehat{B} \equiv \left(\mathbf{B}_M(\Phi)\partial_\alpha \Phi^M + \varepsilon_{\alpha\beta}\widetilde{\mathbf{B}}_M(\Phi)\partial^\beta \Phi^M\right) d\zeta^\alpha \tag{3}$$

is a flat $U(n)$ connection, i.e., if in laconic notation $d\widehat{B} + [\widehat{B}, \widehat{B}] = 0$. The loop operator is in this case invariant under arbitrary continuous deformations of C, as follows from the non-abelian Stoke's theorem. Such defects can therefore be called *topological.* The eigenvalues of topological loops $W(C)$, with C winding around compact space, are charges conserved by the time evolution. The existence of a one-(spectral-) parameter family of flat connections is, for this reason, often tantamount to classical integrability, see [6].

Quantization breaks, in general, the scale invariance of the defect loop even when the bulk theory is conformal. This is because the definition of $W(C)$ requires the introduction of a short-distance cutoff ε. As the cutoff is being removed the couplings run to infrared fixed points, $\mathbf{B}^{(\varepsilon)} \to \mathbf{B}^*$ and $\widetilde{\mathbf{B}}^{(\varepsilon)} \to \widetilde{\mathbf{B}}^*$ as $\varepsilon \to 0$. I will explain later that this renormalization-group flow can be described perturbatively [8] by generalized Dirac–Born–Infeld equations. The fixed-point operators commute with a diagonal conformal algebra. More specifically, if C is the circle of a cylindrical spacetime, and $L_N, \overline{L}_N$ the left- and right-moving Virasoro generators, then

$$[L_N - \overline{L}_{-N}, \mathrm{tr}_V W^*(C)] = 0 \quad \forall N. \tag{4}$$

Topological quantum defects satisfy stronger conditions: they must commute separately with the L_N and with the $\overline{L}_N$.

These facts can be illustrated with the symmetry-preserving defect loops of the WZW model [8]. Consider the following *chiral*, symmetry-preserving defect:

$$\mathscr{O}_r(C) = \chi_r(\mathrm{P}e^{i\oint_C \lambda J^a t^a}), \tag{5}$$

where J^a are the left-moving Kac–Moody currents, t^a the generators of the global group G, and χ_r the character of the G-representation, r, carried by the state-space of the defect. In the classical theory $\mathscr{O}_r(C)$ is topological for all values of the parameter λ. But upon quantization, the spectral parameter runs from the UV fixed point $\lambda^* = 0$ to an IR fixed point $\lambda^* \simeq 1/k$, where k is the level of the Kac–Moody algebra (and $k \gg 1$ for perturbation theory to be valid). It is interesting here to note [8] that one can regularize (5) while preserving the following symmetries: (a) chirality, i.e., $[\mathcal{O}_r^\varepsilon(C),\bar{J}_N^a]=0$ for all right-moving Kac–Moody (and Virasoro) generators, (b) translations on the cylinder, i.e., $[\mathcal{O}_r^\varepsilon(C),L_0\pm\bar{L}_0]=0$, and (c) global G_{left}-invariance. These imply, among other things, that the RG flow can be restricted to the single parameter λ, and that the IR fixed-point loop operator is topological. This fixed-point operator is the quantum-monodromy matrix of the WZW model [4]. It can be constructed explicitly, to all orders in the $1/k$ expansion, as a central element of the enveloping algebra of the Kac–Moody algebra [3,32].

The above renormalization-group flow describes, for $G=SU(2)$, the screening of a magnetic impurity interacting with the left-moving spin current in a quantum wire. This is the celebrated Kondo problem[1] [48] which can be solved exactly by the Bethe ansatz [5,46]. It was first rephrased in the language of conformal field theory by Affleck [1]. Close to the spirit of our discussion here is also the work of Bazhanov et al. [11–13], who proposed to study quantum loop operators in minimal models using conformal (as opposed to integrable lattice-model) techniques. Topological loop operators were first introduced and analyzed in CFT by Petkova and Zuber [40]. Working directly in the CFT makes it possible to use the powerful (geometric and algebraic) tools that were developed for the study of D-branes.

2 Interfaces as Spectrum-Generating Symmetries

Conformal defects in a sigma model with target $\mathcal{M}$ can be mapped to conformal boundaries in a model with target $\mathcal{M}\otimes\mathcal{M}$ by the folding trick [10,39], i.e., by folding space so that all bulk fields live on the same side of the defect. Conformal boundaries can, in turn, be described either as geometric D-branes [41], or algebraically as conformal boundary states on the cylinder [17,42]. In the latter description space is taken to be a compact circle, and the boundary state is a (generally entangled) state of the two decoupled copies of the conformal theory:

$$\|\mathcal{B}\rangle\rangle = \sum \mathcal{B}_{\alpha_1\tilde{\alpha}_1\alpha_2\tilde{\alpha}_2}\,|\alpha_1,\tilde{\alpha}_1\rangle\otimes|\alpha_2,\tilde{\alpha}_2\rangle. \tag{6}$$

Here α_j ($\tilde{\alpha}_j$) labels the state of the left- (right-) movers in the jth copy. Unfolding reverses the sign of time for one copy, and thus transforms the corresponding states by hermitean conjugation. This converts $\|\mathcal{B}\rangle\rangle$ to a formal operator, $\mathcal{O}$, which acts on the Hilbert space $\mathcal{H}$ of the conformal field theory. The fixed-point operators of the previous section are all, in principle, unfolded boundary states.

This discussion can be extended readily to the case where the theories on the left and on the right of the defect are different, CFT1 $\neq$ CFT2. Such defects should be, more properly, called *interfaces* or domain walls. They can be described similarly by a boundary state of CFT1$\otimes$ CFT2, or by the corresponding unfolded operator

[1] Strictly-speaking, in the Kondo setup the magnetic impurity interacts with the s-wave conduction electrons of a 3D metal. This is mathematically identical to the problem discussed here.

$\mathcal{O}_{21} : \mathcal{H}_1 \to \mathcal{H}_2$. Conformal interfaces correspond to operators that intertwine the action of the diagonal Virasoro algebra,

$$(L_N^{(2)} - \overline{L}_{-N}^{(2)})\mathcal{O}_{21} = \mathcal{O}_{21}(L_N^{(1)} - \overline{L}_{-N}^{(1)}), \tag{7}$$

while topological interfaces intertwine separately the action of the left- and right-movers. In the string-theory literature conformal interfaces were first studied as holographic duals [10, 18, 20, 37] to codimension-one anti-de Sitter branes [9, 36]. Note that conformal boundaries are special conformal interfaces for which CFT2 is the trivial theory, i.e., a theory with no massless degrees of freedom. If $\mathcal{O}_{1\emptyset}$ is the corresponding operator (where the empty symbol denotes the trivial theory) then conformal invariance implies that $(L_N^{(1)} - \overline{L}_{-N}^{(1)})\mathcal{O}_{1\emptyset} = 0$.

Let me now come to the main point of this talk. Consider a closed-string background described by the worldsheet theory CFT1, and let $\mathcal{O}_{1\emptyset}$ correspond to a D-brane in this background. Take the worldsheet to be the unit disk, or equivalently the semi-infinite cylinder, with the boundary described by the above D-brane. Consider also a conformal interface $\mathcal{O}_{21}$, where CFT2 describes another admissible closed-string background. Now insert this interface at infinity and push it to the boundary of the cylinder, as in Fig. 1. The operation is, in general, singular except when $\mathcal{O}_{21}$ is a topological interface in which case it can be displaced freely. Let us assume, more generally, that this fusion operation can be somehow defined and yields a boundary state of CFT2 which we denote by $\mathcal{O}_{21} \circ \mathcal{O}_{1\emptyset}$. We assume that the Virasoro generators commute past the fusion symbol. It follows then from (7) that the new boundary state is conformal whenever the old one was. Since conformal invariance is equivalent to the classical string equations, one concludes that $\mathcal{O}_{21}$ acts as a spectrum-generating symmetry of classical string theory. Conformal interfaces could, in other words, play a similar role as the Ehlers–Geroch transformations [19, 27] of General Relativity.

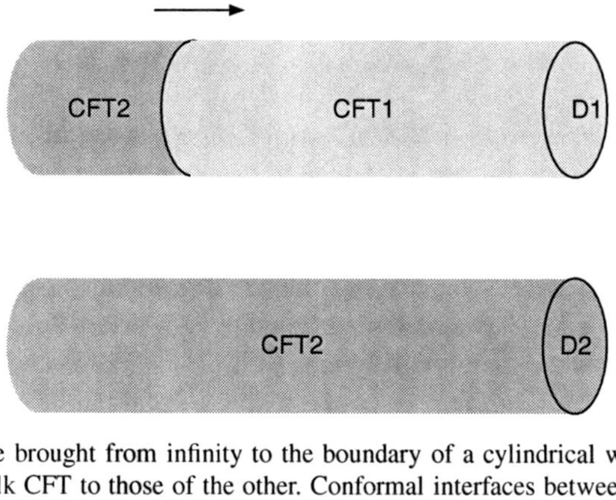

Fig. 1 An interface brought from infinity to the boundary of a cylindrical worldsheet maps the D-branes of one bulk CFT to those of the other. Conformal interfaces between two theories with the same central charge act thus as spectrum-generating symmetries of classical string theory. In many worked-out examples these include and extend the perturbative dualities, and other classical symmetries, of the open- and closed-string action

Bringing an interface to the boundary is a special case of the more general process of *fusion*, i.e., of juxtaposing and then bringing two interfaces together on the string worldsheet. This is of course only possible when the CFT on the right side of the first interface coincides with the CFT on the left side of the second. Furthermore, two interfaces can only be added when their left and right CFTs are identical. Since fusion and addition cannot be defined for arbitrary elements, the set of all conformal interfaces is neither an algebra nor a group. By abuse of language, I will nevertheless refer to it as the "interface algebra."[2]

The first thing to note is that the interface "algebra" is non-trivial even if restricted only to elements with non-singular fusion. These include all the topological interfaces, for which fusion is the regular product of the corresponding operators, $\mathscr{O}_A \circ \mathscr{O}_B = \mathscr{O}_A \mathscr{O}_B$. The simplest topological defects are those whose internal state is decoupled from the dynamics in the bulk. They correspond to multiples of the identity operator, $\mathscr{O} = n\mathbf{1}$ with n a natural number. Their action on any D-brane endows this latter with Chan–Paton multiplicity. Less trivial are the topological defects which generate symmetries of the CFT, as well as the topological interfaces that generate perturbative T-dualities. These were first studied, for several examples, in two beautiful papers by Fröhlich et al. [23,24]. The fact that all perturbative string symmetries can be realized through the action of local defects is not a priori obvious (and needs still to be generally established). Other interesting examples are the minimal-model topological defects, shown to generate universal boundary flows [22,28]. A different set of conformal interfaces whose fusion is non-singular are those that preserve at least $N = (2,2)$ supersymmetry [14, 15]. Some of these descend from supersymmetric gauge theories in higher dimensions [29, 33–35]. Such interfaces were, in particular, used to generate the monodromy transformations of supersymmetric D-branes transported around singular points in the Calabi–Yau moduli space [16]. As these and other examples demonstrate, the interface "algebra" is very rich even if restricted to interfaces with non-singular fusion.

Extending the structure to arbitrary interfaces is, nevertheless, an interesting problem. Firstly, the algebras (without quotation marks) of non-topological defects would provide, if they could be defined, large extensions of the automorphism groups of various CFTs. Furthermore, while topological interfaces are rare – they may only join CFTs that have isomorphic Virasoro representations – the conformal ones are on the contrary common. A useful quantity is the reflection coefficient, $\mathscr{R}$, [44] which vanishes in the topological case. To see that conformal interfaces are not rare, consider the nth multiple of the identity defect which is mapped, after folding, to n diagonally-embedded middle-dimensional branes in $\mathscr{M} \times \mathscr{M}$ [10]. A generic Hamiltonian of the form (2), with the tachyon potential $\mathbf{T}$ set to zero, corresponds to arbitrary geometric and gauge-field perturbations of these diagonal branes. Any solution of the (non-abelian, α' corrected) Dirac–Born–Infeld equations for these branes gives therefore rise to a conformal defect [8]. Likewise, any

[2] The correct term for the interfaces is "functors." For a more accurate mathematical terminology the reader should consult, for instance, ref. [26].

non-factorizable D-brane of CFT1⊗CFT2 unfolds to a non-trivial (i.e., not purely reflecting) interface between the two conformal field theories.

For most of these interfaces the product of the corresponding operators is singular, so the fusion needs to be appropriately defined. A first step in this direction was taken, in the context of a free-scalar theory, in [7]. The rough idea is to define the fusion product as the renormalization-group fixed point to which the system of the two interfaces flows when their separation, ε, goes to zero. A systematic way of doing this, consistent with the distributive property of fusion,[3] has not yet been worked out for interacting theories. For free fields, on the other hand, the story is simpler. The short-distance singularities are in this case expected to be of the general form

$$\mathscr{O}_A\, e^{-\varepsilon(L_0+\overline{L}_0)}\, \mathscr{O}_B \;\simeq\; \sum_C (e^{2\pi/\varepsilon})^{d_{AB}^{C}}\, N_{AB}^{C}\, \mathscr{O}_C, \tag{8}$$

where $\varepsilon \simeq 0$ is the separation of the two (circular) interfaces on the cylinder, $L_0+\overline{L}_0$ is the translation operator in the middle CFT, the d_{AB}^{C} are (non-universal) constants, and the N_{AB}^{C} are integer multiplicities. The singular coefficients in the above expression are Boltzmann factors for divergent Casimir energies. The latter must be proportional to $1/\varepsilon$ which is the only scale in the problem (other than the inverse temperature normalized to $\beta = 2\pi$).

By analogy with the operator-product expansion and the Verlinde algebra [45] we may extract from expression (8) the fusion rule

$$\mathscr{O}_A \circ \mathscr{O}_B \;=\; \sum_C N_{AB}^{C}\, \mathscr{O}_C. \tag{9}$$

The following iterative argument shows that this definition respects the conformal symmetry: first multiply the left-hand-side of (8) with the most singular inverse Boltzmann factor (the one with the largest d_{AB}^{C}) and take the limit $\varepsilon \to 0$ so as to extract the leading term of the product. Since $[L_N - \overline{L}_{-N}, e^{-\varepsilon(L_0+\overline{L}_0)}] \simeq o(\varepsilon)$ the result commutes with the diagonal Virasoro algebra. Next subtract the leading term from the left-hand-side of (8), and mutliply by the inverse Boltzmann factor with the second-largest d_{AB}^{C}. This picks up the subleading term which, thanks to the above argument and the conformal symmetry of the leading term, commutes also with the diagonal Virasoro algebra. Continuing this iterative reasoning proves that the right-hand-side of (9) is conformal as claimed.

3 The $c = 1$ CFT and a Black Hole Analogy

A simple context in which to illustrate the above ideas is the $c = 1$ conformal theory of a periodically-identified free scalar field, $\phi = \phi + 2\pi R$. Consider the interfaces that preserve a $U(1) \times U(1)$ symmetry, i.e., those described by linear

[3] I thank Maxim Kontsevich for stressing this point.

gluing conditions for the field ϕ. They correspond, after folding, to combinations of D1-branes and of magnetized D2-branes on the orthogonal two-torus whose radii, R_1 and R_2, are the radii on either side of the interface. The D1-branes are characterized by their winding numbers, k_1 and k_2, and by the Wilson line and periodic position moduli α and β. The magnetized D2-branes are obtained from the D1-branes by T-dualizing one of the two directions of the torus – they have therefore the same number of discrete and of continuous moduli.

Let us focus here on the D1-branes. The corresponding boundary states read

$$\| \mathrm{D1}, \vartheta \rangle\rangle = g^{(+)} \prod_{n=1}^{\infty} (e^{S^{(+)}_{ij} a^i_n \tilde{a}^j_n})^{\dagger} \sum_{N,M=-\infty}^{\infty} e^{iN\alpha - iM\beta} |k_2 N, k_1 M\rangle \otimes |-k_1 N, k_2 M\rangle, \quad (10)$$

where a^j_n and $\tilde{a}^j_n$ are the left- and right-moving annihilation operators of the field ϕ_j (for $j = 1,2$) and the dagger denotes hermitean conjugation. The ground states $|m, \tilde{m}\rangle$ of the scalar fields are characterized by a momentum (m) and a winding number ($\tilde{m}$). The states in the above tensor product correspond to ϕ_1 and ϕ_2. Furthermore

$$S^{(+)} = \mathscr{U}^T(\vartheta) \begin{pmatrix} -1 & 0 \\ 0 & 1 \end{pmatrix} \mathscr{U}(\vartheta) = \begin{pmatrix} -\cos 2\vartheta & -\sin 2\vartheta \\ -\sin 2\vartheta & \cos 2\vartheta \end{pmatrix}, \quad (11)$$

where $\mathscr{U}(\vartheta)$ is a rotation matrix and $\vartheta = \arctan(k_2 R_2 / k_1 R_1)$ is the angle between the D1-brane and the ϕ_1 direction. Finally, the normalization constant is the g-factor [2] of the boundary state. It is given by

$$g^{(+)} = \frac{\ell}{\sqrt{2V}} = \sqrt{\frac{k_1^2 R_1^2 + k_2^2 R_2^2}{2R_1 R_2}} = \sqrt{\frac{k_1 k_2}{\sin 2\vartheta}}, \quad (12)$$

where ℓ is the length of the D1-brane, V the volume of the two-torus, and the last rewriting follows from straightforward trigonometry. The logarithm of the g factor is the invariant entropy of the interface.

Inspection of the expression (10) shows that the non-zero modes of the fields ϕ_j are only sensitive to the angle ϑ, which also determines the reflection coefficient of the interface. For fixed k_1 and k_2 the g factor is minimal when $\vartheta = \pm\pi/4$, in which case the reflection $\mathscr{R} = 0$ and the interface is topological. Note that this requirement fixes the ratio of the two bulk moduli: $R_1/R_2 = |k_2/k_1|$. When $|k_1| = |k_2| = 1$ the two radii are equal and the invariant entropy is zero. The corresponding topological defects generate the automorphisms of the CFT, i.e., sign flip of the field ϕ and separate translations of its left- and right-moving pieces. The identity defect corresponds to the diagonal D1-brane, with $k_1 = k_2 = 1$ and $\alpha = \beta = 0$. A T-duality along ϕ_1 maps this topological defect to a D2-brane with one unit of magnetic flux. The corresponding interface operator is the generator of the radius-inverting T-duality transformation. All other topological interfaces have positive entropy, $\log g = \log\sqrt{|k_1 k_2|} > 0$. One may conjecture that the following statement is more generally true: the entropy of all topological interfaces is non-negative, and it vanishes only for T-duality transformations and for CFT automorphisms.

The interfaces given by (10)–(12) exist for all values of the bulk radii R_1 and R_2. By choosing the radii to be equal we obtain a large set of conformal defects whose algebra is an extension of the automorphism group of the CFT. For a detailed derivation of this algebra see [7]. The fusion rule for the discrete defect moduli turns out to be multiplicative,

$$[k_1, k_2; s] \circ [k_1', k_2'; s'] \;=\; [k_1 k_1', k_2 k_2'; ss'], \tag{13}$$

where $[k_1, k_2; s]$ denotes a defect with integer moduli k_1, k_2, s, where $s = +, -$ according to whether the folded defect is a D1-brane or a magnetized D2-brane. The above fusion rule continues to hold for general interfaces, i.e., when the radii on either side are not the same. Let me also give the composition rule for the angle ϑ in this general case (assuming $s = s' = +$):

$$\tan(\vartheta \circ \vartheta') = \tan\vartheta \tan\vartheta', \tag{14}$$

where $\vartheta \circ \vartheta'$ denotes the angle of the fusion product. The composition rule (13) was first derived, for the topological interfaces, in [25]. In this case the tangents in the last equation are ± 1 and all operator products are non-singular.

There exist some intriguing similarities [7] between the above conformal interfaces and supergravity black holes. The counterpart of BPS black holes are the topological interfaces, which (a) minimize the free energy for fixed values of the discrete charges, (b) fix through an "attractor mechanism" [21] a combination of the bulk moduli, and (c) are marginally stable against dissociation – the inverse process of fusion. The interface "algebra" is, in this sense, reminiscent of an earlier effort by Harvey and Moore [30] to define an extended symmetry algebra for string theory. Their symmetry generators were vertex operators for supersymmetric states of the compactified string. One noteworthy difference is that the additively-conserved charges in our case are logarithms of natural numbers, rather than taking values in a charge lattice as in [30]. Whether these observations have any deeper meaning remains to be seen. Another direction worth exploring is a possible relation of the above ideas with efforts to formulate string theory in a "doubled geometry," see for instance [31]. The doubling of spacetime after folding suggests that this may provide the natural setting in which to formulate the defect algebras.

Time to conclude: conformal interfaces and defects are examples of extended operators, which are a rich and still only partially-explored chapter of quantum field theory. They describe a variety of critical phenomena in low-dimensional condensed-matter systems which, for lack of time, I did not discuss. They can be, furthermore, both added and juxtaposed or fused. When this latter operation can be defined, the conformal interfaces form interesting algebraic structures which could shed new light on the symmetries of string theory. For all these reasons they deserve to be studied more.

Acknowledgments I thank Ilka Brunner, Jürg Fröhlich and Samuel Monnier for very pleasant collaborations during the last couple of years, on different aspects of this talk. Many thanks also to Eric D'Hoker, Mike Douglas, Sergei Gukov, Chris Hull and Maxim Kontsevich for useful conversations and comments. Claudio Bunster supervised my senior thesis and helped me publish my first scientific article – always a source of considerable pride for a student. My gratitude, after all these years, remains intact.

References

1. I. Affleck, Acta Phys. Polon. B **26** (1995) 1869 [arXiv:cond-mat/9512099].
2. I. Affleck and A. W. W. Ludwig, Phys. Rev. Lett. **67** (1991) 161.
3. A. Alekseev and S. Monnier, JHEP **0708** (2007) 039 [arXiv:hep-th/0702174].
4. A. Alekseev and S. L. Shatashvili, Commun. Math. Phys. **133** (1990) 353.
5. N. Andrei, Phys. Rev. Lett. **45** (1980) 379.
6. O. Babelon, D. Bernard and M. Talon, *Introduction to Classical Integrable Systems* (Cambridge University Press, Cambridge, 2003).
7. C. Bachas and I. Brunner, JHEP **0802** (2008) 085 [arXiv:0712.0076 [hep-th]].
8. C. Bachas and M. Gaberdiel, JHEP **0411** (2004) 065 [arXiv:hep-th/0411067].
9. C. Bachas and M. Petropoulos, JHEP **0102** (2001) 025 [arXiv:hep-th/0012234].
10. C. Bachas, J. de Boer, R. Dijkgraaf and H. Ooguri, JHEP **0206** (2002) 027 [arXiv:hep-th/0111210].
11. V. V. Bazhanov, S. L. Lukyanov and A. B. Zamolodchikov, Commun. Math. Phys. **177** (1996) 381 [arXiv:hep-th/9412229].
12. V. V. Bazhanov, S. L. Lukyanov and A. B. Zamolodchikov, Commun. Math. Phys. **190** (1997) 247 [arXiv:hep-th/9604044].
13. V. V. Bazhanov, S. L. Lukyanov and A. B. Zamolodchikov, Commun. Math. Phys. **200** (1999) 297 [arXiv:hep-th/9805008].
14. I. Brunner and D. Roggenkamp, JHEP **0708** (2007) 093 [arXiv:0707.0922 [hep-th]].
15. I. Brunner and D. Roggenkamp, JHEP **0804** (2008) 001 [arXiv:0712.0188 [hep-th]].
16. I. Brunner, H. Jockers and D. Roggenkamp, arXiv:0806.4734 [hep-th] (unpublished).
17. C. G. Callan, C. Lovelace, C. R. Nappi and S. A. Yost, Nucl. Phys. B **293** (1987) 83.
18. O. DeWolfe, D. Z. Freedman and H. Ooguri, Phys. Rev. D **66** (2002) 025009 [arXiv:hep-th/0111135].
19. J. Ehlers, *Konstruktionen und Charakterisierung von Lsungen der Einsteinschen Gravitationsfeldgleichungen*, Dissertation, Hamburg University, Hamburg, Germany (1957).
20. J. Erdmenger, Z. Guralnik and I. Kirsch, Phys. Rev. D **66** (2002) 025020 [arXiv:hep-th/0203020].
21. S. Ferrara, R. Kallosh and A. Strominger, Phys. Rev. D **52** (1995) 5412 [arXiv:hep-th/9508072].
22. S. Fredenhagen, Nucl. Phys. B **660** (2003) 436 [arXiv:hep-th/0301229].
23. J. Frohlich, J. Fuchs, I. Runkel and C. Schweigert, Phys. Rev. Lett. **93** (2004) 070601 [arXiv:cond-mat/0404051].
24. J. Frohlich, J. Fuchs, I. Runkel and C. Schweigert, Nucl. Phys. B **763** (2007) 354 [arXiv:hep-th/0607247].
25. J. Fuchs, M. R. Gaberdiel, I. Runkel and C. Schweigert, J. Phys. A **40** (2007) 11403 [arXiv:0705.3129 [hep-th]].
26. J. Fuchs, I. Runkel and C. Schweigert, arXiv:math/0701223 (unpublished).
27. R. Geroch, J. Math. Phys. **12** (1971) 918.
28. K. Graham and G. M. T. Watts, JHEP **0404** (2004) 019 [arXiv:hep-th/0306167].
29. S. Gukov and E. Witten, arXiv:hep-th/0612073 (unpublished).
30. J. A. Harvey and G. W. Moore, Commun. Math. Phys. **197** (1998) 489 [arXiv:hep-th/ 9609017].

31. C. M. Hull, JHEP **0707** (2007) 080 [arXiv:hep-th/0605149].
32. V. G. Kac, Proc. Natl. Acad. Sci. U. S. A. **81** (1984) 645.
33. A. Kapustin, arXiv:hep-th/0612119 (unpublished).
34. A. Kapustin and N. Saulina, arXiv:0710.2097 [hep-th] (unpublished).
35. A. Kapustin and E. Witten, arXiv:hep-th/0604151 (unpublished).
36. A. Karch and L. Randall, JHEP **0105** (2001) 008 [arXiv:hep-th/0011156].
37. A. Karch and L. Randall, JHEP **0106** (2001) 063 [arXiv:hep-th/0105132].
38. J. M. Maldacena, Phys. Rev. Lett. **80** (1998) 4859 [arXiv:hep-th/9803002].
39. M. Oshikawa and I. Affleck, Nucl. Phys. B **495** (1997) 533 [arXiv:cond-mat/9612187].
40. V. B. Petkova and J. B. Zuber, Phys. Lett. B **504** (2001) 157 [arXiv:hep-th/0011021].
41. J. Polchinski, Phys. Rev. Lett. **75** (1995) 4724 [arXiv:hep-th/9510017].
42. J. Polchinski and Y. Cai, Nucl. Phys. B **296** (1988) 91.
43. S. J. Rey and J. T. Yee, Eur. Phys. J. C **22** (2001) 379 [arXiv:hep-th/9803001].
44. T. Quella, I. Runkel and G. M. T. Watts, JHEP **0704** (2007) 095 [arXiv:hep-th/0611296].
45. E. P. Verlinde, Nucl. Phys. B **300** (1988) 360.
46. P. B. Wiegmann, JETP Lett. **31** (1980) 392.
47. K. G. Wilson, Phys. Rev. D **10** (1974) 2445.
48. K. G. Wilson, Rev. Mod. Phys. **47** (1975) 773.

Eddington–Born–Infeld Action and the Dark Side of General Relativity

Máximo Bañados

Abstract We review a recent proposal to describe dark matter and dark energy based on an Eddington–Born–Infeld action. The theory is successful in describing the evolution of the expansion factor as well as galactic flat rotation curves. Fluctuations and the CMB spectra are currently under study. This paper in written in honor of Claudio Bunster on the occasion of his 60th birthday.

Marc Henneaux remarked in his lecture at Claudio's Fest (Valdivia Chile, January 2008) that working with him one learns to be brave. Claudio's tuition has been particularly important for me over the last 2 years. I have been working on an idea that looked crazy at first sight and still looks pretty mad today. I have not had the chance to talk to Claudio about these ideas and I have not sent him the papers I have written [2, 3] because I know he will not read them! In this short contribution in his honor I attempt to shoot two birds with one stone, hoping that the bird will not be me.

I shall review a project whose aim is to provide a candidate for dark matter and dark energy, and whose seed relies on the study of general relativity around the degenerate field $g_{\mu\nu} = 0$. The project started as a purely formal idea, but it immediately succeeded in reproducing some of the known phenomenological curves associated to dark matter and dark energy. I thus decided to purse the idea to the end. In this contribution I shall concentrate on the original motivation based on studying general relativity near $g_{\mu\nu} = 0$ [3]. This controversial motivation is now not needed because an action implementing most of the ideas is available [2]. However, I believe that exploring physics at $g_{\mu\nu} = 0$ is an attractive idea, certainly not new, and perhaps necessary to understand the origin of the Universe.

The first step comes from the first order vielbein formulation of general relativity. An intriguing solution to the equations of motion,

$$\varepsilon_{abcd} R^{ab} e^c = 0 \tag{1}$$

$$\varepsilon_{abcd} T^a e^c = 0, \tag{2}$$

M. Bañados
P. Universidad Católica de Chile, Av. Vicuña Mackenna 4860, Santiago, Chile
e-mail: mbandos@uc.cl

M. Henneaux, J. Zanelli (eds.), *Quantum Mechanics of Fundamental Systems: The Quest for Beauty and Simplicity*, DOI 10.1007/978-0-387-87499-9_4,

is

$$e^a = 0. \tag{3}$$

The existence of this solution has not gone unnoticed, e.g., [9,10,13]. Its most salient property is that it preserves the full set of diffeomorphisms and for this reason it is often called the 'unbroken' ground state of general relativity. This point was stressed in [9] were a symmetry breaking transition from $e^a = 0$ to $e^a \neq 0$, the big-bang, was suggested. Topological transitions in this formulation were studied in [10].

Now, a key aspect of this solution is the fact that the spin connection is left undetermined. The above equations of motion are supposed to determine both a^a and w^{ab}. However, since $e^a = 0$ kills both (1) and (2), the spin connection becomes a random field.[1] The first step towards accepting $e^a = 0$ as a solution is to understand the nature of the spin connection at that point.

If one first solves the algebraic equation for the torsion expressing $w \sim e^{-1}\partial e$ as a function of e^a the same problem reappears in a different way. One may try to recover the solution $e^a = 0$ by a limit $e^a \to 0$. The connection $w \simeq e^{-1}\partial e$ has the structure $\frac{0}{0}$ and can be anything.[2] This is equivalent to the statement that w becomes random at $e^a = 0$. One does learn something with this exercise though. The structure $\frac{0}{0}$ appears provided both e^a and its derivative vanish at all points. This tells us that the limit $e^a \to 0$ cannot be associated to time evolution (where $e^a = 0$ would occur at some particular time t_0). Trying to understand the big-bang as a transition from zero metric into non-zero metric raises complicated issues on the role of causality. Before the metric is created there is no causality at all [9].

The problem we shall attack in this paper is the arbitrariness of the connection at $e^a = 0$. Our approach will consist on adding a new interaction to the action such that, as $e^a \to 0$, the spin connection does not go random but continues to be controlled by second order field equations.

Interestingly there are no too many terms one could add. For reasons we shall explain in a moment, it is necessary to go back to the metric formulation. Consider the following Palatini action including a new term that depends only on the connection,

$$I[g,\Gamma] = \int \left[\sqrt{g}(g^{\mu\nu}R_{\mu\nu}(\Gamma) + \Lambda) + \kappa\sqrt{|R_{\mu\nu}(\Gamma)|}\right]. \tag{4}$$

Here $R_{\mu\nu}$ is the Ricci tensor, which only depends on the connection, not the metric. The new term is known as Eddington theory. The constant κ is a coupling constant which in principle should be small enough such that this action is not in

---

[1] This is related to another feature of $e^a = 0$. The leading term of the action $I[e,w] \sim \int \varepsilon_{abcd} R^{ab} e^c e^d$ is cubic with respect to $w = e = 0$, $\int \varepsilon_{abcd} dw^{ab} e^c e^d$. Thus, around $e^a = 0$ there is no quadratic term to expand, and no linearized theory can be defined. Of course the action can be expanded around the 'broken' solution $e^a_\mu = \delta^a_\mu$ with a well-defined linearized theory, but the interactions become non-renormalizable. In three dimensions this problem does not occur because the action has one less power of e^a and the quantum theory can be explored much further [13].

[2] Note that in particular the limit may be a smooth differentiable function. In that case the curvature is well-defined. In particular $R_{\mu\nu}$ exists at the limit while $R = g^{\mu\nu}R_{\mu\nu}$ does not. Not surprisingly, metric invariants are not good objects to characterize the $g = 0$ phase.

contradiction with well-known experiments. It is interesting to note the uniqueness of this term. In the absence of a metric, Eddington's functional is the only density with the correct weight to respect diffeomorphism invariance. Note that Eddington's action cannot be defined in the first order tetrad formalism. The $SO(3,1)$ curvature $R^a{}_{b\mu\nu}(w)$ cannot be traced to produce a two index object, without using e^a_μ. This is in contrast with the $GL(4)$ curvature $R^\alpha{}_{\beta\mu\nu}(\Gamma)$ whose trace $R_{\beta\nu}(\Gamma)$ is a tensor and independent from the metric.

The attractive feature of the action (4) is that if the metric was not present, then the first two terms are not present and the dynamics is governed by Eddington's action.[3] In this sense we have produced an action whose dynamics is well-defined even if the metric is switched off.

However, it is now a simple exercise to prove that the action (4) does not produce any interesting new effects. Actually, this was already known to Eddington. What happens is that the Einstein–Hilbert action with a cosmological term is dual to Eddington's action [8]. In other words, the Eddington term in (4) only renormalizes Newton's constant.

This can be seen as follows. Consider the Palatini action for gravity with a cosmological constant,

$$I_P[g,\Gamma] = \int \sqrt{g}(g^{\mu\nu}R_{\mu\nu}(\Gamma) - 2\Lambda) \tag{5}$$

It is well known that upon eliminating the connection using its own equation of motion one arrives at the usual second order Hilbert action

$$I_H[g] = \int \sqrt{g}(R(g) - 2\Lambda). \tag{6}$$

It is less well-known but also true [8] that, if $\Lambda \neq 0$, then the metric can also be eliminated by using its own equations.[4] The variation $\frac{\delta I}{\delta g^{\mu\nu}} = 0$ yields,

$$g_{\mu\nu} = \frac{1}{\Lambda}R_{\mu\nu}(\Gamma). \tag{7}$$

Since this is an algebraic relation for $g_{\mu\nu}$ it is legal to replace it back in the the action obtaining Eddington's functional

$$I_E[\Gamma] = \frac{2}{\Lambda}\int \sqrt{\det(R_{\mu\nu})}. \tag{8}$$

---

[3] We borrow here the prescription from the tetrad formalism: $\varepsilon_{abcd}R^{ab}e^c e^d \sim \sqrt{g}g^{\mu\nu}R_{\mu\nu}$ vanishes if $e^a \sim g_{\mu\nu} \to 0$. Another way to see this is by noticing that the volume element $\sqrt{g}$ scales faster than the metric inverse $g^{\mu\nu}$, at least for $d > 2$.

[4] This duality is of course well know to Claudio, and in fact the first time I heard about it was from him.

In the terminology of dualities, the action (5) is called the Parent action, while the Einstein–Hilbert action (6) and Eddington's action (8) are its daughters. I_H and I_E are said to be dual to each other, and in many respects they are equivalent [6, 8].

Summarizing, the action (4) can be understood as general relativity interacting with its own dual field. By a set of duality transformations we can transform the whole action (4) into standard general relativity with a new coupling constant. (Starting from (4) one eliminate the metric and to get Eddington action twice. Then apply a new transformation to get back to Einstein–Hilbert.)

An important note of caution is in order here. The equivalence between the Einstein–Hilbert and Eddington actions is true provided $g_{\mu\nu}$ is not degenerate. For degenerate fields they do represent different dynamics, and in fact, only Eddington's action is well-defined in that case. The reason we shall not consider these cases is that, at the end of the day, we are interested in non-degenerate metrics anyway. Our guiding principle is to uncover what sort of modifications would be necessary to incorporate $g_{\mu\nu} = 0$ as an allowed state. But, the physics phenomena we shall be interested do require a non-degenerate metric.

We shall now recall an analogy with condensed matter physics, suggesting a different interpretation for Eddington's action, which will truly depart from pure general relativity.

Within standard general relativity, we have observed that if the metric is removed then the spin connection goes random. This looks very similar to a set of spins, at $T > T_c$, in the presence of an external magnetic field. As the field is removed the spins go random. However, if the temperature is below its critical value $T < T_c$, then the external field can be removed and the spins retain their ordered state. The crucial property of spins which makes this possible is their self interaction. Eddington's action has a similar role. For the action (4), as the metric is removed, the connection continues to be described by a well-defined set of equations.[5]

Now, spins also exhibit the opposite phenomena, namely spontaneous "magnetization." If $T < T_c$ then the random disordered state is unstable and decays spontaneously into a broken ordered state. No external field is needed to trigger this phenomena. Let us imagine that the connection in general relativity exhibits a similar phenomena. That is, *without* introducing a metric, we assume that a connection can exists and be described by some well-defined equations. We shall call this connection $C^{\mu}{}_{\alpha\beta}$. This field is fully independent from the metric. Obviously, the only action consistent with general covariance is again Eddington's theory,

$$I[C] = \int \sqrt{|K_{\mu\nu}(C)|} \tag{9}$$

[5] At this point we treat the metric or tetrad as an external field which can be switched on and off, as a mathematician would do. On a first approximation one does not look at the Maxwell equations governing the external field but simply assume that it can be controlled at will. We have assumed the same with the metric, treating it as an external field. A full action governing the coupled system will be displayed below.

where $K_{\mu\nu}(C)$ is the (traced) curvature associated to the connection C. No metric is needed for this construction. What we have in mind is the existence of a connection $C^{\mu}{}_{\nu\alpha}$ that existed before the Universe, as a Riemannian manifold, was created. This has to be interpreted with care because without $g_{\mu\nu}$ there are no causality relations.

At this point we have departed from the treatment suggested in [9]. In that reference the field which acquires an spontaneous non-zero expectation value was the metric itself.

We shall now turn on the metric and consider the whole self consistent system. The metric $g_{\mu\nu}$ generates its own connection, the Christoffel symbol $\Gamma(g)$. Thus, our theory will contain two independent connections. One, called C, is generated spontaneously. The other, called $\Gamma(g)$ is driven by the external field $g_{\mu\nu}$.

The action describing the coupled system is notably simple. The pair g,Γ are of course described by the standard Einstein–Hilbert action either in first or second order form. On the other hand the field C is described by Eddington's action.

Now, to make things more interesting we shall couple both connections through the metric. Again, there are no two many couplings one can write. An attractive possibility is the Einstein–Hilbert–Eddington–Born–Infeld action [2],

$$I = \int \sqrt{g}(R - 2\Lambda) + \frac{2}{\alpha l^2}\sqrt{|l^2 K_{\mu\nu} - g_{\mu\nu}|}. \tag{10}$$

This action has several interesting properties. First, note that as $g \to 0$ we recover Eddington's action for the field C. (We shall not need this limit in what follows.) Second, (10) is a tensor Born–Infeld theory analogous to the scalar $\sqrt{\det(g_{\mu\nu} + \partial_\mu\phi\partial_\nu\phi)}$ and vector $\sqrt{\det(g_{\mu\nu} + F_{\mu\nu})}$ Born–Infeld theories. Observe that the equations of motion for the whole action are of second order. This is different from the gravitational BI action written in [7] where extra terms had to be added in order to eliminate the ghost. Finally, the action (10) contains dynamical properties which makes it an attractive candidate for dark matter and dark energy [2]. Both appear in a unified way, just like the Chapligyn formulation [11]. It is also curious to observe that the Chapligyn gas can be derived from a scalar Born–Infeld theory.

More specifically, for Friedman models, it follows that the Eddington field C behaves like matter for early times and as dark energy for late times: its equation of state $w = p/\rho$ evolve from $w = 0$ near the big bang to $w = -1$ for late times. One can also analyze the dynamics of objects moving around spherically symmetric sources. The Eddington field in this case yields asymptotically flat rotation curves and thus again provides a candidate for dark matter.

The reader may wonder what does $g_{\mu\nu} = 0$ have to do with dark matter and dark energy. Could one had predicted this (suggested) relationship? Our initial motivation to look at $g_{\mu\nu} = 0$ came from the following analogy. A particle at rest has an energy mc^2. The most direct manifestation of this energy is through gravity, and in fact mc^2 is a source of curvature. From a Newtonian point of view, the energy of a particle at rest is zero. Can we ask the same question in general relativity? Could flat space have an energy-density associated to it? In the standard choice for zero point of energy, flat space has zero energy. The very definition of energy in general relativity requires knowledge of boundary conditions and certainly $g_{\mu\nu} = 0$ falls outside all examples. However at the level of energy density, namely, the Einstein tensor $G_{\mu\nu}(g)$, one may wonder about its value at $g_{\mu\nu} = 0$. Interestingly this tensor depends only on $g^{-1}\partial g$ and thus the limit can be defined in a way that $G_{\mu\nu}(0)$ becomes finite. By a mixture of Bianchi identities and some reasonable assumptions its value can be computed and yields a contribution to Einstein equations similar to those expected from dark matter [3].

The ideas presented in this contribution are highly speculative and need formalization. The analogy with spin systems is the most challenging and difficult problem. Other applications like fluctuations and the CMB spectra are presently under analysis [4]. The action (10) has several interesting formal properties under duality transformations. This theory can be written as a bigravity theory, which have been under great scrutiny in the past and also recently [1]. A detailed analysis will be reported in [5]. Rotation curves for several galaxies has been analyzed in [12], with interesting results.

Acknowledgments Some of the ideas presented here have been developed in collaboration with A. Gomberoff and D. Rodriguez. I would like to thank them for their permission to include here some unpublished material. I would also like to thank S. Theisen for many useful conversations and for bringing to my attention [8]. P. Ferreira and C. Skordis have been crucial to keep this idea alive.

References

1. N. Arkani-Hamed, H. Georgi and M. D. Schwartz, Ann. Phys. **305**, 96 (2003) [arXiv:hep-th/0210184].
2. M. Bañados, Class. Quant. Grav. **24**, 5911 (2007) [arXiv:hep-th/0701169].
3. M. Bañados, Phys. Rev. **D77**, 123534 (2008) arXiv:0801.4103 [hep-th].
4. M. Bañados, P. Ferreira and C. Skordis, arXiv:0811.1272[astro-ph] (unpublished).
5. M. Bañados, A. Gomberoff and D. Rodrigues (unpublished).
6. N. Boulanger, S. Cnockaert and M. Henneaux, JHEP **0306**, 060 (2003) [arXiv:hep-th/0306023].
7. S. Deser and G. W. Gibbons, Class. Quant. Grav. **15**, L35 (1998) [arXiv:hep-th/9803049].
8. E. S. Fradkin and A. A. Tseytlin, Ann. Phys. **162**, 31 (1985).
9. S. B. Giddings, Phys. Lett. B **268**, 17 (1991).
10. G. T. Horowitz, Class. Quant. Grav. **8**, 587 (1991).
11. A. Y. Kamenshchik, U. Moschella and V. Pasquier, Phys. Lett. B **511**, 265 (2001) [arXiv:gr-qc/0103004].
12. A. Reisenegger and N. Rojas (unpublished).
13. E. Witten, Nucl. Phys. B **311**, 46 (1988).

Light-Cone Field Theory, Maximal Supersymmetric Theories and $E_{7(7)}$ in Light-Cone Superspace

Lars Brink

Abstract In this lecture I describe the light-cone formulation of quantum field theories especially the maximally supersymmetric ones. This is a formalism in which we keep only the physical degrees of freedom for both bosons and fermions. I show how $N = 4$ Yang–Mills Theory and $N = 8$ supergravity come out very naturally and that they look very much alike. I finally show how to implement the $E_{7(7)}$ symmetry for the supergravity theory. The new feature in this formulation is that all fields of the supermultiplet including the graviton transform under $E_{7(7)}$.

1 Introduction

When we study supersymmetric theories, the maximally supersymmetric ones, $(\mathcal{N} = 8, d = 4)$ supergravity [1, 2] and $(\mathcal{N} = 4, d = 4)$ Yang–Mills Theory [3, 4] or their 11-dimensional or ten-dimensional versions always show up. The $\mathcal{N} = 1$ supergravity in 11 dimensions [5], is the largest supersymmetric local field theory with maximum helicity two (on reduction to $d = 4$). This theory has gained renewed prominence since its recognition as the infrared limit of M-Theory [6, 7]. Although M-Theory casts well-defined shadows on lower-dimensional manifolds, its actual structure remains a mystery. We must therefore glean all we can from the $\mathcal{N} = 1$ supergravity theory or its dimensionally reduced versions before tackling M-Theory. $\mathcal{N} = 1$ supergravity is ultraviolet divergent in $d = 11$ but this divergence is presumably tamed by M-Theory and the hope is that an understanding of this divergent structure, will give us a window into the workings of M-Theory. Similarly the $\mathcal{N} = 1$ Yang–Mills Theory in ten dimensions [3, 4], which is the low-energy limit of the open string theory in ten dimensions has been shown to play an important

L. Brink
Chalmers University of Technology, Göteborg, Sweden
e-mail: lars.brink@chalmers.se

M. Henneaux, J. Zanelli (eds.), *Quantum Mechanics of Fundamental Systems: The Quest for Beauty and Simplicity*, DOI 10.1007/978-0-387-87499-9_5,

role in the AdS/CFT duality [8]. The four-dimensional version $\mathcal{N}=4$ Yang–Mills Theory is also very special since it is free of ultraviolet divergences [9, 10].

In the normal covariant treatments of these two theories they look quite different, one being a reparametrization invariant gravity theory, while the other is a Yang–Mills gauge theory. However, in the light-cone formulations, the so-called LC_2 formulations where all auxiliary degrees of freedom have been eliminated, [11, 12], the two superfields describing the field content of the two theories are particularly simple and and very much alike. Indeed these superfields can be regarded as master fields for a series of theories. Since they are natural partners in string theory this similarity must be quite important and much of my research in recent years has been to use this similarity and to try to use it to learn more about these theories and the underlying string theory. They are, of course, very well studied during a long time but they have consistently shown themselves to be more interesting than what meets the eye.

Writing the two theories in the LC_2 formulation has a price. We loose a lot of information from the geometry and the only guideline will be the non-linearly realized superPoincaré algebra. However, it is important to view these very important theories from different angles and for certain question this formulation is the most adequate one.

In this talk I will start from the beginning to build up light-cone field theories and then go over to the supersymmetric ones. I will start the analysis in four dimensions of space–time and then 'oxidize' them to higher dimensions keeping the specific form of the superfields. Finally I will show how the $E_{7(7)}$ symmetry is implemented in this formulation.

2 Light-Frame Formulation of Field Theories

In his famous paper of 1949 Dirac [13] argued that for a relativistically invariant theory any direction within the light-cone can be the evolution parameter, the "time". In particular we can use one of the light-cone directions. For this discussion we will use $x^+ = \frac{1}{\sqrt{2}}(x^0 + x^3)$ as the time. The coordinates and the derivatives that we will use will then be

$$x^{\pm} = \frac{1}{\sqrt{2}}(x^0 \pm x^3); \qquad \partial^{\pm} = \frac{1}{\sqrt{2}}(-\partial_0 \pm \partial_3); \tag{1}$$

$$x = \frac{1}{\sqrt{2}}(x_1 + ix_2); \qquad \bar{\partial} = \frac{1}{\sqrt{2}}(\partial_1 - i\partial_2); \tag{2}$$

$$\bar{x} = \frac{1}{\sqrt{2}}(x_1 - ix_2); \qquad \partial = \frac{1}{\sqrt{2}}(\partial_1 + i\partial_2), \tag{3}$$

so that

$$\partial^+ x^- = \partial^- x^+ = -1; \qquad \bar{\partial} x = \partial \bar{x} = +1. \tag{4}$$

The derivatives are, of course, related to the momenta through the usual formula $p^\mu = -i\partial^\mu$ and we use the light-cone decomposition also for p^μ. We will only consider massless theories so we solve the condition $p^2 = 0$. We then find

$$p^- = \frac{p\bar{p}}{p^+}. \tag{5}$$

The generator p^- is really the Hamiltonian conjugated to the light cone time x^+ and we see that the translation generators of the Poincare algebra are written with just three operators. We will use Dirac's vocabulary that generators that involve the "time" are called dynamical (or Hamiltonians) and the others kinematical. Using light-cone notation and the complex one from above for the transverse directions, the most general form of the generators of the full Poincaré algebra at $x^+ = 0$ is then given by the four momenta

$$p^- = -i\frac{\partial\bar{\partial}}{\partial^+}, \qquad p^+ = -i\partial^+, \qquad p = -i\partial, \qquad \bar{p} = -i\bar{\partial}, \tag{6}$$

the kinematical transverse space rotation

$$j = j^{12} = x\bar{\partial} - \bar{x}\partial + \lambda, \tag{7}$$

the other kinematical generators

$$j^+ = ix\partial^+, \qquad \bar{j}^+ = i\bar{x}\partial^+, \tag{8}$$

and

$$j^{+-} = ix^-\partial^+, \tag{9}$$

as well as the dynamical boosts

$$j^- = ix\frac{\partial\bar{\partial}}{\partial^+} - ix^-\partial + i\lambda\frac{\partial}{\partial^+}, \tag{10}$$

$$\bar{j}^- = i\bar{x}\frac{\partial\bar{\partial}}{\partial^+} - ix^-\bar{\partial} + i\lambda\frac{\bar{\partial}}{\partial^+}. \tag{11}$$

There is one degree of freedom in the algebra, namely the parameter λ which is the helicity. At this stage it is arbitrary and checking the corresponding spin one finds, of course, that it is $|\lambda|$. Hence the algebra covers all possible free field theories. We can let the generators act on a complex field $\phi(x)$ with helicity λ, with its complex conjugate having the opposite helicity. This is the "first-quantized" version. We can also consider the fields as operators having the commutation relation.

$$[\partial^+\bar{\phi}(x), \phi(x')] = -\frac{i}{2}\delta(x-x'), \tag{12}$$

where hence the momentum field conjugate to ϕ is $\partial^+\bar{\phi}$.

We then introduce the "second-quantized" representation O in terms of the "first-quantized" representation o as $O = 2i\int d^4x\partial^+\bar{\phi}(x)\ o\ \phi(x)$. We then find that the commutator between two of the generators J_1 and J_2 is

$$[J_1, J_2] = 2i\int d^4x\partial^+\bar{\phi}(x)[j_1, j_2]\phi(x). \tag{13}$$

We can understand that P^- truly is the Hamiltonian using (5)

$$P^- = 2\int d^4x\partial^+\bar{\phi}(x)\frac{\partial\bar{\partial}}{\partial^+}\phi(x). \tag{14}$$

Legendre transforming to the Lagrangian using the field momenta from (12) we get the action

$$\begin{aligned} S &= \int d^4x\left[\partial^+\bar{\phi}(x)\partial^-\phi(x) + \partial^+\phi(x)\partial^-\bar{\phi}(x) - 2\partial^+\bar{\phi}(x)\frac{\partial\bar{\partial}}{\partial^+}\phi(x)\right] \\ &= \int d^4x\partial^+\bar{\phi}(x)\Box\phi(x). \end{aligned} \tag{15}$$

It is remarkable that there is a unique form of the kinematic term for any spin-λ field. We should remember though that to specify the theory we have to give all Poincaré generators, since the action via the Hamiltonian is just one of those generators. They will show what spin the field describes.

In this representation it is straightforward to try to add interaction terms to the Hamiltonian. This was done in [12]. Every dynamical generator will have interaction terms. The procedure is very painstaking and there are as far as I know no other way than trial and error to find the non-linear representation. On the other hand, once such a representation is found it represents a possible relativistically invariant interacting field theory. The result is that for every integer λ there exists a possible three-point interaction. For λ even, the unique solutions are

$$\begin{aligned} S = \int d^4x\Big\{&\bar{\phi}(x)\Box\phi(x) \\ &+g\Big[\sum_{n=0}^{\lambda}(-1)^n\binom{\lambda}{n}\bar{\phi}(x)\partial^{+\lambda}\big(\frac{\bar{\partial}^{\lambda-n}}{\partial^{+\lambda-n}}\phi(x)\frac{\bar{\partial}^{\lambda}}{\partial^{+\lambda}}\phi(x)\big) + c.c.\Big]\Big\} \\ &+O(g^2). \end{aligned} \tag{16}$$

For λ odd, the field $\phi(x)$ must be in the adjoint representation of an external group $\phi^a(x)$ and we have to introduce the fully antisymmetric structure constants f^{abc} in the interaction terms to find a possible term. (It is really by checking the four-point coupling that we find that the field has to be representation of a Lie group.) The results is

$$S = \int d^4x \Big\{ \bar{\phi}^a(x)\Box\phi^a(x)$$
$$+ g f^{abc}\Big[\sum_{n=0}^{\lambda}(-1)^n \binom{\lambda}{n} \bar{\phi}^a(x)\partial^{+\lambda}\big(\frac{\bar{\partial}^{\lambda-n}}{\partial^{+\lambda-n}}\phi^b(x)\frac{\bar{\partial}^{\lambda}}{\partial^{+\lambda}}\phi^c(x)\big) + c.c.\Big]\Big\}$$
$$+ O(g^2). \tag{17}$$

We note the non-locality in the interaction term in terms of inverses of ∂^+. The easiest way to understand it is to Fourier transform to momentum space. In the calculations it is really defined by the rule $\frac{1}{\partial^+}\partial^+ f(x^+) = f(x^+)$. When performing a calculation one has to specify exactly the situation of the pole in ∂^+. In an sense this is a remainder of the gauge invariance.

We can now check for special values of λ.

- $\lambda = 0$

The dimension of the coupling constant g is 1 (in mass units) and this is the usual ϕ^3-theory. This theory is superrenormalizable but not physical since it does not have a stable vacuum having a potential with no minimum.

- $\lambda = 1$

The dimension of the coupling constant g is 0 and this theory is nothing but non-abelian gauge theory in a specific gauge. If we go on we know that we need a four-point coupling to fully close the algebra. Note that the action has no local symmetry and the gauge group only appears as the external symmetry group.

- $\lambda = 2$

The dimension of the coupling constant g is -1 and this theory is the beginning series of a gravity theory. It is clear from the dimensions of the coupling constant that interaction terms to arbitrary order can be constructed without serious non-localities. The four-point function related to Einstein's theory is known [14]. Going beyond the four-point coupling is probably too difficult, unless powerful computer methods could be devised. We expect several solutions, of course, since we know that the Hilbert action is but the simplest of all actions consistent with the equivalence principle. Note that the action above, which is a fully gauge fixed Hilbert action expanded in the fluctuations around the Minkowski metric, has no local symmetry, no covariance and knows nothing about curved spaces. It is probably useless for discussions about global properties of space and time but can be useful in the study of quantum corrections; to understand the finiteness properties of the quantum theory.

- $\lambda > 2$

The dimension of the coupling constant g is <-1 and these theories are theories for higher spins. Again they are non-renormalizable in the naive sense like the spin-2 theory above. There are strong reason to believe that these theories cannot be Poincaré invariant one by one when we go to higher orders in the coupling constant, but the result above is an indication that certain sums of such theories interacting with each other could possibly be invariant theories.

We can also find interacting solutions for λ half-integer. We can, of course, not have a three-point coupling. We will in fact not be able to find self-interacting theories but have to consider the coupling of the half-integer spin field to an integer spin field. We then find that we can couple a spin-1/2 field to a spin-1 or a spin-0 field to recover in the first case a non-abelian gauge field coupled to a spin-1/2 field $\psi^i(x)$ in a representation characterized by i of the external group such that we can have a coupling $\bar{\psi}_i \psi^j \phi^a C^i_{ja}$, with C^i_{ja} the Clebsch–Gordan coefficient. It is interesting to note that it is only in the interacting theory that we can prove the spin-statistics theorem [15]. The formalism demands the spin-1/2 field to be of odd Grassmann type and the integer spin fields to be even. Note that there is no spinor space. The spin-1/2 field is a complex (Grassmann odd) field with no space–time index. Its equation of motion looks just like the one for a bosonic field. (Remember the free equation the follows from (5).) However, the dimension of the field $\psi(x)$ is different from the one of the bosonic field, so the free action is

$$S = \int d^4x \partial^+ \bar{\psi}(x) \frac{\Box}{\partial^+} \psi(x). \tag{18}$$

The fact that we do not need to use spinors is very special for $d = 4$, since the transverse symmetry which is covariantly realized is $SO(2) \approx U(1)$, which does not distinguish spinor representations.

We have hence seen that we can find all known unitary relativistic field theories as representations of the Poincaré algebra, and we see their uniqueness and also what kind of possibilities there are for higher spin fields. In a gauge invariant formulation one can attempt to add in new terms that are gauge invariant. Invariably they lead to problems with unitarity. We do not see those terms here since the theories are unitary by construction.

It should be said here that we could have derived the expressions above by starting with a gauge-covariant action and implement the light-cone gauge by choosing the A^+-component of the vector field to be zero and then solve for the A^--component.

3 Light-Frame Formulation of Supersymmetric Field Theories

The known extension of the Poincaré algebra is to make it into a supersymmetry algebra. This will lead to a restriction on relativistic dynamics. It is true that the world does not look supersymmetric as such, but a good working hypothesis is that at some stage supersymmetry is indeed a symmetry of the world.

The standard covariant supersymmetry generator Q_α is a spinor with the anti-commutator

$$\{Q_\alpha, \bar{Q}_\beta\} = \gamma^\mu_{\alpha\beta} P_\mu. \tag{19}$$

The spinor Q_α is four-component. It satisfies the Majorana condition which makes it real in a certain representation of the γ-matrices. In the light-cone frame the spinor splits up into two two-component spinor that can be rewritten as two complex operators, which we call $Q_+ = -\frac{1}{2}\gamma_+\gamma_- Q$ and $Q_- = -\frac{1}{2}\gamma_-\gamma_+ Q$. From the Clifford algebra $\{\gamma^\mu, \gamma^\nu\} = 2\eta^{\mu\nu}$ with $\eta = diag(-1,1,1,1)$ we see that $Q = Q_+ + Q_-$, and that the products $-\frac{1}{2}\gamma_+\gamma_-$ and $-\frac{1}{2}\gamma_-\gamma_+$ are projection operators. We can linearly combine the two components of the spinors into complex entities with no indices. We can also augment by letting the Q's transform as the representation **N** under $SU(N)$. The light-cone supersymmetry algebra is then

$$\{Q^m_+, \bar{Q}_{+n}\} = -\sqrt{2}\delta^m_n P^+ \tag{20}$$

$$\{Q^m_-, \bar{Q}_{-n}\} = -\sqrt{2}\delta^m_n P^- \tag{21}$$

$$\{Q^m_+, \bar{Q}_{-n}\} = -\sqrt{2}\delta^m_n P, \tag{22}$$

where all other anticommutators are zero, except for the complex conjugate of the last one. The indices m,n run from 1 to N.

The superPoincaré algebra can now be represented on a superspace with coordinates $x^\pm, x, \bar{x}, \theta^m, \bar{\theta}_n$, where the coordinates θ^m and $\bar{\theta}_n$ are complex conjugates, Grassmann odd and transform as **N** and $\bar{\mathbf{N}}$ under $SU(N)$. We will denote their derivatives as

$$\bar{\partial}_m \equiv \frac{\partial}{\partial \theta^m}; \qquad \partial^m \equiv \frac{\partial}{\partial \bar{\theta}_m}. \tag{23}$$

The Q's are then represented as (We use the notation with lower case letters for operators that act on the field.)

$$q^m_+ = -\partial^m + \frac{i}{\sqrt{2}}\theta^m \partial^+; \qquad \bar{q}_{+n} = \bar{\partial}_n - \frac{i}{\sqrt{2}}\bar{\theta}_n \partial^+, \tag{24}$$

and the dynamical ones as

$$q^m_- = \frac{\bar{\partial}}{\partial^+} q^m_+, \qquad \bar{q}_{-m} = \frac{\partial}{\partial^+}\bar{q}_{+m}. \tag{25}$$

On this space we can also represent "chiral" derivatives anticommuting with the supercharges Q.

$$d^m = -\partial^m - \frac{i}{\sqrt{2}}\theta^m \partial^+; \qquad \bar{d}_n = \bar{\partial}_n + \frac{i}{\sqrt{2}}\bar{\theta}_n \partial^+, \tag{26}$$

which satisfy the anticommutation relations

$$\{d^m, \bar{d}_n\} = -i\sqrt{2}\,\delta^m{}_n\,\partial^+. \tag{27}$$

To find an irreducible representation we have to impose the chiral constraints

$$d^m\,\phi = 0; \qquad \bar{d}_m\,\bar{\phi} = 0, \tag{28}$$

on a complex superfield $\phi(x^{\pm}, x, \bar{x}, \theta^m, \bar{\theta}_n)$. The solution is then that

$$\phi = \phi(x^+, y^- = x^- - \frac{i}{\sqrt{2}}\,\theta^m\,\bar{\theta}_m, x, \bar{x},\, \theta^m). \tag{29}$$

We now have to add in θ-terms into the Lorentz generators to complete the representation of the free algebra. The result is for $\lambda = 0$

$$j \;=\; x\bar{\partial} - \bar{x}\partial + S^{12}, \tag{30}$$

where the little group helicity generator is

$$S^{12} \;=\; \frac{1}{2}\,(\,\theta^p\,\bar{\partial}_p \,-\, \bar{\theta}_p\,\partial^p\,) - \frac{i}{4\sqrt{2}\,\partial^+}\,(d^p\,\bar{d}_p - \bar{d}_p\,d^p\,). \tag{31}$$

It ensures that the chirality constraints are preserved

$$[\,j, d^m\,] \;=\; [\,j, \bar{d}_m\,] \;=\; 0. \tag{32}$$

The other kinematical generators are

$$j^+ \;=\; ix\partial^+, \qquad \bar{j}^+ \;=\; i\bar{x}\partial^+. \tag{33}$$

The rest of the generators must be specified separately for chiral and antichiral fields. Acting on ϕ, we have

$$j^{+-} \;=\; ix^-\,\partial^+ - \frac{i}{2}\,(\,\theta^p\bar{\partial}_p + \bar{\theta}_p\,\partial^p\,), \tag{34}$$

chosen so as to preserve the chiral combination

$$[\,j^{+-}, y^-\,] \;=\; -\,iy^-, \tag{35}$$

and such that its commutators with the chiral derivatives

$$[\,j^{+-}, d^m\,] \;=\; \frac{i}{2}\,d^m, \qquad [\,j^{+-}, \bar{d}_m\,] \;=\; \frac{i}{2}\,\bar{d}_m, \tag{36}$$

preserve chirality. Similarly the dynamical boosts are

$$
\begin{aligned}
j^- &= ix\frac{\partial\bar{\partial}}{\partial^+} - ix^-\partial + i\Big(\theta^p\bar{\partial}_p + \frac{i}{4\sqrt{2}\partial^+}(d^p\bar{d}_p - \bar{d}_p d^p)\Big)\frac{\partial}{\partial^+},\\
\bar{j}^- &= i\bar{x}\frac{\partial\bar{\partial}}{\partial^+} - ix^-\bar{\partial} + i\Big(\bar{\theta}_p\partial^p + \frac{i}{4\sqrt{2}\partial^+}(d^p\bar{d}_p - \bar{d}_p d^p)\Big)\frac{\bar{\partial}}{\partial^+}.
\end{aligned}
\tag{37}
$$

They do not commute with the chiral derivatives,

$$
[j^-, d^m] = \frac{i}{2}d^m\frac{\partial}{\partial^+}, \qquad [j^-, \bar{d}_m] = \frac{i}{2}\bar{d}_m\frac{\partial}{\partial^+}, \tag{38}
$$

but do not change the chirality of the fields on which they act. They satisfy the Poincaré algebra, in particular

$$
[j^-, \bar{j}^+] = -ij^{+-} - j, \qquad [j^-, j^{+-}] = ij^-. \tag{39}
$$

We can now follow the same path as we did in the last section to go over to a "second-quantized" version in terms of integrals over the superfield and then add interaction terms to the dynamical generators and try to close the algebra. In this way we can construct all the known supersymmetric field theories as different representations of various supersymmetry algebras with different values of λ and N. It is particularly interesting to study the cases $N = 4 \times integer$. For those values one can impose a further condition on the superfield ϕ namely the "inside out" condition

$$
\begin{aligned}
&\bar{d}_{m_1}\bar{d}_{m_2}\cdots\bar{d}_{m_{N/2-1}}\bar{d}_{m_{N/2}}\phi\\
&= \frac{1}{N/2!}\varepsilon_{m_1 m2\cdots m_{N/2}\cdots m_{N-1}m_N}\, d^{m_{N/2+1}}d^{m_{N/2+2}}\cdots d^{m_{N-1}}d^{m_N}\bar{\phi}.
\end{aligned}
\tag{40}
$$

We can now construct three-point interaction terms for any $\frac{N}{4}$ even in the dynamical generators. This is certainly a tedious exercise based on writing the most general terms in the interaction terms and then check the full algebra. The resulting action is [12]

$$
\begin{aligned}
S = &\int d^4x d^N\theta d^N\bar{\theta}\\
&\times\Big\{\bar{\phi}(x,\theta)\frac{\Box}{\partial^{+\frac{N}{2}}}\phi(x,\theta)\\
&+\frac{4g}{3}\Big[\sum_{n=0}^{\frac{N}{4}}(-1)^n\binom{\frac{N}{4}}{n}\frac{1}{\partial^{+N/2}}\bar{\phi}(x,\theta)\bar{\partial}^{\frac{N}{4}-n}\partial^{+n}\phi(x,\theta)\bar{\partial}^n\partial^{+\frac{N}{4}-n}\phi(x,\theta)\\
&\qquad+c.c.\Big]\Big\} + O(g^2).
\end{aligned}
\tag{41}
$$

When $\frac{N}{4}$ is odd, again the superfield has to transform as the adjoint representation of an external group with structure constants f^{abc}. The corresponding action is then

$$
\begin{aligned}
S = & \int d^4x d^N\theta d^N\bar{\theta} \\
& \times \Big\{ \bar{\phi}^a(x,\theta) \frac{\Box}{\partial^{+\frac{N}{2}}} \phi^a(x,\theta) \\
& + \frac{4g}{3} f^{abc} \Big[\sum_{n=0}^{\frac{N}{4}} (-1)^n \binom{\frac{N}{4}}{n} \frac{1}{\partial^{+N/2}} \bar{\phi}^a(x,\theta) \bar{\partial}^{\frac{N}{4}-n} \partial^{+n} \phi^b(x,\theta) \bar{\partial}^n \partial^{+\frac{N}{4}-n} \phi^c(x,\theta) \\
& + c.c. \Big] \Big\} + O(g^2).
\end{aligned} \tag{42}
$$

We note that we can construct theories with higher spin if $\frac{N}{4} > 2$. These are then very special combinations of the theories constructed in the previous section, with better quantum properties, since we know by experience that the more supersymmetry there is the better are the quantum properties.

3.1 Maximally Supersymmetric Yang–Mills Theory

The case $N = 4$ is especially interesting [3,4]. All the physical degrees of freedom are present in the superfield which can be expanded as

$$
\begin{aligned}
\phi(y) = & \frac{1}{\partial^+} A(y) + \frac{i}{\sqrt{2}} \theta^m \theta^n \overline{C}_{mn}(y) + \frac{1}{12} \theta^m \theta^n \theta^p \theta^q \varepsilon_{mnpq} \partial^+ \bar{A}(y) \\
& + \frac{i}{\partial^+} \theta^m \bar{\chi}_m(y) + \frac{\sqrt{2}}{6} \theta^m \theta^n \theta^p \varepsilon_{mnpq} \chi^q(y).
\end{aligned} \tag{43}
$$

The fields A and $\bar{A}$ constitute the two helicities of a vector field while the antisymmetric $SU(4)$ bi-spinors C_{mn} represent six scalar fields since they satisfy

$$
\overline{C}_{mn} = \frac{1}{2} \varepsilon_{mnpq} C^{pq}. \tag{44}
$$

The fermion fields are denoted by χ^m and $\bar{\chi}_m$. All have adjoint indices (not shown here), and are local fields in the modified light-cone coordinates. This is the maximal supersymmetric Yang–Mills theory. The full action is known [11]

$$
\begin{aligned}
\mathscr{S} = -\int d^4x \int d^4\theta\, d^4\bar{\theta} \Big\{ & \bar{\phi}^a \frac{\Box}{\partial^{+2}} \phi^a + \frac{4g}{3} f^{abc} \Big(\frac{1}{\partial^+} \bar{\phi}^a \phi^b \bar{\partial} \phi^c + \text{c.c.} \Big) \\
& - g^2 f^{abc} f^{ade} \Big(\frac{1}{\partial^+} (\phi^b \partial^+ \phi^c) \frac{1}{\partial^+} (\bar{\phi}^d \partial^+ \bar{\phi}^e) \\
& + \frac{1}{2} \phi^b \bar{\phi}^c \phi^d \bar{\phi}^e \Big) \Big\}.
\end{aligned} \tag{45}
$$

With this action it was shown [9] that the perturbation expansion is finite. There is no need for renormalization and the theory is very special. It is one of the cornerstones of modern particle physics. From the point of this lecture it appears as a very special representation of the superPoincaré algebra.

3.2 *Maximal Supergravity*

The next case is $N = 8$ [12]. In this case the superfield can be expanded as

$$\begin{aligned}\varphi(y) &= \frac{1}{\partial^{+2}}h(y) + i\theta^m \frac{1}{\partial^{+2}}\bar{\psi}_m(y) + i\theta^{mn}\frac{1}{\partial^+}\bar{A}_{mn}(y) \\ &\quad - \theta^{mnp}\frac{1}{\partial^+}\bar{\chi}_{mnp}(y) - \theta^{mnpr}C_{mnpr}(y) + i\tilde{\theta}^{(5)}_{mnp}\chi^{mnp}(y) \\ &\quad + i\tilde{\theta}^{(6)}_{mn}\partial^+ A^{mn}(y) + \tilde{\theta}^{(7)}_m \partial^+\chi^m(y) + \tilde{\theta}^{(8)}\partial^{+2}\bar{h}(y),\end{aligned} \tag{46}$$

where

$$\theta^{m_1\cdots m_n} \equiv \frac{1}{n!}\theta^{m_1}\cdots\theta^{m_n}, \quad \tilde{\theta}^{(n)}_{n_1\cdots n_{8-n}} \equiv \frac{1}{n!}\theta^{m_1\cdots m_n}\varepsilon_{m_1\cdots m_n n_1\cdots n_{8-n}}. \tag{47}$$

The helicity in the field goes from 2 to -2 and the theory has a spectrum comprised of a metric, 28 vector fields, 70 scalar fields, 56 spin one-half fields and 8 spin three-half fields. This theory is the maximal supergravity theory in $d = 4$. The action can be simplified [16] to

$$S = \int d^4x d^8\theta d^8\bar{\theta}\left\{\bar{\varphi}(x,\theta)\frac{\Box}{\partial^{+4}}\varphi(x,\theta) + \frac{3}{2}g\frac{1}{\partial^{+2}}\bar{\varphi}\;\bar{\partial}\varphi\;\bar{\partial}\varphi + c.c.\Big]\right\} + O(g^2). \tag{48}$$

The four-point coupling was finally found a few years ago [17]. It is however quite complicated and reflects the fact that the ∂^+-derivatives can be sprinkled out in very many ways. It is remarkable though, that the actions for the maximally supersymmetric Yang–Mills theory and supergravity theory are so similar. In some sense the supergravity theory is just an extension of the Yang–Mills one. In the modern particle physics these two theories are very intimately connected even though the direct physical consequences of them look quite different.

3.3 *The Hamiltonian as a Quadratic Form*

The two theories share also another unique property. We note that the free Hamiltonian for the $(\mathcal{N} = 4, d = 4)$ Yang–Mills Theory

$$H^0 = \int d^4x d^4\theta\, d^4\bar{\theta}\,\bar{\phi}^a\frac{2\partial\bar{\partial}}{\partial^{+2}}\phi^a, \tag{49}$$

can be rewritten as a quadratic form

$$H^0 = \frac{1}{2\sqrt{2}}(\mathscr{W}_0, \mathscr{W}_0), \tag{50}$$

using the inner product notation

$$(\phi, \xi) \equiv 2i \int d^4x d^4\theta\, d^4\bar{\theta}\, \bar{\phi}\, \frac{1}{\partial^+}\xi, \tag{51}$$

where ϕ and ξ are chiral superfields and

$$\mathscr{W}_0^a = \frac{\partial}{\partial^+}\bar{q}_+\phi^a, \tag{52}$$

is a fermionic superfield, the *free* dynamical supersymmetry variation of the superfield ($SU(4)$ spinor indices are summed over). The proof is straightforward, and requires integration by parts and the use of of the inside-out property of the superfields.

Also the fully interacting Hamiltonian [18] can be expressed as a quadratic form

$$H = \frac{1}{2\sqrt{2}}(\mathscr{W}^a, \mathscr{W}^a), \tag{53}$$

where now

$$\mathscr{W}^a = \frac{\partial}{\partial^+}\bar{q}_+\phi^a - g f^{abc}\frac{1}{\partial^+}(\bar{d}\phi^b\partial^+\phi^c), \tag{54}$$

is the complete (classical) dynamical supersymmetry variation. The power of supersymmetry allows for this simple rewriting of the fully interacting Hamiltonian.

The same can be shown to be true also for the $(\mathscr{N} = 8, d = 4)$ supergravity [17] and was an important clue to find the four-point coupling. We have not found any other theory with this property which again renders the two models to behave very similarly and to have unique properties.

Note that this is not the same as the statement that the Hamiltonian satisfies

$$H = -2\sqrt{2}\{Q_-^m, \bar{Q}_{-m}\}. \tag{55}$$

4 Light-Frame Formulations of Higher Dimensional Theories

The procedure to find representations of the Poincaré algebra that we have followed in the previous section can, of course, be extended to field theories in dimensions of space–time higher than four. The covariant subalgebra which will be linearly realized is then $SO(d-2)$, so the physical fields will be representations of this algebra and hence characterized by these representations like we used helicity to distinguish the physical fields in four dimensions. If we just implement Poincaré invariance as in Sect. 2 we can, in principle, find all the possible field theories. However, the procedure gets easily tedious and furthermore there are few interesting quantum field theories in higher dimensions because of the renormalization problems. The only ones that are discussed are supersymmetric field theories since they are connected to the superstring Theory. The ones that we have been interested in are the ones which lead to interesting field theories when compactified to four dimensions, so let us concentrate on those. The ones I will discuss here are ten-dimensional SuperYang–Mills and 11-dimensional supergravity which under compactification leads to the maximal theories discussed above.

4.1 Ten-Dimensional SuperYang–Mill Theory

The physical degrees of freedom of this theory are $\mathbf{8_v}$ and an $\mathbf{8_s}$. If we insist that the superfield should be a representation of the transverse $SO(8)$ it must be in one of the representations above. Since the natural spinor coordinate will also be an $\mathbf{8_s}$, such a superfield must include 8×2^8 components and must hence be very strongly restricted. Such a formalism has been developed [19], but it is not clear that the formalism is useful. Also it is not easily generalizable to the 11-dimensional case. Instead I will describe a recent procedure developed in [20].

The idea is to use the same superfield as in four dimensions. In order to do that we have to sacrifice the explicit covariance under $SO(8)$ and use the decomposition

$$SO(8) \supset SO(2) \times SO(6). \tag{56}$$

Since $SO(6) \sim SU(4)$ we can identify the $SU(4)$ as the external symmetry group in the superfield equation (46). The remaining symmetry $SO(8)/(SO(6) \times SO(2))$ will transform among the components of the superfield. First of all, the transverse light-cone space variables need be generalized to eight. We stick to the representations used in the superfield, and introduce the six extra coordinates and their derivatives as antisymmetric bi-spinors

$$x^{m4} = \frac{1}{\sqrt{2}}(x_{m+3} + i x_{m+6}), \qquad \partial^{m4} = \frac{1}{\sqrt{2}}(\partial_{m+3} + i\partial_{m+6}), \tag{57}$$

for $m \neq 4$, and their complex conjugates

$$\bar{x}_{pq} = \frac{1}{2}\varepsilon_{pqmn} x^{mn}; \qquad \bar{\partial}_{pq} = \frac{1}{2}\varepsilon_{pqmn}\partial^{mn}. \tag{58}$$

Their derivatives satisfy

$$\bar{\partial}_{mn} x^{pq} = (\delta_m{}^p \delta_n{}^q - \delta_m{}^q \delta_n{}^p); \qquad \partial^{mn} \bar{x}_{pq} = (\delta^m{}_p \delta^n{}_q - \delta^m{}_q \delta^n{}_p), \tag{59}$$

and

$$\partial^{mn}\, x^{pq} \;=\; \frac{1}{2}\,\varepsilon^{pqrs}\,\partial^{mn}\,\bar{x}_{rs} \;=\; \varepsilon^{mnpq}. \tag{60}$$

There are then no modifications to be made to the chiral superfield, except for the dependence on the extra coordinates

$$A(y) \;=\; A(x,\bar{x},x^{mn},\bar{x}_{mn},y^{-}),\;\; \text{etc.} \tag{61}$$

These extra variables will be acted on by new operators that generate the higher-dimensional symmetries.

4.2 The SuperPoincaré Algebra in Ten Dimensions

The SuperPoincaré algebra needs to be generalized from the form in four dimensions. One starts with the construction of the $SO(8)$ little group using the decomposition $SO(8) \supset SO(2) \times SO(6)$. The $SO(2)$ generator is the same; the $SO(6) \sim SU(4)$ generators are given by

$$\begin{aligned} j^{m}{}_{n} \;=\; & \frac{1}{2}\,(x^{mp}\,\bar{\partial}_{pn} - \bar{x}_{pn}\,\partial^{mp}\,) - \theta^{m}\,\bar{\partial}_{n} + \bar{\theta}_{n}\,\partial^{m} + \frac{1}{4}\,(\,\theta^{p}\,\bar{\partial}_{p} - \bar{\theta}_{p}\,\partial^{p}\,)\,\delta^{m}{}_{n} \\ & + \frac{i}{2\sqrt{2}\,\partial^{+}}\,(d^{m}\,\bar{d}_{n} - \bar{d}_{n}\,d^{m}\,) + \frac{i}{8\sqrt{2}\,\partial^{+}}\,(d^{p}\,\bar{d}_{p} - \bar{d}_{p}\,d^{p}\,)\,\delta^{m}{}_{n}. \end{aligned} \tag{62}$$

Note that we use the same spinors as in four dimensions because of the decomposition $SO(8) \supset SO(2) \times SO(6)$, where $SO(6) \sim SU(4)$. The extra terms with the d and $\bar{d}$ operators are not necessary for closure of the algebra. However they insure that the generators commute with the chiral derivatives. They satisfy the commutation relations

$$\left[\,j\,,\,j^{m}{}_{n}\,\right] \;=\; 0, \qquad \left[\,j^{m}{}_{n}\,,\,j^{p}{}_{q}\,\right] \;=\; \delta^{m}{}_{q}\,j^{p}{}_{n} - \delta^{p}{}_{n}\,j^{m}{}_{q}. \tag{63}$$

The remaining $SO(8)$ generators lie in the coset $SO(8)/(SO(2) \times SO(6))$

$$\begin{aligned} j^{pq} &= x\,\partial^{pq} - x^{pq}\,\partial + \frac{i}{\sqrt{2}}\,\partial^{+}\,\theta^{p}\,\theta^{q} - i\,\sqrt{2}\,\frac{1}{\partial^{+}}\,\partial^{p}\,\partial^{q} + \frac{i}{\sqrt{2}\,\partial^{+}}\,d^{p}\,d^{q}, \\ \bar{j}_{mn} &= \bar{x}\,\bar{\partial}_{mn} - \bar{x}_{mn}\,\bar{\partial} + \frac{i}{\sqrt{2}}\,\partial^{+}\,\bar{\theta}_{m}\,\bar{\theta}_{n} - i\,\sqrt{2}\,\frac{1}{\partial^{+}}\,\bar{\partial}_{m}\,\bar{\partial}_{n} + \frac{i}{\sqrt{2}\,\partial^{+}}\,\bar{d}_{m}\,\bar{d}_{n}. \end{aligned} \tag{64}$$

All $SO(8)$ transformations are specially constructed so as not to mix chiral and antichiral superfields,

$$[\,j^{mn}\,,\,\bar{d}_{p}\,] \;=\; 0; \qquad [\,\bar{j}_{mn}\,,\,d^{p}\,] \;=\; 0, \tag{65}$$

and satisfy the $SO(8)$ commutation relations

$$\Big[j, j^{mn}\Big] = j^{mn}, \qquad \Big[j, \bar{j}_{mn}\Big] = -\bar{j}_{mn},$$
$$\Big[j^m{}_n, j^{pq}\Big] = \delta^q{}_n\, j^{mp} - \delta^p{}_n\, j^{mq}, \qquad \Big[j^m{}_n, \bar{j}_{pq}\Big] = \delta^m{}_q\, \bar{j}_{np} - \delta^m{}_p\, \bar{j}_{nq},$$
$$\Big[j^{mn}, \bar{j}_{pq}\Big] = \delta^m{}_p j^n{}_q + \delta^n{}_q j^m{}_p - \delta^n{}_p j^m{}_q - \delta^m{}_q j^n{}_p - (\delta^m{}_p\delta^n{}_q - \delta^n{}_p\delta^m{}_q)\, j.$$

Rotations between the 1 or 2 and 4 through 9 directions induce on the chiral fields the changes

$$\delta\phi = \Big(\frac{1}{2}\,\omega_{mn}\, j^{mn} + \frac{1}{2}\,\bar{\omega}^{mn}\,\bar{j}_{mn}\Big)\,\phi, \tag{66}$$

where complex conjugation is like duality

$$\bar{\omega}_{pq} = \frac{1}{2}\,\varepsilon_{mnpq}\,\omega^{mn}. \tag{67}$$

For example, a rotation in the 1–4 plane through an angle θ corresponds to taking $\theta = \omega_{14} = \omega_{23}$ ($= \omega^{23} = \omega^{14}$ by reality), all other components being zero. Finally, we verify that the kinematical supersymmetries are duly rotated by these generators

$$[\,j^{mn}, \bar{q}_{+\,p}] = \delta^n{}_p\, q^m_+ - \delta^m{}_p\, q^n_+; \qquad [\,\bar{j}_{mn}, q^p_+] = \delta_n{}^p\, \bar{q}_{+m} - \delta_m{}^p\, \bar{q}_{+n}. \tag{68}$$

We now use the $SO(8)$ generators to construct the SuperPoincaré generators

$$j^+ = ix\partial^+; \qquad \bar{j}^+ = i\bar{x}\partial^+$$
$$j^{+\,mn} = ix^{mn}\,\partial^+; \qquad \bar{j}^+{}_{mn} = i\bar{x}_{mn}\,\partial^+. \tag{69}$$

The dynamical boosts are now

$$j^- = ix\frac{\partial\bar{\partial} + \frac{1}{4}\bar{\partial}_{pq}\,\partial^{pq}}{\partial^+} - ix^-\,\partial + i\frac{\partial}{\partial^+}\Big\{\,\theta^m\,\bar{\partial}_m + \frac{i}{4\sqrt{2}\,\partial^+}\,(d^p\,\bar{d}_p - \bar{d}_p d^p)\,\Big\}$$
$$-\frac{1}{4}\,\frac{\bar{\partial}_{pq}}{\partial^+}\Big\{\,\frac{\partial^+}{\sqrt{2}}\,\theta^p\,\theta^q - \frac{\sqrt{2}}{\partial^+}\,\partial^p\,\partial^q + \frac{1}{\sqrt{2}\partial^+} d^p\,d^q\,\Big\}, \tag{70}$$

and its conjugate

$$\bar{j}^- = i\bar{x}\frac{\partial\bar{\partial} + \frac{1}{4}\bar{\partial}_{pq}\,\partial^{pq}}{\partial^+} - ix^-\,\bar{\partial} + i\frac{\bar{\partial}}{\partial^+}\Big\{\,\bar{\theta}_m\,\partial^m + \frac{i}{4\sqrt{2}\,\partial^+}\,(d^p\,\bar{d}_p - \bar{d}_p d^p)\,\Big\}$$
$$-\frac{1}{4}\,\frac{\partial^{pq}}{\partial^+}\Big\{\,\frac{\partial^+}{\sqrt{2}}\,\bar{\theta}_p\,\bar{\theta}_q - \frac{\sqrt{2}}{\partial^+}\,\bar{\partial}_p\,\bar{\partial}_q + \frac{1}{\sqrt{2}\partial^+}\bar{d}_p\,\bar{d}_q\,\Big\}. \tag{71}$$

The others are obtained by using the $SO(8)/(SO(2)\times SO(6))$ rotations

$$j^{-mn} = [j^-, j^{mn}]; \qquad \bar{j}^-{}_{mn} = [\bar{j}^-, \bar{j}_{mn}]. \tag{72}$$

We do not show their explicit forms as they are too cumbersome. The four supersymmetries in four dimensions turn into one supersymmetry in ten dimensions. In our notation, the kinematical supersymmetries q^n_+ and $\bar{q}_{+n}$, are assembled into one $SO(8)$ spinor. The dynamical supersymmetries are obtained by boosting

$$i[\bar{j}^-, q^m_+] \equiv \mathcal{Q}^m, \qquad i[j^-, \bar{q}_{+m}] \equiv \overline{\mathcal{Q}}_m, \tag{73}$$

where

$$\begin{aligned} \mathcal{Q}^m &= \frac{\bar{\partial}}{\partial^+} q_+{}^m + \frac{1}{2}\frac{\partial^{mn}}{\partial^+}\bar{q}_{+n}, \\ \overline{\mathcal{Q}}_m &= \frac{\partial}{\partial^+}\bar{q}_{+m} + \frac{1}{2}\frac{\bar{\partial}_{mn}}{\partial^+} q^{\,n}_+. \end{aligned} \tag{74}$$

They satisfy the supersymmetry algebra

$$\{\mathcal{Q}^m, \overline{\mathcal{Q}}_n\} = i\sqrt{2}\,\delta^m_n \frac{1}{\partial^+}\left(\partial\bar{\partial} + \frac{1}{4}\bar{\partial}_{pq}\partial^{pq}\right), \tag{75}$$

and can be obtained from one another by $SO(8)$ rotations, as

$$\frac{1}{2}\varepsilon_{pqmn}\,[\,j^{pq}, \mathcal{Q}^m] = 4\overline{\mathcal{Q}}_n, \tag{76}$$

while

$$[\bar{j}_{pq}, \mathcal{Q}^m] = 0. \tag{77}$$

Note also that

$$\{\mathcal{Q}^m, q^n_+\} = \frac{i}{\sqrt{2}}\partial^{mn}, \tag{78}$$

4.3 The Generalized Derivatives

The cubic interaction in the $N = 4$ Lagrangian contains explicitly the derivative operators ∂ and $\bar{\partial}$. To achieve covariance in ten dimensions, these must be generalized. Remarkably the only change we have to is to introduce the following operator

$$\overline{\nabla} \equiv \bar{\partial} + \frac{i}{4\sqrt{2}\partial^+}\bar{d}_p\bar{d}_q\partial^{pq}, \tag{79}$$

which naturally incorporates the rest of the derivatives ∂^{pq}, with α as an arbitrary parameter. After some algebra, we find that $\overline{\nabla}$ is covariant under $SO(8)$ transformations. We define its rotated partner as

$$\nabla^{mn} \equiv \left[\overline{\nabla}, j^{mn}\right], \tag{80}$$

where

$$\nabla^{mn} = \partial^{mn} - \frac{i}{4\sqrt{2}\partial^+} \bar{d}_r \bar{d}_s \varepsilon^{mnrs} \partial. \tag{81}$$

If we apply to it to the inverse transformation, it goes back to the original form

$$\left[\bar{j}_{pq}, \nabla^{mn}\right] = (\delta_p{}^m \delta_q{}^n - \delta_q{}^m \delta_p{}^n) \overline{\nabla}, \tag{82}$$

and these operators transform under $SO(8)/(SO(2) \times SO(6))$, and $SO(2) \times SO(6)$ as the components of an 8-vector.

We introduce the conjugate operator ∇ by requiring that

$$\nabla \bar{\phi} \equiv \overline{(\overline{\nabla} \phi)}, \tag{83}$$

with

$$\nabla \equiv \partial + \frac{i}{4\sqrt{2}\partial^+} d^p d^q \bar{\partial}_{pq}. \tag{84}$$

Define

$$\overline{\nabla}_{mn} \equiv \left[\nabla, \bar{j}_{mn}\right], \tag{85}$$

which is given by

$$\overline{\nabla}_{mn} = \bar{\partial}_{mn} - \frac{i}{4\sqrt{2}\partial^+} d^r d^s \varepsilon_{mnrs} \bar{\partial}. \tag{86}$$

We then verify that

$$\left[j^{mn}, \overline{\nabla}_{pq}\right] = (\delta_p{}^m \delta_q{}^n - \delta_q{}^m \delta_p{}^n) \nabla. \tag{87}$$

The kinetic term is trivially made $SO(8)$-invariant by including the six extra transverse derivatives in the d'Alembertian. The quartic interactions are obviously invariant since they do not contain any transverse derivative operators. Hence we need only consider the cubic vertex. In the paper [20] it is shown that to achieve covariance in ten dimensions, it suffices indeed to replace the transverse ∂ and $\bar{\partial}$ by ∇ and $\overline{\nabla}$, respectively. This is done by checking the invariance under the little group $SO(8)$. Together with the result from four dimensions this is enough to warrant invariance under the full superPoincaré group in ten dimensions. The full action is then

$$\begin{aligned}\mathscr{S} = -\int d^4x \int d^4\theta\, d^4\bar{\theta} \Big\{ &\bar{\phi}^a \frac{\Box}{\partial^{+2}} \phi^a + \frac{4g}{3} f^{abc} \Big(\frac{1}{\partial^+} \bar{\phi}^a \phi^b \overline{\nabla} \phi^c + \text{c.c.} \Big) \\ &- g^2 f^{abc} f^{ade} \Big(\frac{1}{\partial^+} (\phi^b \partial^+ \phi^c) \frac{1}{\partial^+} (\bar{\phi}^d \partial^+ \bar{\phi}^e) \\ &+ \frac{1}{2} \phi^b \bar{\phi}^c \phi^d \bar{\phi}^e \Big) \Big\}. \end{aligned} \tag{88}$$

This action is suitable in order to investigate the perturbative properties of the theory. It is, of course, non-renormalizable but has still remarkable properties that Nature might use. One can also study possible higher symmetries of this action.

4.4 Eleven-Dimensional Supergravity

$N=1$ Supergravity in 11 dimensions, contains three different massless fields, two bosonic (gravity and a three-form) and one Rarita–Schwinger spinor. Its physical degrees of freedom are classified in terms of the transverse little group, $SO(9)$, with the graviton $G^{(MN)}$, transforming as a symmetric second-rank tensor, the three-form $B^{[MNP]}$ as an anti-symmetric third-rank tensor and the Rarita–Schwinger field as a spinor-vector, Ψ^M ($M,N,\ldots$ are $SO(9)$ indices). This theory on reduction to four dimensions leads to the maximally supersymmetric $N=8$ theory.

In order to use the formalism and especially the superfield equation (46) developed in four dimensions for the maximally supersymmetric $N=8$ theory we have to decompose

$$SO(9) \supset SO(2) \times SO(7). \tag{89}$$

The $SO(7)$ symmetry can in fact be upgraded to an $SU(8)$ symmetry. However, it is important to remember that it is really the $SO(7)$ which is relevant when we "oxidize" the theory to $d=11$ and the coordinates θ^m and $\bar{\theta}_n$ used in the four-dimensional case will now be interpreted as spinors under $SO(7)\times SO(2)$. To distinguish this we will change the notation m, n to α, β for the spinors and use the notation a, b for the vector indices of $SO(7)$.

The first step is to generalize the transverse variables to nine. In the Yang–Mills case, the compactified $SO(6)$ was easily described by $SU(4)$ parameters and we made use of the convenient bi-spinor notation. In the present case, the compactified $SO(7)$ has no equivalent unitary group so we simply introduce additional real coordinates, x^a and their derivatives ∂^a(where a runs from 4 through 10). The chiral superfield remains unaltered, except for the added dependence on the extra coordinates

$$h(y) = h(x,\bar{x},x^a,y^-), \text{ etc.} \tag{90}$$

These extra variables will be acted on by new operators that will restore the higher-dimensional symmetries.

4.5 The SuperPoincaré Algebra in 11 Dimensions

The SuperPoincaré algebra needs to be generalized from its four-dimensional version. The $SO(2)$ generators stay the same and we propose generators of the coset $SO(9)/(SO(2)\times SO(7))$, of the form,

$$j^a = -i(x\partial^a - x^a\partial) + \frac{i}{2\sqrt{2}}\partial^+\,\theta^\alpha\,(\gamma^a)_{\alpha\beta}\,\theta^\beta - \frac{i}{\sqrt{2}\,\partial^+}\,\partial^\alpha\,(\gamma^a)_{\alpha\beta}\,\partial^\beta + \frac{i}{2\sqrt{2}\,\partial^+}\,d^\alpha\,(\gamma^a)_{\alpha\beta}\,d^\beta \tag{91}$$

$$\bar{j}^{\,b} = -i(\bar{x}\partial^b - x^b\bar{\partial}) + \frac{i}{2\sqrt{2}}\partial^+\,\bar{\theta}_\alpha\,(\gamma^b)^{\alpha\beta}\,\bar{\theta}_\beta - \frac{i}{\sqrt{2}\,\partial^+}\,\bar{\partial}_\alpha\,(\gamma^b)^{\alpha\beta}\,\bar{\partial}_\beta + \frac{i}{2\sqrt{2}\,\partial^+}\,\bar{d}_\alpha\,(\gamma^b)^{\alpha\beta}\,\bar{d}_\beta \tag{92}$$

which satisfy the $SO(9)$ commutation relations,

$$\begin{aligned}
\left[j, j^a\right] &= j^a, \qquad \left[j, \bar{j}^b\right] = -\bar{j}^b \\
\left[j^{cd}, j^a\right] &= \delta^{ca}\,j^d - \delta^{da}\,j^c \\
\left[j^a, \bar{j}^b\right] &= i\,j^{ab} + \delta^{ab}\,j,
\end{aligned} \tag{93}$$

where j is the same as before, and the $SO(7)$ generators read,

$$j^{ab} = -i(x^a\partial^b - x^b\partial^b) + \theta^\alpha\,(\gamma^a)^{\alpha\beta}\,(\gamma^b)^{\beta\sigma}\,\bar{\partial}_\sigma + \bar{\theta}_\alpha\,(\gamma^a)^{\alpha\beta}\,(\gamma^b)^{\beta\sigma}\,\partial^\sigma - \frac{1}{\sqrt{2}\,\partial^+}\,d^\alpha\,(\gamma^a)^{\alpha\beta}\,(\gamma^b)^{\beta\sigma}\,\bar{d}_\sigma. \tag{94}$$

The full $SO(9)$ transverse algebra is generated by j, j^{ab}, j^a and $\bar{j}^b$. All rotations are specially constructed to preserve chirality. For example,

$$[j^a, \bar{d}_\alpha] = 0; \qquad [\bar{j}^b, d^\alpha] = 0. \tag{95}$$

The remaining kinematical generators do not get modified,

$$j^+ = j^+, \qquad j^{+-} = j^{+-}, \tag{96}$$

while new kinematical generators appear,

$$j^{+a} = i x^a\,\partial^+; \qquad \bar{j}^{+b} = i\bar{x}^b\,\partial^+. \tag{97}$$

We generalize the linear part of the dynamical boosts to,

$$j^- = ix\frac{\partial\bar{\partial} + \frac{1}{2}\partial^a\partial^a}{\partial^+} - ix^-\partial + i\frac{\partial}{\partial^+}\left\{\theta^\alpha\,\bar{\partial}_\alpha + \frac{i}{4\sqrt{2}\,\partial^+}(d^\alpha\,\bar{d}_\alpha - \bar{d}_\alpha\,d^\alpha)\right\} - \frac{1}{4}\frac{\partial^a}{\partial^+}\left\{\partial^+\,\theta^\alpha\,(\gamma^a)_{\alpha\beta}\,\theta^\beta - \frac{2}{\partial^+}\,\partial^\alpha\,(\gamma^a)_{\alpha\beta}\,\partial^\beta + \frac{1}{\partial^+}\,d^\alpha\,(\gamma^a)_{\alpha\beta}\,d^\beta\right\}. \tag{98}$$

The other boosts may be obtained by using the $SO(9)/(SO(2)\times SO(7))$ rotations,

$$j^{-a} = [j^-, j^a]; \qquad \bar{j}^{-b} = [\bar{j}^-, \bar{j}^b]. \tag{99}$$

We do not show their explicit forms as they are too cumbersome. The dynamical supersymmetries are obtained by boosting

$$\begin{aligned}
[j^-, \bar{q}_{+\eta}] &\equiv \overline{\mathcal{Q}}_\eta = -i\,\frac{\partial}{\partial^+}\bar{q}_{+\eta} - \frac{i}{\sqrt{2}}(\gamma^b)_{\eta\rho}\,q_+^{\rho}\,\frac{\partial^b}{\partial^+},\\
[\bar{j}^-, q_+^{\alpha}] &\equiv \mathcal{Q}^\alpha = i\,\frac{\bar{\partial}}{\partial^+}q_+{}^{\alpha} + \frac{i}{\sqrt{2}}(\gamma^a)^{\alpha\beta}\,\bar{q}_{+\beta}\,\frac{\partial^a}{\partial^+}.
\end{aligned} \tag{100}$$

They satisfy,

$$\{\mathcal{Q}^\alpha, q_+^\eta\} = -(\gamma^a)^{\alpha\eta}\,\partial^a, \tag{101}$$

and the supersymmetry algebra,

$$\{\mathcal{Q}^\alpha, \overline{\mathcal{Q}}_\eta\} = i\sqrt{2}\,\delta^\alpha{}_\eta\,\frac{1}{\partial^+}\left(\partial\bar{\partial} + \frac{1}{2}\partial^a\partial^a\right). \tag{102}$$

Having constructed the free $N = 1$ SuperPoincaré generators in 11 dimensions which act on the chiral superfield, we turn to building the interacting theory.

4.6 The Generalized Derivatives

The cubic interaction in the $N = 8$ Lagrangian explicitly contains the transverse derivative operators ∂ and $\bar{\partial}$. To achieve covariance in 11 dimensions, we proceed to generalize these operators as we did for $N = 4$ Yang–Mills. We found the generalized derivative

$$\overline{\nabla} = \bar{\partial} - \frac{\sqrt{2}}{16}\,\bar{d}_\alpha\,(\gamma^a)^{\alpha\beta}\,\bar{d}_\beta\,\frac{\partial^a}{\partial^+}, \tag{103}$$

which naturally incorporates the coset derivatives ∂^m. We use the coset generators to produce its rotated partner $\overline{\nabla}$ by,

$$[\overline{\nabla}, j^a] \equiv \nabla^a = -i\partial^a - \frac{i\sqrt{2}}{16}\,\bar{d}_\alpha\,(\gamma^a)^{\alpha\beta}\,\bar{d}_\beta\,\frac{\partial}{\partial^+}. \tag{104}$$

It remains to verify that the original derivative operator is reproduced by undoing this rotation; indeed we find the required closure,

$$[\nabla^a, \bar{j}^b] = \delta^{ab}\,\overline{\nabla}$$

The new derivative ($\overline{\nabla}$, ∇^a), thus transforms as a 9-vector under the little group in 11 dimensions. W. We define the conjugate derivative ∇, by requiring that

$$\nabla\,\bar{\varphi} \;\equiv\; \overline{(\overline{\nabla}\,\varphi)}. \tag{105}$$

This tells us that,

$$\nabla \;\equiv\; \partial - \frac{\sqrt{2}}{16}\,d^{\alpha}\,(\gamma^{b})^{\alpha\beta}\,d^{\beta}\,\frac{\partial^{b}}{\partial^{+}} \tag{106}$$

This construction is akin to that for the $N=4$ Yang–Mills theory, but this time it applies to the "oxidation" of the ($N=8$, $d=4$) theory to ($N=1$, $d=11$) supergravity. This points to remarkable algebraic similarities between the two theories, with possibly profound physical consequences. It remains to show that the simple replacement of the transverse derivatives $\partial, \bar{\partial}$ by $\nabla, \overline{\nabla}$ in the ($N=8$, $d=4$) interacting theory yields the fully covariant Lagrangian in 11 dimensions.

This can be done by checking the invariance under the little group $SO(9)$. This is a very tedious exercise which was done in paper [16]. Indeed it is possible to show that the three-point coupling is invariant and that the 11-dimensional supergravity theory can be written as

$$S=\int d^{10}x d^{8}\theta d^{8}\bar{\theta}\left\{\bar{\varphi}(x,\theta)\frac{\Box}{\partial^{+4}}\varphi(x,\theta)+\,\frac{3}{2}g\left[\frac{1}{\partial^{+2}}\,\overline{\varphi}\;\overline{\nabla}\varphi\,\overline{\nabla}\varphi+c.c.\right]\right\}+O(g^{2}). \tag{107}$$

We have not yet shown that this also works for the four-point coupling. With it one can study various properties of this theory, such as the one-loop graphs. They will diverge but there might be ways to add more fields to get convergent answer. This is one long term goal of this project. One can also study the symmetries of the action. It is clear that the action is quite unique and has a profound rôle in modern particle physics and any symmetry that can be found for this action is a genuine physical symmetry. This theory is also the low-energy limit of the mystic M-theory which is supposed to be the underlying theory to all string theories. This theory is shrouded in mystery and any attempt to better understand the supergravity theory can help us eventually understand M-theory.

5 Exceptional Symmetries in Maximal Supergravity

In their original work on $\mathcal{N}=8$ supergravity [1,2] Cremmer and Julia found an on-shell, $E_{7(7)}$ duality symmetry. It is therefore natural to ask if this symmetry can be exploited to bring simplicity to the quartic and higher-order interactions of $\mathcal{N}=8$ supergravity. In a recent paper [21], as a first step in this direction, we have shown how to exploit this symmetry to construct the light-cone Hamiltonian to order κ^2. Our resulting expression is remarkably simpler than that formulation of the same

Hamiltonian with over ninety terms [17]. In this process we also got the $E_{7(7)}$ transformations to lowest order for all the fields in the theory. After a brief review o f$E_{7(7)}$ duality in the covariant formalism with the scalar and field strengths alone, I will express the action of $E_{7(7)}$ in the LC_2 formalism. The explicit non-linear $E_{7(7)}$ action on the scalars and vector potentials to lowest non-trivial order in κ are derived in this gauge; they stand as the starting point for our analysis.

5.1 Covariant $\mathscr{N}=8$ Supergravity

$\mathscr{N}=8$ Supergravity contains a graviton $h_{\mu\nu}$ and its 8 gravitinos ψ^i_μ interacting with matter composed of 28 vectors $A^{[ij]}_\mu$, 56 spinors $\chi^{[ijk]}$, and 70 scalars $C^{[ijkl]}$, labelled with $SO(8)$ indices, $i,j,k,l=1,2,\ldots,8$. The much larger Cremmer–Julia $E_{7(7)}$ symmetry acts on the scalars and the field strengths, and we begin with the manifestly $SO(8)$ symmetric order-κ^2 $\mathscr{N}=8$ supergravity Lagrangian [22] with those fields only. The scalar part is given by

$$\mathscr{L}_S = -\frac{1}{48}\left\{\partial_\mu C^{ijkl}\,\partial^\mu\overline{C}^{ijkl} + \frac{\kappa^2}{2}C^{ijkl}\,\overline{C}^{klmn}\,\partial_\mu C^{mnpq}\,\partial^\mu\overline{C}^{pqij} + \mathscr{O}(\kappa^3)\right\}, \quad (108)$$

where the scalar fields satisfy

$$C^{ijkl} = \frac{1}{4!}\,\varepsilon^{ijklmnpq}\,\overline{C}^{mnpq}. \quad (109)$$

The Lagrangian with the field strengths is given by

$$\mathscr{L}_V = -\frac{1}{8}\,\mathscr{F}^{ij}_{\mu\nu}\,\mathscr{G}^{\mu\nu ij} + c.c. \quad (110)$$

written in terms of the self-dual complex field strengths

$$\mathscr{F}^{\mu\nu ij} = \frac{1}{2}F^{\mu\nu ij} + \frac{i}{2}\widetilde{F}^{\mu\nu ij}, \quad (111)$$

and

$$\mathscr{G}^{\mu\nu ij} = \mathscr{F}^{\mu\nu ij} + \kappa\overline{C}^{ijkl}\,\mathscr{F}^{\mu\nu kl} + \frac{\kappa^2}{2}\overline{C}^{ijkl}\overline{C}^{klmn}\,\mathscr{F}^{\mu\nu mn} + O(\kappa^2), \quad (112)$$

is linear in the field strengths.

The electro-magnetic duality transformations exchange equations of motion

$$\partial_\mu\left(\mathscr{G}^{\mu\nu ij} + \overline{\mathscr{G}}^{\mu\nu ij}\right) = 0, \quad (113)$$

for Bianchi identities

$$\partial_\mu\left(\mathscr{F}^{\mu\nu ij}-\overline{\mathscr{F}}^{\mu\nu ij}\right) = 0. \tag{114}$$

These equations are manifestly $SO(8)$ covariant. We can elevate this symmetry to $SU(8)$ [23] on the complex field strengths by demanding

$$\delta\left(\mathscr{G}^{\mu\nu ij}+\mathscr{F}^{\mu\nu ij}\right) = \left(R^{ik}+iS^{ik}\right)\left(\mathscr{G}^{\mu\nu kj}+\mathscr{F}^{\mu\nu kj}\right)-(i\leftrightarrow j), \tag{115}$$

transforming as **28**, while the other combinations $(\mathscr{G}^{\mu\nu ij}-\mathscr{F}^{\mu\nu ij})$ transform as the complex conjugate $\overline{\mathbf{28}}$,

$$\delta\left(\mathscr{G}^{\mu\nu ij}-\mathscr{F}^{\mu\nu ij}\right) = \left(R^{ik}-iS^{ik}\right)\left(\mathscr{G}^{\mu\nu kj}-\mathscr{F}^{\mu\nu kj}\right)-(i\leftrightarrow j). \tag{116}$$

where R^{ij} are the 28 *real* antisymmetric rotation tensors which generate $SO(8)$, and S^{ij} are 35 *real* symmetric traceless matrices in the coset $SU(8)/SO(8)$. The $SU(8)/SO(8)$ coset transformations δ' on the complex field strengths

$$\delta'\mathscr{F}^{\mu\nu ij} = iS^{ik}\mathscr{G}^{\mu\nu kj}-(i\leftrightarrow j), \qquad \delta'\mathscr{G}^{\mu\nu ij} = iS^{ik}\mathscr{F}^{\mu\nu kj}-(i\leftrightarrow j), \tag{117}$$

are the duality transformations which map the equations of motion into the Bianchi identities and vice versa

$$\delta'\left\{\partial_\mu(\mathscr{G}^{\mu\nu ij}+\overline{\mathscr{G}}^{\mu\nu ij})\right\} = iS^{ik}\,\partial_\mu(\mathscr{F}^{\mu\nu kj}-\overline{\mathscr{F}}^{\mu\nu kj}) - (i\leftrightarrow j). \tag{118}$$

The $SU(8)/SO(8)$ transformations are only symmetries of the equations of motion and the Bianchi identities, but not of the Lagrangian.

Consistency of the coset variation of this expression with the two variations of (117) requires that the scalar fields transform linearly under the full $SU(8)$, that is

$$\delta'\overline{C}^{ijkl} = -iS^{im}\overline{C}^{mjkl} - (i\leftrightarrow j)-(i\leftrightarrow k)-(i\leftrightarrow l), \tag{119}$$

i.e., as a **70**. This is an exact equation with no order κ corrections. It follows that the scalar Lagrangian (108) is $SU(8)$ invariant. On the other hand, the complex field strengths have more complicated non-linear coset transformation

$$\delta'\mathscr{F}^{\mu\nu ij} = iS^{im}\left(\mathscr{F}^{\mu\nu mj}+\kappa\overline{C}^{mjkl}\mathscr{F}^{\mu\nu kl}+\frac{\kappa^2}{2}\overline{C}^{mjkl}\overline{C}^{klpq}\mathscr{F}^{\mu\nu pq}+O(\kappa^3)\right) \\ -(i\leftrightarrow j). \tag{120}$$

The terms on the right-hand-side transform differently order by order in κ: $\mathscr{F}^{\mu\nu mj}\sim\mathbf{28}$, while $\overline{C}^{mjkl}\mathscr{F}^{\mu\nu kl}\sim\overline{\mathbf{28}}$, and the order κ^2 term has even more complicated coset transformations. Yet, one can check that the commutator of two such variations closes on $SO(8)$ transformation, as required. The extension to $SU(8)$ duality on the field strengths is meaningful only in the interacting case when $\kappa \neq 0$, since $\mathscr{G}^{\mu\nu ij} - \mathscr{F}^{\mu\nu ij} = \mathscr{O}(\kappa)$.

Cremmer and Julia extended the duality symmetries to the non-compact $E_{7(7)}$. Assemble the complex field strengths in one column vector with 56 complex entries [24].

$$Z^{\mu\nu} = \begin{pmatrix} \mathscr{G}^{\mu\nu ij} + \mathscr{F}^{\mu\nu ij} \\ \mathscr{G}^{\mu\nu ij} - \mathscr{F}^{\mu\nu ij} \end{pmatrix} \equiv \begin{pmatrix} X^{\mu\nu ab} \\ Y^{\mu\nu}{}_{ab} \end{pmatrix}, \tag{121}$$

where a, b are $SU(8)$ indices, with upper(lower) antisymmetric indices for $\mathbf{28}(\overline{\mathbf{28}})$. Its two components

$$X^{\mu\nu ab} = 2\mathscr{F}^{\mu\nu ij} + \kappa\overline{C}^{ijkl}\mathscr{F}^{\mu\nu kl} + \frac{\kappa^2}{2}\overline{C}^{ijkl}\overline{C}^{klmn}\mathscr{F}^{\mu\nu mn} + O(\kappa^3), \tag{122}$$

$$Y^{\mu\nu}{}_{ab} = \kappa\overline{C}^{ijkl}\mathscr{F}^{\mu\nu kl} + \frac{\kappa^2}{2}\overline{C}^{ijkl}\overline{C}^{klmn}\mathscr{F}^{\mu\nu mn} + O(\kappa^3), \tag{123}$$

are *not independent*, but related by

$$Y^{\mu\nu}{}_{ab} - \frac{\kappa}{2}\overline{C}_{abcd}X^{\mu\nu cd} + O(\kappa^2) = 0. \tag{124}$$

The equations of motion (113) and Bianchi identities (114) can be written in terms of $Z^{\mu\nu}$

$$\partial_\mu \left(Z^{\mu\nu} + \widetilde{Z}^{\mu\nu} \right) = 0, \tag{125}$$

where

$$\widetilde{Z}^{\mu\nu} \equiv \begin{pmatrix} 0 & \mathbf{1} \\ \mathbf{1} & 0 \end{pmatrix} \overline{Z}^{\mu\nu} = \begin{pmatrix} \overline{\mathscr{G}}^{\mu\nu ij} - \overline{\mathscr{F}}^{\mu\nu ij} \\ \overline{\mathscr{G}}^{\mu\nu ij} + \overline{\mathscr{F}}^{\mu\nu ij} \end{pmatrix} = \begin{pmatrix} \overline{Y}^{\mu\nu\, ab} \\ \overline{X}^{\mu\nu}{}_{ab} \end{pmatrix}.$$

The upper component of (125) is the sum of the equations of motion and the Bianchi identities, and the lower component the difference. It follows that the duality transformations are those which act the same way on both $Z^{\mu\nu}$ and $\widetilde{Z}^{\mu\nu}$. Explicitly, under the coset transformation denoted by δ

$$\delta X^{\mu\nu ab} = \Xi^{abcd}Y^{\mu\nu}{}_{cd}, \tag{126}$$

$$\delta Y^{\mu\nu}{}_{ab} = \overline{\Xi}_{abcd}X^{\mu\nu cd}, \tag{127}$$

transform $\mathbf{28}$ into $\overline{\mathbf{28}}$ and vice versa. It can be checked that such transformations with real Ξ^{abcd} leave both equations of motion and Bianchi identities invariant, while those with pure imaginary Ξ^{abcd} are duality transformations which interchange the two. The transformations must respect the constraint (124) between the upper and lower components of $Z^{\mu\nu}$

$$\delta Y^{\mu\nu}{}_{ab} = \frac{\kappa}{2}\delta\left(\overline{C}_{abcd}X^{\mu\nu ab}\right) + O(\kappa^2),$$

that is

$$\Xi_{abcd}X^{\mu\nu cd} = \frac{\kappa}{2}\,\delta\overline{C}_{abcd}X^{\mu\nu cd} + \frac{\kappa}{2}\overline{C}_{abef}\Xi^{efmn}(\frac{\kappa}{2}\overline{C}_{mncd}X^{\mu\nu cd}) + O(\kappa^2).$$

It follows that the scalars must transform non-linearly as

$$\delta\overline{C}_{abcd} = \frac{2}{\kappa}\overline{\Xi}_{abcd} - \frac{\kappa}{2}\overline{C}_{ef[ab}\overline{C}_{cd]mn}\Xi^{efmn} + O(\kappa^3), \tag{128}$$

where the indices inside the square brackets are antisymmetrized. Since the scalars satisfy the self duality condition (109), so must Ξ^{abcd}

$$\Xi^{abcd} = \frac{1}{4!}\,\varepsilon^{abcdefgh}\,\overline{\Xi}_{efgh}, \tag{129}$$

which restricts Ξ^{abcd} to 70 real parameters. It also means that the extra term in (128) is self-dual. Repeated use of (127) yields the commutator

$$[\delta_1, \delta_2]X^{\mu\nu ab} = \left(\Xi^{abef}_{(2)}\,\overline{\Xi}_{(1)efcd} - \Xi^{abef}_{(1)}\,\overline{\Xi}_{(2)efcd}\right)X^{\mu\nu cd}.$$

We can show [25] that the duality requirement (129) on the parameters of this commutator yields exactly the 63 parameters of $SU(8)$, resulting in a 133-parameter group, the non-compact e since the e$/SU(8)$ transformations are not unitary. The $E_{7(7)}/SU(8)$ transformations of the complex field strengths follow

$$\begin{aligned}\delta\,\mathscr{F}^{\mu\nu ij} = &-\overline{\Xi}^{ijkl}\,\mathscr{F}^{\mu\nu kl} + \frac{\kappa}{2}\left(\Xi^{ijkl} - \overline{\Xi}^{ijkl}\right)\overline{C}^{klmn}\,\mathscr{F}^{\mu\nu mn}\\ &+\frac{\kappa^2}{4}\left(\Xi^{ijkl} - \overline{\Xi}^{ijkl}\right)\overline{C}^{klmn}\overline{C}^{mnpq}\,\mathscr{F}^{\mu\nu pq} + O(\kappa^3).\end{aligned} \tag{130}$$

As we mentioned before, this equation is meaningful only when $\kappa \neq 0$. While the scalar part of the Lagrangian $\mathscr{L}_S$ is $E_{7(7)}$-invariant, the vector Lagrangian $\mathscr{L}_V$ is not. Invariance is attained only after invoking the equations of motion.

5.2 $E_{7(7)}$ *Invariance on the Light-Cone*

The Abelian field strengths are written in terms of the potentials A^{ij}_{μ} through

$$F^{\mu\nu ij} = \partial^{\mu}A^{\nu ij} - \partial^{\nu}A^{\mu ij}.$$

In the LC_2 formalism we choose the gauge conditions

$$A^{+ij} = \frac{1}{\sqrt{2}}\left(A^0 + A^3\right)^{ij} = 0, \tag{131}$$

and invert the equations of motion to express A^{-ij} in terms of the remaining variables in the theory, the physical transverse components of the *complex* vector potentials

$$\bar{A}^{ij} = \frac{1}{\sqrt{2}}(A^1 + iA^2)^{ij}; \qquad A^{ij} = \frac{1}{\sqrt{2}}(A^1 - iA^2)^{ij}.$$

A lengthy but straightforward computation yields

$$\begin{aligned}
A^{-ij} &\equiv \frac{1}{\sqrt{2}}\left(A^0 - A^3\right)^{ij} \\
&= \frac{\partial}{\partial^+}A^{ij} + \frac{\bar{\partial}}{\partial^+}\bar{A}^{ij} - \kappa\frac{1}{\partial^+}\left(\overline{C}^{ijkl}\partial A^{kl}\right) - \kappa\frac{1}{\partial^+}\left(C^{ijkl}\bar{\partial}\bar{A}^{kl}\right) \\
&\quad + \kappa\frac{\partial}{\partial^{+2}}(\overline{C}^{ijkl}\partial^+ A^{kl}) + \kappa\frac{\bar{\partial}}{\partial^{+2}}(C^{ijkl}\partial^+ \bar{A}^{kl}) \\
&\quad + \frac{\kappa^2}{2}\frac{1}{\partial^+}\Big[C^{ijkl}\overline{C}^{klmn}\partial A^{mn} + \overline{C}^{ijkl}C^{klmn}\bar{\partial}\bar{A}^{mn} \\
&\quad - (C^{ijkl} + \overline{C}^{ijkl})\frac{\partial}{\partial^+}(\overline{C}^{klmn}\partial^+ A^{mn}) - (C^{ijkl} + \overline{C}^{ijkl})\frac{\bar{\partial}}{\partial^+}(C^{klmn}\partial^+ \bar{A}^{mn}) \\
&\quad + \frac{\partial}{\partial^+}(\overline{C}^{ijkl}\overline{C}^{klmn}\partial^+ A^{mn}) + \frac{\bar{\partial}}{\partial^+}(C^{ijkl}C^{klmn}\partial^+\bar{A}^{mn})\Big] + O(\kappa^3), \qquad (132)
\end{aligned}$$

where

$$\bar{\partial} = \frac{1}{\sqrt{2}}(\partial_1 - i\partial_2), \qquad \partial = \frac{1}{\sqrt{2}}(\partial_1 + i\partial_2), \qquad \partial^+ = \frac{1}{\sqrt{2}}(-\partial_0 + \partial_3)$$

(The occurrence of the non-local operator $\frac{1}{\partial^+}$ is abundant in the LC_2 formalism. It is a harmless non-locality along the light-cone which is well understood.)

This enables us to find the LC_2 complex field strengths $\mathscr{F}^{+-ij}$

$$\begin{aligned}
\mathscr{F}^{+-ij} &= \frac{1}{2}\left(\partial^+ A^{-ij} + \partial A^{ij} - \bar{\partial}\bar{A}^{ij}\right) \\
&= \partial A^{ij} - \frac{\kappa}{2}\overline{C}^{ijkl}\partial A^{kl} - \frac{\kappa}{2}C^{ijkl}\bar{\partial}\bar{A}^{kl} + \frac{\kappa}{2}\frac{\partial}{\partial^+}(\overline{C}^{ijkl}\partial^+ A^{kl}) + \frac{\kappa}{2}\frac{\bar{\partial}}{\partial^+}(C^{ijkl}\partial^+\bar{A}^{kl}) \\
&\quad + \frac{\kappa^2}{4}\Big[C^{ijkl}\overline{C}^{klmn}\partial A^{mn} + \overline{C}^{ijkl}C^{klmn}\bar{\partial}\bar{A}^{mn} \\
&\qquad - (C^{ijkl} + \overline{C}^{ijkl})\left(\frac{\partial}{\partial^+}(\overline{C}^{klmn}\partial^+ A^{mn}) + \frac{\bar{\partial}}{\partial^+}(C^{klmn}\partial^+\bar{A}^{mn})\right) \\
&\qquad + \frac{\partial}{\partial^+}\left(\overline{C}^{ijkl}\overline{C}^{klmn}\partial^+ A^{mn}\right) + \frac{\bar{\partial}}{\partial^+}\left(C^{ijkl}C^{klmn}\partial^+\bar{A}^{mn}\right)\Big] + \cdots. \qquad (133)
\end{aligned}$$

By varying this expression and using (120), we arrive at the non-linear transformation of the physical vector potentials under $SU(8)/SO(8)$

$$\delta' A^{ij} \;=\; iS^{im}\left(A^{mj} + \kappa\frac{1}{\partial^+}\left(\overline{C}^{mjkl}\partial^+ A^{kl}\right) + O(\kappa^3)\right) \;-\; (i \leftrightarrow j). \qquad (134)$$

As in the covariant case, the terms on the right-hand-side do not share the same coset transformations.

Similarly, the coset $e/SU(8)$ transformations of the vector potentials are obtained by substituting $\mathscr{F}^{+-ij}$ in (130) with (133). Remembering that the scalars transform non-linearly under $E_{7(7)}/SU(8)$ (128) , we find for the vector potentials

$$\delta A^{ij} \;=\; -\,\overline{\Xi}^{ijkl}A^{kl} + \frac{\kappa}{2}(\Xi^{ijkl} - \overline{\Xi}^{ijkl})\frac{1}{\partial^+}\left(\overline{C}^{klmn}\partial^+ A^{mn}\right) + \mathscr{O}(\kappa^3), \qquad (135)$$

which preserve helicity, and exist as long as $\kappa \neq 0$.

5.3 The Vector and Scalar LC_2 Hamiltonians

The vector Lagrangian (110) in the LC_2 gauge, obtained by setting $A^{+ij} = 0$ and replacing A^{-ij} using the equations of motion, is given by

$$\begin{aligned}
\mathscr{L}_V =\; & \bar{A}^{ij}(-\partial^+\partial^- + \partial\bar{\partial})A^{ij} \\
& + \frac{\kappa}{2}\Big[\partial^+ A^{ij}\overline{C}^{ijmn}\,\partial^-\bar{A}^{mn} + \partial A^{ij}\Big(\overline{C}^{ijkl}\partial A^{kl} - \frac{\partial}{\partial^+}(\overline{C}^{ijkl}\partial^+ A^{kl}) \\
& \qquad\qquad - \frac{\bar{\partial}}{\partial^+}(C^{ijkl}\partial^+\bar{A}^{kl})\Big) + c.c.\Big] \\
& - \frac{\kappa^2}{2}\frac{\bar{\partial}}{\partial^+}(\partial^+\bar{A}^{ij}C^{ijkl})\frac{\partial}{\partial^+}(\overline{C}^{klmn}\partial^+ A^{mn}) \\
& + \frac{\kappa^2}{2}\Big[-\frac{1}{2}\partial A^{ij}\overline{C}^{ijkl}C^{klmn}\bar{\partial}\bar{A}^{mn} + \partial A^{ij}\overline{C}^{ijkl}\frac{\bar{\partial}}{\partial^+}(C^{klmn}\partial^+\bar{A}^{mn}) + c.c.\Big] \\
& + \frac{\kappa^2}{2}\Big[\partial A^{ij}\overline{C}^{ijkl}\frac{\partial}{\partial^+}(\overline{C}^{klmn}\partial^+ A^{mn}) - \frac{1}{2}\frac{\partial}{\partial^+}(\partial^+ A^{ij}\overline{C}^{ijkl})\frac{\partial}{\partial^+}(\overline{C}^{klmn}\partial^+ A^{mn}) + c.c.\Big] \\
& + \frac{\kappa^2}{4}\Big[\partial^-\bar{A}^{ij}\overline{C}^{ijkl}\overline{C}^{klmn}\partial^+ A^{mn} - \frac{\partial}{\partial^+}(\partial^+ A^{ij}\overline{C}^{ijkl}\overline{C}^{klmn})(\partial A^{mn} + \bar{\partial}\bar{A}^{mn}) + c.c.\Big] \\
& + O(\kappa^3),
\end{aligned} \qquad (136)$$

while the scalar supergravity Lagrangian (108) becomes

$$\begin{aligned}
\mathscr{L}_S = & -\frac{1}{24}C^{ijkl}(\partial^+\partial^- - \partial\bar{\partial})\overline{C}^{ijkl} \\
& + \frac{\kappa^2}{96}C^{ijkl}\overline{C}^{klmn}\,(\partial^+ C^{mnpq}\,\partial^-\overline{C}^{pqij} + \partial^- C^{mnpq}\,\partial^+\overline{C}^{pqij} \\
& \qquad - \partial C^{mnpq}\,\bar{\partial}\overline{C}^{pqij} - \bar{\partial}C^{mnpq}\,\partial\overline{C}^{pqij}) + O(\kappa^3).
\end{aligned} \qquad (137)$$

Both contain the light-cone time derivative ∂^- in their interactions. In order to have a Hamiltonian without this derivative we eliminate it by the field redefinitions

$$\begin{aligned} C^{ijkl} &= D^{ijkl} - \frac{\kappa^2}{4}\frac{1}{\partial^+}\left(D^{pq[ij}\partial^+ D^{kl]mn}\overline{D}_{pqmn}\right) \\ &+\frac{3\,\kappa^2}{2\partial^+}\left(\partial^+ B^{[ij}\frac{1}{\partial^+}(D^{kl]mn}\partial^+\overline{B}_{mn})\right) \\ &+\frac{3\,\kappa^2}{2\cdot 4!\,\partial^+}\varepsilon^{ijklrstu}\left(\partial^+\overline{B}_{rs}\frac{1}{\partial^+}(\overline{D}_{tumn}\partial^+ B^{mn})\right) + O(\kappa^3), \end{aligned} \tag{138}$$

$$A^{ij} = B^{ij} - \frac{\kappa}{2}\frac{1}{\partial^+}\left(\overline{D}_{ijkl}\partial^+ B^{kl}\right) + \frac{\kappa^2}{8}D^{ijkl}\frac{1}{\partial^+}\left(\partial^+ B^{mn}\overline{D}_{mnkl}\right) + O(\kappa^3). \tag{139}$$

This procedure leads to the unique Hamiltonian of the theory in component form.

The new vector potentials, B^{ij} now transform *linearly* under $SU(8)$,

$$\delta' B^{ij} = iS^{ik}B^{kj} - (i \leftrightarrow j),$$

so that $i, j, \ldots$ are now true $SU(8)$ indices; in particular, their lowering produces the "barred" representation. The $E_{7(7)}/SU(8)$ variations of the redefinitions yield the transformation properties of the new fields

$$\delta B^{ij} = -\frac{\kappa}{4}\overline{\Xi}_{mnkl}D^{ijkl}B^{mn} + \frac{\kappa}{4}\,\Xi^{ijkl}\frac{1}{\partial^+}\left(\overline{D}_{mnkl}\partial^+ B^{mn}\right) + O(\kappa^3), \tag{140}$$

$$\begin{aligned} \delta D^{ijkl} &= \frac{2}{\kappa}\,\Xi^{ijkl} - \frac{\kappa}{2}\overline{\Xi}_{mnpq}\frac{1}{\partial^+}\left(D^{mn[kl}\partial^+ D^{ij]pq}\right) \\ &+\frac{\kappa}{2}\,\Xi^{pq[ij}\frac{1}{\partial^+}\left(\partial^+ D^{kl]mn}\overline{D}_{pqmn}\right) \\ &-3\,\kappa\left(\frac{\Xi^{mn[kl}}{\partial^+}\left(\partial^+ B^{ij]}\overline{B}_{mn}\right) + \varepsilon^{ijklrstu}\frac{\overline{\Xi}_{tumn}}{4!\,\partial^+}\left(B^{mn}\partial^+\overline{B}_{rs}\right)\right) + O(\kappa^3). \end{aligned} \tag{141}$$

We note that the $E_{7(7)}/SU(8)$ variation of the scalars contains terms quadratic in the gauge fields. This mixing does not occur in the covariant formalism. Complicated as they may seem, these variations are still incomplete since they do not include the other fields of the theory. We will use the supersymmetry of $\mathcal{N} = 8$ supergravity to generalize the transformations (140) and (141) to include them.

5.4 $E_{7(7)}$ Light-Cone Superspace

We will now implement the $E_{7(7)}$ transformations on the superfield equation (46). We remind ourselves that the kinematical supersymmetry transformations of the physical fields are

$$\delta_s h \,=\, 0, \qquad \delta_s \overline{h} \,=\, -\,i\frac{\sqrt{2}}{4}\,\overline{\varepsilon}_m\,\psi^m,$$

$$\begin{aligned}
\delta_s\,\psi^m &= 2\sqrt{2}\,\overline{\varepsilon}_n \partial^+ B^{mn}, & \delta_s\,\overline{\psi}_m &= -\sqrt{2}\,\overline{\varepsilon}_m\,\partial^+ h,\\
\delta_s\,B^{mn} &= -\,3i\sqrt{2}\,\overline{\varepsilon}_p\,\chi^{mnp}, & \delta_s\,\overline{B}_{mn} &= -\,2i\sqrt{2}\,\overline{\varepsilon}_{[m}\overline{\psi}_{n]},\\
\delta_s\,\chi^{lmn} &= -\,\frac{\sqrt{2}}{3!}\,\overline{\varepsilon}_k\,\partial^+ D^{klmn}, & \delta_s\,\overline{\chi}_{mnp} &= -\,3\sqrt{2}\,\varepsilon_{[p}\,\partial^+\,\overline{B}_{mn]},
\end{aligned}$$

and finally

$$\delta_s\,\overline{D}_{klmn} \,=\, -\,4i\sqrt{2}\,\overline{\varepsilon}_{[n}\,\overline{\chi}_{klm]}.$$

The quadratic operators

$$T^i{}_j \,=\, -\,\frac{i}{\sqrt{2}\,\partial^+}\left(q^i\bar{q}_j - \frac{1}{8}\delta^i{}_j q^k\bar{q}_k\right), \tag{142}$$

which satisfy the $SU(8)$ algebra

$$[\,T^i{}_j,\,T^k{}_l\,] \,=\, \delta^k{}_j\,T^i{}_l - \delta^i{}_l\,T^k{}_j,$$

also act linearly on the chiral superfield

$$\delta_{SU_8}\,\varphi(y) \,=\, \omega^j{}_i\,T^i{}_j\,\varphi(y).$$

We can now include the other fields of the theory by demanding that the e$/SU(8)$ transformations commute with the kinematical supersymmetries, that is

$$[\,\delta_s,\,\delta\,]\,\varphi(y) \,=\, 0. \tag{143}$$

We begin by applying this equation to the vector potential. By carefully choosing the parameters of both supersymmetry and of $E_{7(7)}/SU(8)$, we arrive at the generalization of (140) to order κ

$$\begin{aligned}
\delta\,\overline{B}_{ij} = -\,\kappa\,\Xi^{klmn}\bigg(&\frac{1}{4}\overline{D}_{ijkl}\,\overline{B}_{mn} + \frac{1}{4!}\frac{1}{\partial^+}\overline{D}_{klmn}\partial^+\overline{B}_{ij} - \frac{1}{4!}\varepsilon_{ijklmnrs}\frac{1}{\partial^+}B^{rs}\partial^+ h\\
&+\frac{i}{3!}\frac{1}{\partial^+}\overline{\chi}_{klm}\,\overline{\chi}_{ijn} - \frac{i}{3!}\,\varepsilon_{ijklmrst}\,\frac{1}{\partial^+}\chi^{rst}\overline{\psi}_n\bigg)\\
+\,\kappa\,\overline{\Xi}_{ijkl}\,\frac{1}{\partial^+}\bigg(&\frac{1}{4}D^{klmn}\,\partial^+\overline{B}_{mn} - \frac{1}{\partial^+}B^{kl}\,\partial^{+2}\,h\\
&+\frac{i}{4(3!)^2}\overline{\chi}_{mnp}\overline{\chi}_{rst}\varepsilon^{klmnprst} - 3\,i\,\frac{1}{\partial^+}\chi^{kln}\partial^+\overline{\psi}_n\bigg)\\
+\,O(\kappa^3),&
\end{aligned} \tag{144}$$

as well as to the $E_{7(7)}/SU(8)$ transformations of the gravitinos since commutativity implies

$$\delta_s\,\delta\,\overline{B}_{ij} \;=\; -2i\sqrt{2}\overline{\varepsilon}_{[i}\,\delta\,\overline{\psi}_{j]}.$$

The result is

$$\begin{aligned}\delta\,\overline{\psi}_i = -\,\kappa\,\Xi^{mnpq}\Big(&\frac{1}{4!}\frac{1}{\partial^+}\overline{D}_{mnpq}\partial^+\overline{\psi}_i+\frac{1}{3!}\overline{D}_{mnpi}\overline{\psi}_q\\ &-\frac{1}{4!}\varepsilon_{mnpqirst}\frac{1}{\partial^+}\chi^{rst}\partial^+h+\frac{1}{4}\overline{\chi}_{imn}\overline{B}_{pq}+\frac{1}{3!}\frac{1}{\partial^+}\overline{\chi}_{mnp}\partial^+\overline{B}_{iq}\Big)\\ &+O(\kappa^3). \qquad (145)\end{aligned}$$

Applying commutativity on the gravitinos yields the $E_{7(7)}/SU(8)$ transformation of the graviton

$$\delta_s\,\delta\,\overline{\psi}_i \;=\; -\sqrt{2}\overline{\varepsilon}_i\,\partial^+\,\delta\,h,$$

with

$$\delta h \;=\; -\,\kappa\,\Xi^{ijkl}\left(\frac{1}{8}\overline{B}_{ij}\overline{B}_{kl}+\frac{1}{4!}\frac{1}{\partial^+}\overline{D}_{ijkl}\partial^+h+\frac{i}{6}\frac{1}{\partial^+}\overline{\chi}_{ijk}\overline{\psi}_l\right)+O(\kappa^3). \quad (146)$$

All these transformations are non-linear. Similar equations can be derived for the 56 spinors and 70 scalars.

The inhomogeneous $E_{7(7)}/SU(8)$ transformations of order κ^{-1} of the scalar fields can be expressed in superfield language, that is

$$\delta^{(-1)}\,\varphi \;=\; -\frac{2}{\kappa}\,\theta^{ijkl}\,\overline{\Xi}_{ijkl},$$

which is chiral since $E_{7(7)}$ is a global symmetry: $\partial^+\overline{\Xi}_{ijkl}=0$. The order κ transformations of the superfield itself take a particularly simple form. We need only require that its variation be chiral, with the tensor structure

$$\kappa\,\Xi^{ijkl}(\cdots)_{ijkl}.$$

Assuming that the lower indices are carried by the antichiral derivatives $\overline{d}_n$ leads to the unique form of the transformation to first order in κ

$$\frac{\kappa}{4!}\,\Xi^{ijkl}\,\frac{1}{\partial^{+2}}\left(\overline{d}_{ijkl}\frac{1}{\partial^+}\varphi\,\partial^{+3}\varphi-4\overline{d}_{ijk}\varphi\,\overline{d}_l\partial^{+2}\varphi+3\overline{d}_{ij}\partial^+\varphi\,\overline{d}_{kl}\partial^+\varphi\right),$$

where $\bar{d}_{k\ldots l}$ is a shorthand notation for $\bar{d}_k\cdots\bar{d}_l$. Including the inhomogeneous term, the e$/SU(8)$ transformation can be written in a more compact way by introducing a coherent state-like representation

$$\delta\varphi = -\frac{2}{\kappa}\,\theta^{ijkl}\,\Xi_{ijkl} + \frac{\kappa}{4!}\,\Xi^{ijkl}\left(\frac{\partial}{\partial\eta}\right)_{ijkl}\frac{1}{\partial^{+2}}\left(e^{\eta\hat{\bar{d}}}\partial^{+3}\varphi\, e^{-\eta\hat{\bar{d}}}\partial^{+3}\varphi\right)\Bigg|_{\eta=0}$$
$$+O(\kappa^3), \tag{147}$$

where

$$\eta\hat{\bar{d}} = \eta^m\frac{\bar{d}_m}{\partial^+}, \quad \text{and} \quad \left(\frac{\partial}{\partial\eta}\right)_{ijkl} \equiv \frac{\partial}{\partial\eta^i}\frac{\partial}{\partial\eta^j}\frac{\partial}{\partial\eta^k}\frac{\partial}{\partial\eta^l}.$$

We note that these $E_{7(7)}/SU(8)$ transformations do close properly to an $SU(8)$ transformation on the superfield

$$[\,\delta_1,\delta_2\,]\,\varphi \;=\; \delta_{SU(8)}\,\varphi.$$

It is chiral by construction $d^n\,\delta\varphi \;=\; 0$, with the power of the first inverse derivative set by comparing with the graviton transformation. Hence, *all* physical fields, including the graviton transform under $E_{7(7)}$ and can be read off from this equation. It will be interesting to see what constraints this puts on the geometry.

We can now extend the method to the dynamical supersymmetries, and determine the form of the interactions implied by the $E_{7(7)}$ symmetry.

5.5 Superspace Action

The $\mathcal{N}=8$ supergravity action in superspace was first obtained in [26] and its LC_2 form is derived in [12] to order κ, using algebraic consistency and simplified further in [16]. It is remarkably simple:

$$S = -\frac{1}{64}\int d^4x\int d^8\theta\, d^8\overline{\theta}\,\left\{-\overline{\varphi}\,\frac{\Box}{\partial^{+4}}\,\varphi - 2\kappa\left(\frac{1}{\partial^{+2}}\overline{\varphi}\,\overline{\partial}\varphi\,\overline{\partial}\varphi + c.c.\right) + O(\kappa^2)\right\}, \tag{148}$$

where $\Box \equiv 2\,(\partial\bar{\partial} - \partial^+\partial^-)$. The light-cone superfield Hamiltonian density is then written as

$$\mathcal{H} \;=\; 2\overline{\varphi}\,\frac{\partial\bar{\partial}}{\partial^{+4}}\,\varphi + 2\,\kappa\left(\frac{1}{\partial^{+2}}\overline{\varphi}\,\overline{\partial}\varphi\,\overline{\partial}\varphi + c.c.\right) + \mathcal{O}(\kappa^2). \tag{149}$$

It can be derived from the action of the dynamical supersymmetries on the chiral superfield

$$\delta_s^{dyn}\,\varphi = \delta_s^{dyn\,(0)}\,\varphi + \delta_s^{dyn\,(1)}\,\varphi + \delta_s^{dyn\,(2)}\,\varphi + O(\kappa^3),$$
$$= \varepsilon^m\left\{\frac{\partial}{\partial^+}\,\bar{q}_m\,\varphi + \kappa\,\frac{1}{\partial^+}\left(\bar{\partial}\,\bar{d}_m\,\varphi\,\partial^{+2}\,\varphi - \partial^+\,\bar{d}_m\,\varphi\,\partial^+\,\bar{\partial}\,\varphi\right) + O(\kappa^2)\right\}. \tag{150}$$

We now require that the $E_{7(7)}/SU(8)$ commutes with the dynamical supersymmetries

$$[\delta, \delta_s^{dyn}]\,\varphi \;=\; 0. \tag{151}$$

This commutativity is valid only on the chiral superfield. For example, $[\delta_1, \delta_s]$ $\delta_2\varphi \neq 0$, due to the non-linearity of the e transformation. This helps us understand how the Jacobi identity

$$([\delta_1, [\delta_2, \delta_s]] + [\delta_2, [\delta_s, \delta_1]] + [\delta_s, [\delta_1, \delta_2]])\,\varphi \;=\; 0,$$

is algebraically consistent. In the last term the commutator of the two $E_{7(7)}/SU(8)$ transformations, $[\delta_1, \delta_2]$, yields an $SU(8)$ under which the supersymmetry transforms. This is precisely compensated by contributions from the first two terms.

Although the dynamical supersymmetry to order κ is already known, we re-derive $\delta_s^{dyn\,(1)}\,\varphi$ from the commutativity between the dynamical supersymmetries and $E_{7(7)}/SU(8)$ transformations.

The inhomogeneous $E_{7(7)}$ transformations link interaction terms with different order in κ. To zeroth order, one finds

$$[\delta^{(-1)}, \delta_s^{dyn\,(1)}]\,\varphi \;=\; \delta^{(-1)}\,\delta_s^{dyn\,(1)}\varphi \;=\; 0, \tag{152}$$

since $\delta_s^{dyn\,(1)}\,\delta^{(-1)}\varphi \;=\; 0$. To find $\delta_s^{dyn\,(1)}\varphi$ that satisfies both the above equation and the SuperPoincaré algebra, one may start with a general form that satisfies all the commutation relations with the kinematical SuperPoincaré generators,

$$\delta_s^{dyn\,(1)}\varphi \propto \frac{\partial}{\partial a}\frac{\partial}{\partial b}\frac{1}{\partial^{+(m+n+1)}}\left(e^{a\hat{\bar{\partial}}}e^{b\,\varepsilon\hat{\bar{q}}}\partial^{+(2+m)}\varphi\, e^{-a\hat{\bar{\partial}}}e^{-b\,\varepsilon\hat{\bar{q}}}\partial^{+(2+n)}\varphi\right)\Big|_{a=b=0},$$

where $\hat{\bar{\partial}} = \frac{\bar{\partial}}{\partial^+}$, $\varepsilon\hat{\bar{q}} = \varepsilon^m\frac{\bar{q}_m}{\partial^+}$. It is not difficult to see that this form with non-negative m, n satisfies (152). The number of powers of ∂^+ can be determined by checking the commutation relation between two dynamical generators δ_{p^-} (Hamiltonian variation which is derived from the supersymmetry algebra) and δ_{j^-} (the boost which can also be obtained through $[\delta_{j^-}, \delta_{\bar{q}}]\,\varphi \;=\; \delta_s^{dyn}\varphi$), yielding that the commutator between δ_{j^-} and δ_{p^-} vanishes only when $m = n = 0$, which leads to the same form as (150) written in a coherent-like form

$$\delta_s^{dyn\,(1)}\varphi = \frac{\kappa}{2}\frac{\partial}{\partial a}\frac{\partial}{\partial b}\frac{1}{\partial^+}\left[e^{a\hat{\bar{\partial}}}e^{b\,\varepsilon\hat{\bar{q}}}\partial^{+2}\varphi\, e^{-a\hat{\bar{\partial}}}e^{-b\,\varepsilon\hat{\bar{q}}}\partial^{+2}\varphi\right]\Big|_{a=b=0}.$$

It is worth noting that this is the solution that has the least number of powers of ∂^+ in the denominator, and thus the least "non-local".

The same reasoning can be applied to higher orders in κ. To order κ, we find that commutativity

$$[\delta^{(-1)}, \delta_s^{dyn\,(2)}]\,\varphi + [\delta^{(1)}, \delta_s^{dyn\,(0)}]\,\varphi \;=\; 0$$

requires

$$\begin{aligned}
&\delta^{(-1)}\,\delta_s^{dyn\,(2)}\varphi \\
&= \frac{\kappa}{4!}\Xi^{ijkl}\frac{1}{\partial^{+3}}\left[-\bar{d}_{ijkl}\frac{\partial}{\partial^+}\varphi\,\partial^{+3}\varepsilon\bar{q}\,\varphi + 4\bar{d}_{ijk}\partial\varphi\,\bar{d}_l\partial^{+2}\varepsilon\bar{q}\,\varphi - 3\bar{d}_{ij}\partial\partial^+\varphi\,\bar{d}_{kl}\partial^+\varepsilon\bar{q}\,\varphi\right.\\
&\qquad - \bar{d}_{ijkl}\frac{\varepsilon\bar{q}}{\partial^+}\varphi\,\partial\partial^{+3}\varphi + 4\bar{d}_{ijk}\varepsilon\bar{q}\,\varphi\,\bar{d}_l\partial\partial^{+2}\varphi - 3\bar{d}_{ij}\partial^+\varepsilon\bar{q}\varphi\,\bar{d}_{kl}\partial\partial^+\varphi\\
&\qquad + \bar{d}_{ijkl}\frac{\partial}{\partial^{+2}}\varepsilon\bar{q}\,\varphi\,\partial^{+4}\varphi - 4\bar{d}_{ijk}\frac{\partial}{\partial^+}\varepsilon\bar{q}\,\varphi\,\bar{d}_l\partial^{+3}\varphi + 3\bar{d}_{ij}\partial\varepsilon\bar{q}\,\varphi\,\bar{d}_{kl}\partial^{+2}\varphi\\
&\qquad \left. + \bar{d}_{ijkl}\,\varphi\,\partial\partial^{+2}\varepsilon\bar{q}\,\varphi - 4\bar{d}_{ijk}\partial^+\varphi\,\bar{d}_l\partial\partial^+\varepsilon\bar{q}\,\varphi + 3\bar{d}_{ij}\partial^{+2}\varphi\,\bar{d}_{kl}\partial\varepsilon\bar{q}\varphi\right],
\end{aligned} \tag{153}$$

where $\varepsilon\bar{q}$ denotes $\varepsilon^m\bar{q}_m$, which can be written in a simpler form by rewriting it in terms of a coherent state-like form:

$$\begin{aligned}
\delta^{(-1)}\,\delta_s^{dyn\,(2)}\varphi &= \frac{\kappa}{2\cdot 4!}\Xi^{ijkl}\frac{\partial}{\partial a}\frac{\partial}{\partial b}\left(\frac{\partial}{\partial\eta}\right)_{ijkl}\\
&\quad\times\frac{1}{\partial^{+3}}\left[\,e^{a\hat{\partial}}e^{b\,\varepsilon\hat{\bar{q}}}e^{\eta\hat{\bar{d}}}\partial^{+4}\varphi\,e^{-a\hat{\partial}}e^{-b\varepsilon\hat{\bar{q}}}e^{-\eta\hat{\bar{d}}}\partial^{+4}\varphi\right]\Bigg|_{a=b=\eta=0}.
\end{aligned} \tag{154}$$

To find $\delta_s^{dyn(2)}\varphi$ that satisfies (153), consider the chiral combination

$$\begin{aligned}
Z_{mnpq} &\equiv \left(\frac{\partial}{\partial\xi}\right)_{mnpq}\left(e^{\xi\hat{\bar{d}}}\partial^{+4}\varphi e^{-\xi\hat{\bar{d}}}\partial^{+4}\varphi\right)\Big|_{\xi=0},\\
&= \bar{d}_{mnpq}\varphi\,\partial^{+4}\varphi - 4\bar{d}_{mnp}\partial^+\varphi\,\bar{d}_q\partial^{+3}\varphi + 3\bar{d}_{mn}\partial^{+2}\varphi\,\bar{d}_{pq}\partial^{+2}\varphi.
\end{aligned} \tag{155}$$

The inhomogeneous e transformation of

$$Z^{ijkl} \equiv \frac{1}{4!}\varepsilon^{ijklmnpq}Z_{mnpq},$$

has the simple form

$$\delta^{(-1)}Z^{ijkl} = \frac{1}{4!}\varepsilon^{ijklmnpq}\,\bar{d}_{mnpq}\delta^{(-1)}\varphi\,\partial^{+4}\varphi = \frac{2}{\kappa}\,\Xi^{ijkl}\,\partial^{+4}\varphi, \tag{156}$$

which leads to the solution

$$\begin{aligned}
\delta_s^{dyn\,(2)}\varphi &= \frac{\kappa^2}{2\cdot 4!}\frac{\partial}{\partial a}\frac{\partial}{\partial b}\left(\frac{\partial}{\partial\eta}\right)_{ijkl}\frac{1}{\partial^{+4}}\\
&\quad\times\left(e^{a\hat{\partial}+b\,\varepsilon\hat{\bar{q}}+\eta\hat{\bar{d}}}\partial^{+5}\varphi\,e^{-a\hat{\partial}-b\,\varepsilon\hat{\bar{q}}-\eta\hat{\bar{d}}}Z^{ijkl}\right)\Bigg|_{a=b=\eta=0},
\end{aligned}$$

where we have fixed the ambiguity discussed earlier by choosing the expression with the least number of ∂^+ in the denominator. Its algebraic consistency should be checked in a future publication. This coherent state-like form is very efficient; Written out explicitly $\delta_s^{dyn\,(2)}\varphi$ consists of 60 terms.

The dynamical supersymmetry is then written in terms of the coherent state-like form,

$$\delta_s^{dyn}\varphi = \frac{\partial}{\partial a}\frac{\partial}{\partial b}\left\{ e^{a\hat{\partial}} e^{b\varepsilon\hat{\bar{q}}} \partial^+\varphi + \frac{\kappa}{2}\frac{1}{\partial^+}\left(e^{a\hat{\partial}+b\varepsilon\hat{\bar{q}}}\partial^{+2}\varphi e^{-a\hat{\partial}-b\varepsilon\hat{\bar{q}}}\partial^{+2}\varphi\right) \right.$$
$$+\frac{\kappa^2}{2\cdot 4!}\left(\frac{\partial}{\partial\eta}\right)_{ijkl}\frac{1}{\partial^{+4}}\left(e^{a\hat{\partial}+b\varepsilon\hat{\bar{q}}+\eta\hat{\bar{d}}}\partial^{+5}\varphi\, e^{-a\hat{\partial}-b\varepsilon\hat{\bar{q}}-\eta\hat{\bar{d}}}Z^{ijkl}\right)$$
$$\left. + O(\kappa^3)\right\}\Bigg|_{a=b=\eta=0} . \qquad (157)$$

We now use the fact, as Ananth et al. [17] have shown, that the $\mathcal{N}=8$ supergravity light-cone Hamiltonian can be written as a quadratic form (to order κ^2),

$$\mathcal{H} = \frac{1}{4\sqrt{2}}\left(\mathcal{W}_m, \mathcal{W}_m\right) \equiv \frac{2i}{4\sqrt{2}}\int d^8\theta\, d^8\bar{\theta}\, d^4x\, \overline{\mathcal{W}}_m\frac{1}{\partial^{+3}}\mathcal{W}_m,$$

where the fermionic superfield $\mathcal{W}_m$ is the dynamical supersymmetry variation of φ

$$\delta_s^{dyn}\varphi \equiv \varepsilon^m\mathcal{W}_m,$$

with

$$\mathcal{W}_m = \mathcal{W}_m^{(0)} + \mathcal{W}_m^{(1)} + \mathcal{W}_m^{(2)} + \cdots .$$

Up to order κ, the Hamiltonian is simply

$$\mathcal{H} = \frac{1}{4\sqrt{2}}\left[\left(\mathcal{W}_m^{(0)}, \mathcal{W}_m^{(0)}\right) + \left(\mathcal{W}_m^{(0)}, \mathcal{W}_m^{(1)}\right) + \left(\mathcal{W}_m^{(1)}, \mathcal{W}_m^{(0)}\right)\right], \qquad (158)$$

while the Hamiltonian of order κ^2 consists of three parts:

$$\mathcal{H}^{\kappa^2} = \frac{1}{4\sqrt{2}}\left[\left(\mathcal{W}_m^{(1)}, \mathcal{W}_m^{(1)}\right) + \left(\mathcal{W}_m^{(0)}, \mathcal{W}_m^{(2)}\right) + \left(\mathcal{W}_m^{(2)}, \mathcal{W}_m^{(0)}\right)\right], \qquad (159)$$

where the first part was computed by Ananth et al. [17]

$$\left(\mathcal{W}_m^{(1)}, \mathcal{W}_m^{(1)}\right) = i\frac{\kappa^2}{2}\frac{\partial}{\partial a}\frac{\partial}{\partial b}\frac{\partial}{\partial r}\frac{\partial}{\partial s}\int d^8\theta\, d^8\bar{\theta}\, d^4x$$
$$\times\frac{1}{\partial^{+5}}\left(e^{a\hat{\partial}+b\hat{q}^m}\partial^{+2}\bar{\varphi}e^{-a\hat{\partial}-b\hat{q}^m}\partial^{+2}\bar{\varphi}\right)$$
$$\times\left(e^{r\hat{\bar{\partial}}+s\hat{q}_m}\partial^{+2}\varphi e^{-r\hat{\bar{\partial}}-s\hat{q}_m}\partial^{+2}\varphi\right)\Bigg|_{a=b=r=s=0}, \qquad (160)$$

and the second and third parts are complex conjugate of each other. It suffices to consider

$$\left(\mathscr{W}_m^{(0)}, \mathscr{W}_m^{(2)}\right) = i\frac{\kappa^2}{4!}\frac{\partial}{\partial a}\frac{\partial}{\partial b}\left(\frac{\partial}{\partial \eta}\right)_{ijkl}\int d^8\theta\, d^8\bar{\theta}\, d^4x \\ \times \frac{\bar{\partial}}{\partial^+} q^m \overline{\varphi}\, \frac{1}{\partial^{+7}}\left(e^{a\hat{\partial}+b\hat{\bar{q}}_m+\eta\hat{\bar{d}}}\partial^{+5}\varphi\, e^{-a\hat{\partial}-b\hat{\bar{q}}_m-\eta\hat{\bar{d}}} Z^{ijkl}\right)\Bigg|_{a=b=\eta=0} . \tag{161}$$

Integration by parts with respect to $\bar{d}$'s and use of the inside-out constraint (40) allow for an efficient rearrangement of terms to yield the final expression

$$\left(\mathscr{W}_m^{(0)}, \mathscr{W}_m^{(2)}\right) = -i\frac{\kappa^2}{4!}\frac{\partial}{\partial a}\frac{\partial}{\partial b}\int d^8\theta\, d^8\bar{\theta}\, d^4x \\ \times \frac{\bar{\partial}}{\partial^{+4}} q^m d^{ijkl}\overline{\varphi}\left(e^{a\hat{\partial}+b\hat{\bar{q}}_m}\partial^+\overline{\varphi}\, e^{-a\hat{\partial}-b\hat{\bar{q}}_m}\frac{1}{\partial^{+4}}Z_{ijkl}\right)\Bigg|_{a=b=0} . \tag{162}$$

Therefore, the Hamiltonian to order κ^2 is written as

$$\mathscr{H}^{\kappa^2} = i\frac{\kappa^2}{4\sqrt{2}}\int d^8\theta\, d^8\bar{\theta}\, d^4x\, \frac{\partial}{\partial a}\frac{\partial}{\partial b} \\ \times\Bigg\{\frac{1}{2}\frac{\partial}{\partial r}\frac{\partial}{\partial s}\frac{1}{\partial^{+5}}\left(e^{a\hat{\partial}+b\hat{q}}\partial^{+2}\overline{\varphi} e^{-a\hat{\partial}-b\hat{q}}\partial^{+2}\overline{\varphi}\right)\left(e^{r\hat{\bar{\partial}}+s\hat{\bar{q}}}\partial^{+2}\varphi e^{-r\hat{\bar{\partial}}-s\hat{\bar{q}}}\partial^{+2}\varphi\right) \\ -\left[\frac{1}{4!}\frac{\bar{\partial}}{\partial^{+4}}q^m d^{ijkl}\overline{\varphi}\left(e^{a\hat{\partial}+b\hat{\bar{q}}_m}\partial^+\overline{\varphi}\, e^{-a\hat{\partial}-b\hat{\bar{q}}_m}\frac{1}{\partial^{+4}}Z_{ijkl}\right)+c.c.\right]\Bigg\}\Bigg|_{a=b=r=s=0} . \tag{163}$$

to be compared with the 96 terms of Ananth et al. [17]!

6 Concluding Remarks

In this lecture I have shown that all the known quantum field theories follow by studying representations of the Poincaré algebra. What we get though is essentially the part of them which is amenable to perturbation theory, i.e., as expansions in a coupling constant. We have learnt in recent years that quantum field theories are very much richer than what meets the eye in a perturbation expansion. The formalism here is not suitable for such studies. It is very hard if possible to study non-perturbative effects such as solitons, magnetic monopoles, branes and various forms of duality. However, the formalism is a complement to other studies, it shows the physical symmetries of the theories and it is very useful for certain studies about finiteness in perturbations expansions, which is one of the crucial tests of a quantum gravity theory.

Acknowledgement It was a privilege to give this lecture at the celebration of Claudio Bunster's birthday celebration. Claudio has been a close friend of mine for very many years and I have always admired his deep understanding of basic physics as well as of human relations. His tireless and brave work for South American science is an example for all scientists.

References

1. E. Cremmer and B. Julia, "The N=8 supergravity theory. 1. The Lagrangian," Phys. Lett. B **80**, 48 (1978).
2. E. Cremmer and B. Julia, "The SO(8) supergravity," Nucl. Phys. B **159**, 141 (1979).
3. L. Brink, J. H. Schwarz and J. Scherk, "Supersymmetric Yang–Mills theories," Nucl. Phys. B **121**, 77 (1977).
4. F. Gliozzi, J. Scherk and D. I. Olive, "Supersymmetry, supergravity theories and the dual spinor model," Nucl. Phys. B **122**, 253 (1977).
5. E. Cremmer, B. Julia and J. Scherk, "Supergravity theory in 11 dimensions," Phys. Lett. B **76**, 409 (1978).
6. C. M. Hull and P. K. Townsend, "Unity of superstring dualities," Nucl. Phys. B **438**, 109 (1995) [arXiv:hep-th/9410167].
7. E. Witten, "String theory dynamics in various dimensions," Nucl. Phys. B **443**, 85 (1995) [arXiv:hep-th/9503124].
8. J. M. Maldacena, "The large N limit of superconformal field theories and supergravity," Adv. Theor. Math. Phys. **2**, 231 (1998) [Int. J. Theor. Phys. **38**, 1113 (1999)] [arXiv:hep-th/9711200].
9. L. Brink, O. Lindgren and B. E. W. Nilsson, "The ultraviolet finiteness of the N = 4 Yang–Mills theory," Phys. Lett. B **123**, 323 (1983).
10. S. Mandelstam, "Light cone superspace and the ultraviolet finiteness of the N=4 model," Nucl. Phys. B **213**, 149 (1983).
11. L. Brink, O. Lindgren and B. E. W. Nilsson, "N = 4 Yang–Mills theory on the light cone," Nucl. Phys. B **212**, 401 (1983).
12. A. K. H. Bengtsson, I. Bengtsson and L. Brink, "Cubic Interaction terms for arbitrary spin" Nucl. Phys. B **227**, 31 (1983).
13. P. A. M. Dirac, "Forms of relativistic dynamics," Rev. Mod. Phys. **21**, 392 (1949).
14. I. Bengtsson, M. Cederwall and O. Lindgren, "Light cone actions for gravity and higher spins: some further results," GOTEBORG-83-55.
15. I. Bengtsson, "A note on the connection between spin and statistics," Phys. Rev. D **31**, 2525 (1985).
16. S. Ananth, L. Brink and P. Ramond, "Eleven-dimensional supergravity in light-cone superspace," JHEP **0505**, 003 (2005) [arXiv:hep-th/0501079].
17. S. Ananth, L. Brink, R. Heise and H. G. Svendsen, "The N = 8 supergravity Hamiltonian as a quadratic form," Nucl. Phys. B **753**, 195 (2006) [arXiv:hep-th/0607019].
18. S. Ananth, L. Brink, S. S. Kim and P. Ramond, "Non-linear realization of PSU(2,2|4) on the light-cone," Nucl. Phys. B **722**, 166 (2005) [arXiv:hep-th/0505234].
19. L. Brink, M. B. Green and J. H. Schwarz, "Ten-Dimensional Supersymmetric Yang–Mills theory with SO(8) – covariant light cone superfields," Nucl. Phys. B **223**, 125 (1983).
20. S. Ananth, L. Brink and P. Ramond, "Oxidizing SuperYang–Mills from (N = 4, d = 4) to (N = 1, d = 10)," JHEP **0407**, 082 (2004) [arXiv:hep-th/0405150].

21. L. Brink, S. S. Kim and P. Ramond, "$E_7(7)$ on the light cone," JHEP **0806**, 034 (2008) [arXiv:0801.2993 [hep-th]].
22. B. de Wit and D. Z. Freedman, "On SO(8) extended supergravity," Nucl. Phys. B **130**, 105 (1977).
23. B. de Wit, "Properties of SO(8) extended supergravity," Nucl. Phys. B **158**, 189 (1979).
24. B. de Wit and H. Nicolai, "N = 8 supergravity," Nucl. Phys. B **208**, 323 (1982).
25. L. Brink, S-S. Kim and P. Ramond, in preparation.
26. L. Brink and P. S. Howe, "The N = 8 supergravity in superspace," Phys. Lett. B **88**, 268 (1979).

Strongly Hyperbolic Extensions of the ADM Hamiltonian

J. David Brown

Abstract The ADM Hamiltonian formulation of general relativity with prescribed lapse and shift is a weakly hyperbolic system of partial differential equations. In general weakly hyperbolic systems are not mathematically well posed. For well posedness, the theory should be reformulated so that the complete system, evolution equations plus gauge conditions, is (at least) strongly hyperbolic. Traditionally, reformulation has been carried out at the level of equations of motion. This typically destroys the variational and Hamiltonian structures of the theory. Here I show that one can extend the ADM formalism to (a) incorporate the gauge conditions as dynamical equations and (b) affect the hyperbolicity of the complete system, all while maintaining a Hamiltonian description. The extended ADM formulation is used to obtain a strongly hyperbolic Hamiltonian description of Einstein's theory that is generally covariant under spatial diffeomorphisms and time reparametrizations, and has physical characteristics. The extended Hamiltonian formulation with $1+\log$ slicing and gamma-driver shift conditions is weakly hyperbolic.

1 Introduction

This paper is dedicated to Claudio Bunster in celebration of his 60th birthday. In a remarkable body of work Claudio showed that we can view the Hamiltonian formulation of general relativity as fundamental. (See in particular refs. [14, 26, 27].) He considered the requirement that the sequence of spatial three-geometries evolved by the Hamiltonian should be interpretable as a four-dimensional spacetime. From this assertion and a few modest assumptions he was able to derive the ADM Hamiltonian [1, 8] of general relativity. A number of deep insights into the nature of gravity and matter came from his analysis, including the role of gauge symmetries in electrodynamics and Yang–Mills theories and the necessity for all matter fields to couple to gravity.

J.D. Brown
Department of Physics, North Carolina State University, Raleigh, NC 27695, USA
e-mail: david_brown@ncsu.edu

M. Henneaux, J. Zanelli (eds.), *Quantum Mechanics of Fundamental Systems: The Quest for Beauty and Simplicity*, DOI 10.1007/978-0-387-87499-9_6,

In general relativity we are faced with the practical problem of predicting the future evolution of strongly gravitating systems, including interacting black holes and neutron stars. Such problems fall into the realm of numerical relativity. Naturally, the first attempts at numerical modeling in general relativity were based on the ADM Hamiltonian equations. By the early 1990s it became clear that the ADM equations were not appropriate for numerical computation because they are not well posed in a mathematical sense. What followed was more than a decade of activity in which the ADM equations were rewritten in a variety of ways. The goal was to produce a well posed system of partial differential equations (PDEs) for Einstein's theory. One strategy for modifying the ADM equations was to add multiples of the constraints to the right-hand sides. Another strategy was to introduce new independent variables defined as combinations of metric tensor components and their spatial derivatives. This later strategy introduces new constraints into the system, namely, the constraints that the definitions of the new variables should hold for all time.

In a practical sense, the effort to re-express the ADM equations has been successful. Currently there are a number of formulations of the Einstein evolution equations that appear to be "good enough," the most popular for numerical work being the BSSN system [2, 24]. BSSN relies on a conformal splitting of the metric and extrinsic curvature. It introduces new independent variables, the "conformal connection functions," defined as the trace (in its lower indices) of the Christoffel symbols built from the conformal metric.

BSSN and the other "modern" formulations of Einstein's theory are very clever. But at a basic level, these formulations are obtained through a manipulation of the equations of motion as a system of PDEs. What is invariably lost is the beautiful Hamiltonian structure found in the ADM formulation. In this paper I present a systematic procedure that can be used to modify the ADM equations in an effort to obtain a good system of PDEs without losing the Hamiltonian structure.

A good system of PDEs is one that is mathematically well posed. As a general rule, a system formulated in space without boundaries must be strongly hyperbolic to be well posed. If boundaries are present, an even stronger notion of hyperbolicity, symmetric hyperbolicity, is needed to prove well posedness. We are interested in extensions of the ADM equations that, like the ADM equations themselves, have first-order time and second-order space derivatives. It turns out that a simple prescription can be given to test for strong hyperbolicity in such systems of PDEs. The justification for this prescription requires a rather deep mathematical analysis, but the prescription itself is fairly easy to apply. In Sect. 2, I discuss hyperbolicity and justify the prescription for well posedness with heuristic arguments.

Another issue that has become apparent from recent numerical work is the benefit, in practice, of incorporating the slicing and coordinate conditions (gauge conditions) as dynamical equations. That is, the lapse function and shift vector are not fixed a priori but are determined along with the other fields through evolution equations of their own. The hyperbolicity of the entire system of PDEs, including the equations for the lapse and shift, must be considered. The issues of gauge conditions and hyperbolicity cannot be separated.

In this paper I show that the Hamiltonian formulation of general relativity can be extended to (a) incorporate dynamical gauge conditions and (b) alter the level of hyperbolicity. In Sect. 3 the ADM formulation is enlarged by the introduction of momentum variables π and ρ_a conjugate to the lapse function α and shift vector β^a. In this way the lapse and shift become dynamical. The new momenta are primary constraints and they appear in the action with undetermined multipliers Λ and Ω^a. The usual Hamiltonian and momentum constraints, $\mathscr{H}$ and $\mathscr{M}_a$, are secondary constraints. This Hamiltonian formulation of Einstein's theory is not new [7, 12], and is not substantially different from the original ADM formulation – like the ADM formulation, it is only weakly hyperbolic. This is shown in Sect. 4.

The Hamiltonian formulation with dynamical lapse and shift is extended in Sect. 5 by allowing the multipliers Λ and Ω^a to depend on the canonical variables. This has two effects. First, it changes the evolution equations for the lapse and shift, yielding gauge conditions that depend on the dynamical variables. Second, it changes the principal parts of the evolution equations and potentially changes the level of hyperbolicity of the system. The hyperbolicity of the extended Hamiltonian formulation is considered in Sect. 6 for a fairly general choice of multipliers that preserves spatial diffeomorphism invariance and time reparametrization invariance. When the multipliers are chosen so that the evolution equations are strongly hyperbolic with physical characteristics, the system is equivalent in its principal parts to the generalized harmonic formulation of gravity [9, 16, 20]. It is also shown that the extended Hamiltonian formulation with $1 + \log$ slicing and the gamma-driver shift condition is only weakly hyperbolic. A few final remarks are presented in Sect. 7.

2 Strong Hyperbolicity for Quasilinear Hamiltonian Systems

Let q_μ, p_μ denote pairs of canonically conjugate fields. We will consider Hamiltonian systems for which Hamilton's equations are a quasilinear system of partial differential equations (PDEs). Thus we assume that the Hamiltonian H is a linear combination of terms that are at most quadratic in the momenta and spatial derivatives of the coordinates. More precisely, H should be a linear combination of terms $p_\mu p_\nu$, $(\partial_a q_\mu)(\partial_b q_\nu)$, $p_\mu(\partial_a q_\nu)$, p_μ, $(\partial_a q_\mu)$, and 1 with coefficients that depend on the q's.[1] (Here, ∂_a denotes the derivative with respect to the spatial coordinates.) One would like to show that Hamilton's equations are well posed as a system of PDEs. The subject of well posedness is a large, active area of research in mathematics and physics. In this section I present a very pedestrian account of the subject in the context of Hamiltonian field theory. Much more rigorous and complete discussions can be found elsewhere. (See, for example, refs. [6, 10, 11, 15, 19, 22, 23, 25].)

A well posed system is one whose solutions depend continuously on the initial data. For a well posed system, two sets of initial data that are close to one another

[1] Note that terms proportional to $\partial_a p_\mu$ are also allowed in H since they are related to terms of the form $p_\mu(\partial_a q_\nu)$ through integration by parts.

will evolve into solutions that remain close for some finite time. A system is not well posed if it supports modes whose growth rates increase without bound with increasing wave number. A concrete example is given below.

In analyzing well posedness we are primary concerned with the evolution in time of high wave number (short wavelength) perturbations of the initial data. For this purpose we can approximate the quasilinear system of PDEs by linearizing about a solution. That is, we expand the Hamiltonian to quadratic order in perturbations, which we denote δq_μ, δp_μ. We then look for Fourier modes of the form $\delta q_\mu = \bar{q}_\mu e^{i\omega t + ik_a x^a}/(i|k|)$, $\delta p_\mu = \bar{p}_\mu e^{i\omega t + ik_a x^a}$ with nonzero wave number k_a. Here, $|k| \equiv \sqrt{h^{ab}k_a k_b}$ is the norm of k_a defined in terms of a convenient metric h^{ab} (which could be the inverse of the physical spatial metric). If the ansatz for δq_μ, δp_μ is substituted into the linearized Hamilton's equations, the system becomes

$$\omega v = (|k|A - iB - C/|k|)v \tag{1}$$

where v is the column vector $v = (\bar{q}_1, \bar{q}_2, \ldots, \bar{p}_1, \bar{p}_2, \ldots)^T$ of Fourier coefficients. Equation (1) shows that the problem of finding perturbative modes with wave number k_a is equivalent to the eigenvalue problem for the matrix $(|k|A - iB - C/|k|)$. The eigenvector is v and the eigenvalue is the frequency ω.

What one is really doing in the construction above is replacing the system of PDEs with a pseudo-differential system. The factor of $i|k|$ in the denominator of δq_μ is, in effect, equivalent to a change of variables in which q_μ is replaced by $q_\mu/\sqrt{h^{ab}\partial_a\partial_b}$. In this way the second order (in space derivatives) PDEs are replaced with an equivalent first order pseudo-differential system [19].

The behavior of ω as $|k|$ becomes large depends on the leading order term A in the matrix $(|k|A - iB - C/|k|)$. The term A is the "principal symbol" of the system. It is constructed from the coefficients of the highest "weight" terms in the Hamiltonian, namely, the terms proportional to $p_\mu p_\nu$, $(\partial_a q_\mu)(\partial_b q_\nu)$, $p_\mu(\partial_a q_\nu)$ and $(\partial_a p_\mu)$. Note that it is not necessary to linearize the equations of motion (or expand the Hamiltonian to quadratic order) in order to find A. In practice we don't actually linearize, we simply identify the coefficients of the highest weight terms in the PDEs to form the matrix A.

If A has real eigenvalues and a complete set of eigenvectors that have smooth dependence on the unit vector $n_a \equiv k_a/|k|$, the system is said to be *strongly hyperbolic*. If A has real eigenvalues but the eigenvectors are not complete, the system is said to be *weakly hyperbolic*. It can be proved that a strongly hyperbolic system of quasilinear PDEs is well-posed [19].

Here is the rough idea. Let S denote the matrix whose rows are the left eigenvectors of A. Assuming strong hyperbolicity, the eigenvectors are complete and S^{-1} exists. The eigenvalue problem (1) can be written as $\omega\hat{v} = (|k|\hat{A} - i\hat{B} - \hat{C}/|k|)\hat{v}$ where $\hat{v} \equiv Sv$, $\hat{A} \equiv SAS^{-1}$, $\hat{B} \equiv SBS^{-1}$, and $\hat{C} \equiv SCS^{-1}$. Note that $\hat{A}$ is diagonal with entries equal to the (real) eigenvalues. Let a dagger (†) denote the Hermitian conjugate (complex conjugate $*$ plus transpose T). Since $\hat{A}^\dagger = \hat{A}$, we find

$$
\begin{aligned}
(\omega-\omega^*)\hat{v}^\dagger\hat{v} &= \hat{v}^\dagger(\omega\hat{v})-(\omega\hat{v})^\dagger\hat{v} \\
&= \hat{v}^\dagger(|k|\hat{A}-i\hat{B}-\hat{C}/|k|)\hat{v}-\hat{v}^\dagger(|k|\hat{A}+i\hat{B}^T-\hat{C}^T/|k|)\hat{v} \\
&= -i\hat{v}^\dagger(M+M^\dagger)\hat{v}
\end{aligned} \tag{2}
$$

where $M \equiv \hat{B}-i\hat{C}/|k|$. The left-hand side includes the factor $(\omega-\omega^*) = 2i\Im\omega = -2i\Re(i\omega)$, so (2) can be written as $2\Re(i\omega) = \hat{v}^\dagger(M+M^\dagger)\hat{v}/\hat{v}^\dagger\hat{v}$. It follows that $\Re(i\omega) \le \tau^{-1}$ where $2\tau^{-1}$ is the maximum over $|k|$ of the matrix norm of $(M+M^\dagger)$. [The matrix norm is the maximum of the real number $\hat{v}^\dagger(M+M^\dagger)\hat{v}/(\hat{v}^\dagger\hat{v})$.] From this argument we see that the growth rate for the mode k_a is bounded; it can grow no faster than $e^{t/\tau}$ where τ is independent of k_a.

Consider a simple example in one spatial dimension with two pairs of canonically conjugate fields, q_1, p_1 and q_2, p_2. Let the Hamiltonian be given by

$$
\begin{aligned}
H = \int dx\Big\{&\frac{1}{2}[(p_1)^2+(p_2)^2+(q_1')^2+(q_2')^2]+2p_2q_1'+2p_1q_2' \\
&+p_1(q_2+q_1)+p_2(q_2-q_1)+q_1q_2\Big\}.
\end{aligned} \tag{3}
$$

Hamilton's equations are

$$\dot{q}_1 = p_1+2q_2'+q_2+q_1, \tag{4}$$
$$\dot{q}_2 = p_2+2q_1'+q_2-q_1, \tag{5}$$
$$\dot{p}_1 = q_1''+2p_2'+p_2-p_1-q_2, \tag{6}$$
$$\dot{p}_2 = q_2''+2p_1'-p_1-p_2-q_1. \tag{7}$$

In this example the PDEs are linear so the linearization step is trivial: $q_\mu \to \delta q_\mu$, $p_\mu \to \delta p_\mu$. Now insert the ansatz $\delta q_\mu = \bar{q}_\mu e^{i\omega t+ikx}/(i|k|)$, $\delta p_\mu = \bar{p}_\mu e^{i\omega t+ikx}$. This yields

$$\omega\bar{q}_1 = |k|(2n\bar{q}_2+\bar{p}_1)-i(\bar{q}_1+\bar{q}_2), \tag{8}$$
$$\omega\bar{q}_2 = |k|(2n\bar{q}_1+\bar{p}_2)-i(\bar{q}_2-\bar{q}_1), \tag{9}$$
$$\omega\bar{p}_1 = |k|(\bar{q}_1+2n\bar{p}_2)-i(\bar{p}_2-\bar{p}_1)+\bar{q}_2/|k|, \tag{10}$$
$$\omega\bar{p}_2 = |k|(\bar{q}_2+2n\bar{p}_1)+i(\bar{p}_1+\bar{p}_2)+\bar{q}_1/|k|, \tag{11}$$

where $n \equiv k/|k|$ is the sign of the wave number k. Collecting the unknowns into a column vector $v = (\bar{q}_1,\bar{q}_2,\bar{p}_1,\bar{p}_2)^T$, we see that these equations become $\omega v = (|k|A-iB-C/|k|)v$ where the matrices are given by

$$
A = \begin{pmatrix} 0 & 2n & 1 & 0 \\ 2n & 0 & 0 & 1 \\ 1 & 0 & 0 & 2n \\ 0 & 1 & 2n & 0 \end{pmatrix}, \quad
B = \begin{pmatrix} 1 & 1 & 0 & 0 \\ -1 & 1 & 0 & 0 \\ 0 & 0 & -1 & 1 \\ 0 & 0 & -1 & -1 \end{pmatrix}, \quad
C = \begin{pmatrix} 0 & 0 & 0 & 0 \\ 0 & 0 & 0 & 0 \\ 0 & -1 & 0 & 0 \\ -1 & 0 & 0 & 0 \end{pmatrix}. \tag{12}
$$

The principal symbol A has real eigenvalues ± 1, ± 3, and a complete set of eigenvectors. Therefore this system is strongly hyperbolic. The modes with wave number k have frequencies $\omega = \pm|k| + \mathcal{O}(1/|k|)$ and $\omega = \pm 3|k| + \mathcal{O}(1/|k|)$. In particular the imaginary parts of the ω's do not grow with increasing $|k|$.

Now replace the terms $2p_2q_1' + 2p_1q_2'$ in the Hamiltonian with $p_2q_1' - p_1q_2'$. The principal symbol becomes

$$A = \begin{pmatrix} 0 & -n & 1 & 0 \\ n & 0 & 0 & 1 \\ 1 & 0 & 0 & n \\ 0 & 1 & -n & 0 \end{pmatrix} \tag{13}$$

while B and C are unchanged. The eigenvalues of A vanish and there are only two independent eigenvectors. Therefore this system is weakly hyperbolic. The modes with wave number k have frequencies $\omega = \pm i\sqrt{2|k|} + \mathcal{O}(1)$ and $\omega = \pm\sqrt{2|k|} + \mathcal{O}(1)$. The modes with frequency $\omega \approx -i\sqrt{2|k|}$ will grow in time at a rate that is unbounded as $|k|$ increases.

The system described by this last example is not well posed. Indeed, consider two initial data sets that differ from one another by terms $q_\mu \sim \bar{q}_\mu e^{ikx}/|k|^2$, $p_\mu \sim \bar{p}_\mu e^{ikx}/|k|$ where $(\bar{q}_1, \bar{q}_2, \bar{p}_1, \bar{p}_2)^T$ is an eigenvector with eigenvalue $\omega = -i\sqrt{2|k|} + \mathcal{O}(1)$. In the limit as $|k| \to \infty$ these terms vanish and the two initial data sets coincide. However, if we evolve these data sets the solutions will differ at finite time t by terms $q_\mu \sim \bar{q}_\mu e^{\sqrt{2|k|}t + ikx}/|k|^2$, $p_\mu \sim \bar{p}_\mu e^{\sqrt{2|k|}t+ikx}/|k|$. These terms do not vanish in the limit $|k| \to \infty$. This system is ill posed because the solution at finite time does not depend continuously on the initial data.

In some cases it may be possible to model a physical system with ill posed PDEs and to gain important physical insights through a formal analysis. Claudio's beautiful work on the (weakly hyperbolic) ADM equations is a perfect example! One can imagine that the initial data are analytic, in which case the Cauchy–Kowalewski theorem guarantees that a solution exists for a finite time. But most data, even smooth data, are not analytic. From a computational point of view, having an ill posed system is unacceptable. Numerical errors will always introduce modes with large wave numbers, with the size of $|k|$ limited only by the details of the numerical implementation. For example, with a finite difference algorithm the maximum $|k|$ is roughly $1/\Delta x$ where Δx is the grid spacing. In practice it does not take long for the numerical solution to become dominated by this highest-wave number mode. As the grid resolution is increased (Δx is decreased), the unwanted highest wave number mode grows even more quickly. For practical numerical studies, we need our system of PDEs to be well posed.

The analysis outlined above leads to the following test for strong hyperbolicity. We begin by constructing the principal symbol A from the principal parts of Hamilton's equations. The principal parts of the $\dot{q}_\mu$ equations are the terms proportional to p_μ and $\partial_a q_\mu$. In these terms we make the replacements $p_\mu \to \bar{p}_\mu$ and $\partial_a q_\mu \to n_a \bar{q}_\mu$. The principal parts of the $\dot{p}_\mu$ equations are the terms proportional to $\partial_a p_\mu$ and $\partial_a \partial_b q_\mu$. In these terms we make the replacements $\partial_a p_\mu \to n_a \bar{p}_\mu$ and $\partial_a \partial_b q_\mu \to n_a n_b \bar{q}_\mu$. The principal symbol A is the matrix formed from the coefficients of the $\bar{q}$'s and $\bar{p}$'s. The next step is to compute the eigenvalues and eigenvectors of A. If A has real eigenvalues and a complete set of eigenvectors that depend smoothly on n_a, the system is strongly hyperbolic and the initial value problem is well posed.

3 ADM with Dynamical Lapse and Shift

The Einstein–Hilbert action is $S = \int d^4x \sqrt{-\mathbf{g}}\mathbf{R}$ where $\mathbf{g}$ is the determinant of the spacetime metric and $\mathbf{R}$ is the spacetime curvature scalar. Units are chosen such that Newton's constant equals $1/(16\pi)$. With the familiar splitting of the spacetime metric into the spatial metric g_{ab}, lapse function α, and shift vector β^a, the action becomes

$$S[g,\alpha,\beta] = \int d^4x\,\alpha\sqrt{g}\left(R + K_{ab}K^{ab} - K^2\right). \tag{14}$$

The extrinsic curvature is defined by

$$K_{ab} \equiv -\frac{1}{2\alpha}\left(\dot{g}_{ab} - 2D_{(a}\beta_{b)}\right), \tag{15}$$

and D_a denotes the spatial covariant derivative. The Hamiltonian can be derived in a straightforward fashion if one recognizes that the action does not contain time derivatives of the lapse and shift. Time derivatives of the spatial metric appear through the combination K_{ab}. Thus, we introduce the momentum

$$P^{ab} \equiv \frac{\partial \mathscr{L}}{\partial \dot{g}_{ab}} = \sqrt{g}\left(Kg^{ab} - K^{ab}\right), \tag{16}$$

where the Lagrangian density $\mathscr{L}$ is the integrand of the action. This definition can be inverted for $\dot{g}_{ab}$ as a function of P^{ab} and used to define the Hamiltonian: $H \equiv \int d^3x \left(P^{ab}\dot{g}_{ab} - \mathscr{L}\right)$. This yields the ADM Hamiltonian

$$H = \int d^3x(\alpha\mathscr{H} + \beta^a\mathscr{M}_a), \tag{17}$$

where

$$\mathscr{H} \equiv \frac{1}{\sqrt{g}}\left(P^{ab}P_{ab} - P^2/2\right) - \sqrt{g}R, \tag{18}$$

$$\mathscr{M}_a \equiv -2D_b P_a^b \tag{19}$$

are the Hamiltonian and momentum constraints.

In the analysis above the lapse and shift are treated as non dynamical fields. They appear in the Hamiltonian form of the action,

$$S[g,P,\alpha,\beta] = \int_{t_i}^{t_f} dt \int d^3x \left\{ P^{ab}\dot{g}_{ab} - \alpha\mathscr{H} - \beta^a\mathscr{M}_a \right\}, \tag{20}$$

as undetermined multipliers. Here t_i and t_f are the initial and final times. Extremization of S with respect to α and β^a yields the constraints $\mathscr{H} = 0$ and $\mathscr{M}_a = 0$.

We can enlarge the ADM formulation to include the lapse and shift as dynamical variables [7, 12]. Consider again the action (14). In addition to the momentum P^{ab} conjugate to the spatial metric, we also define conjugate variables for the lapse and shift:

$$\pi \equiv \frac{\partial \mathscr{L}}{\partial \dot{\alpha}} = 0, \tag{21}$$

$$\rho_a \equiv \frac{\partial \mathscr{L}}{\partial \dot{\beta}^a} = 0. \tag{22}$$

This leads to primary constraints $\pi = 0$ and $\rho_a = 0$. The resulting Hamiltonian is not unique; it is only determined to within the addition of arbitrary multiples of the constraints:

$$H = \int d^3x \left(\alpha\mathscr{H} + \beta^a\mathscr{M}_a + \Lambda\pi + \Omega^a\rho_a\right). \tag{23}$$

The coefficients Λ and Ω^a appear as undetermined multipliers in the action, which now reads

$$\begin{aligned} &S[g,P,\alpha,\pi,\beta,\rho,\Lambda,\Omega] \\ &\quad = \int_{t_i}^{t_f} dt \int d^3x \left\{ P^{ab}\dot{g}_{ab} + \pi\dot{\alpha} + \rho_a\dot{\beta}^a - \alpha\mathscr{H} - \beta^a\mathscr{M}_a - \Lambda\pi - \Omega^a\rho_a \right\}. \end{aligned} \tag{24}$$

The equations of motion, $\delta S = 0$, are[2]

$$\dot{g}_{ab} = \mathscr{L}_\beta g_{ab} + \frac{\alpha}{\sqrt{g}}(2P_{ab} - Pg_{ab}), \tag{25}$$

$$\begin{aligned} \dot{P}^{ab} &= \mathscr{L}_\beta P^{ab} + \frac{\alpha}{\sqrt{g}}(\delta^a_c\delta^b_d - g^{ab}g_{cd}/4)(PP^{cd} - 2P^{ce}P^d_e) \\ &\quad - \alpha\sqrt{g}G^{ab} + \sqrt{g}(D^aD^b\alpha - g^{ab}D_cD^c\alpha) \end{aligned} \tag{26}$$

$$\dot{\alpha} = \Lambda, \tag{27}$$

$$\dot{\pi} = -\mathscr{H}, \tag{28}$$

$$\dot{\beta}^a = \Omega^a, \tag{29}$$

$$\dot{\rho}_a = -\mathscr{M}_a, \tag{30}$$

$$\pi = 0, \tag{31}$$

$$\rho_a = 0, \tag{32}$$

[2] Throughout this paper I ignore the issues that arise when space has boundaries [5, 21].

where G^{ab} denotes the spatial Einstein tensor and $\mathscr{L}_\beta$ is the Lie derivative with respect to β^a.

The equations above must hold for each time $t_i \leq t \leq t_f$. They are equivalent to the Einstein equations supplemented with evolution equations for the lapse function and shift vector. In particular, observe that π and ρ_a must vanish for all time by (31) and (32). It follows that the time derivatives of π and ρ_a must vanish. In turn, (28) and (30) imply that the usual Hamiltonian and momentum constraints are zero. Equations (25) and (26) are the familiar ADM evolution equations, and (27) and (29) supply evolution equations for the lapse and shift.

The evolution equations (25–30) are Hamilton's equations derived from the Hamiltonian (23). The time derivative of any function F of the canonical variables is $\dot{F} = \{F, H\}$ where the fundamental Poisson brackets relations are defined by $\{g_{ab}(x), P^{cd}(x')\} = \delta_{(a}^{(c}\delta_{b)}^{d)}\delta^3(x,x')$, $\{\alpha(x), \pi(x')\} = \delta^3(x,x')$, and $\{\beta^a(x), \rho_b(x')\} = \delta_b^a\delta^3(x,x')$. We can interpret Hamilton's equations as an initial value problem by following Dirac's reasoning for constrained Hamiltonian systems [13]. The initial data are chosen such that the primary constraints π and ρ_a vanish at the initial time. According to (28) and (30), these constraints will remain zero as long as $\mathscr{H}$ and $\mathscr{M}_a$ are constrained to vanish as well. Thus we impose $\mathscr{H} = 0$ and $\mathscr{M}_a = 0$ as secondary constraints. The complete set of constraints, $\pi = 0$, $\rho_a = 0$, $\mathscr{H} = 0$, and $\mathscr{M}_a = 0$ is first class.

4 Hyperbolicity of ADM with Dynamical Lapse and Shift

Hamilton's equations (25–30) are equivalent to the ADM equations plus evolution equations $\dot{\alpha} = \Lambda$, $\dot{\beta}^a = \Omega^a$ for the lapse and shift. Let us analyze the level of hyperbolicity of these PDEs. The principal parts of the $\dot{q}$ equations are the terms that are proportional to p's or first spatial derivatives of q's. The principal parts of the $\dot{p}$ equations are the terms that are proportional to first spatial derivatives of p's or second spatial derivatives of q's. Thus, we find

$$\hat{\partial}_0 g_{ab} \cong 2g_{c(a}\partial_{b)}\beta^c + \frac{\alpha}{\sqrt{g}}(2P_{ab} - Pg_{ab}), \tag{33}$$

$$\begin{aligned}\hat{\partial}_0 P^{ab} \cong\ & \frac{\alpha\sqrt{g}}{2} g^{ac}g^{bd}g^{ef}(\partial_e\partial_f g_{cd} - 2\partial_e\partial_{(c}g_{d)f} + \partial_c\partial_d g_{ef}) \\ & + \frac{\alpha\sqrt{g}}{2} g^{ab}g^{cd}g^{ef}(\partial_c\partial_e g_{df} - \partial_c\partial_d g_{ef}) + \sqrt{g}(g^{ac}g^{bd} - g^{ab}g^{cd})\partial_c\partial_d\alpha,\end{aligned} \tag{34}$$

$$\hat{\partial}_0 \alpha \cong -\beta^a\partial_a\alpha, \tag{35}$$

$$\hat{\partial}_0 \pi \cong \sqrt{g}g^{ab}g^{cd}(\partial_a\partial_c g_{bd} - \partial_a\partial_b g_{cd}) - \beta^a\partial_a\pi, \tag{36}$$

$$\hat{\partial}_0 \beta^a \cong -\beta^b\partial_b\beta^a, \tag{37}$$

$$\hat{\partial}_0 \rho_a \cong 2g_{ac}\partial_b P^{bc} - \beta^b\partial_b\rho_a, \tag{38}$$

where the symbol $\cong$ is used to denote equality up to lower order (non principal) terms. These equations have been expressed in terms of the operator $\hat{\partial}_0 \equiv \partial_t - \beta^a \partial_a$ so that the characteristic speeds (the eigenvalues of the principal symbol) are defined with respect to observers who are at rest in the spacelike slices.

We now construct the eigenvalue problem $\mu v = Av$ for the principal symbol A. The principal symbol is found from the coefficients on the right-hand sides of (33–38). These coefficients are divided by a factor of α so that the characteristic speeds will be expressed in terms of proper time rather than coordinate time. The result is

$$\mu \bar{g}_{ab} = \frac{2}{\alpha} n_{(a}\bar{\beta}_{b)} + \frac{1}{\sqrt{g}}[2\bar{P}_{ab} - g_{ab}(\bar{P}_{nn} + \bar{P}_{AB}\delta^{AB})], \tag{39}$$

$$\mu \bar{P}_{ab} = \frac{\sqrt{g}}{2}[\bar{g}_{ab} - 2n_{(a}\bar{g}_{b)n} + n_a n_b(\bar{g}_{nn} + \bar{g}_{AB}\delta^{AB}) - g_{ab}\bar{g}_{AB}\delta^{AB}] - \frac{\sqrt{g}}{\alpha}(g_{ab} - n_a n_b)\bar{\alpha}, \tag{40}$$

$$\mu \bar{\alpha} = -(\beta \cdot n/\alpha)\bar{\alpha}, \tag{41}$$

$$\mu \bar{\pi} = -\frac{\sqrt{g}}{\alpha}\bar{g}_{AB}\delta^{AB} - (\beta \cdot n/\alpha)\bar{\pi}, \tag{42}$$

$$\mu \bar{\beta}_a = -(\beta \cdot n/\alpha)\bar{\beta}_a, \tag{43}$$

$$\mu \bar{\rho}_a = \frac{2}{\alpha}\bar{P}_{na} - (\beta \cdot n/\alpha)\bar{\rho}_a. \tag{44}$$

where μ is the eigenvalue and $v = (\bar{g}_{ab}, \bar{P}_{ab}, \bar{\alpha}, \bar{\pi}, \bar{\beta}_a, \bar{\rho}_a)^T$ is the eigenvector. In these equations n^a is normalized with respect to the spatial metric, $n^a g_{ab} n^b = 1$, and a subscript n denotes contraction with n^a. We have also introduced an orthonormal diad e^a_A with $A = 1,2$ in the subspace orthogonal to n_a. That is, $n_a e^a_A = 0$ and $e^a_A g_{ab} e^b_B = \delta_{AB}$. A subscript A on a tensor (such as the metric g_{ab} or momentum P_{ab}) denotes contraction with e^a_A.

The eigenvalue problem (39–44) splits into scalar, vector and trace-free tensor blocks with respect to rotations about the normal direction n_a. The scalar block is

$$\mu \bar{g}_{nn} = \frac{2}{\alpha}\bar{\beta}_n + \frac{1}{\sqrt{g}}(\bar{P}_{nn} - \bar{P}_{AB}\delta^{AB}), \tag{45}$$

$$\mu \bar{g}_{AB}\delta^{AB} = -\frac{2}{\sqrt{g}}\bar{P}_{nn}, \tag{46}$$

$$\mu \bar{P}_{nn} = 0, \tag{47}$$

$$\mu \bar{P}_{AB}\delta^{AB} = -\frac{1}{2}\sqrt{g}\bar{g}_{AB}\delta^{AB} - \frac{2}{\alpha}\sqrt{g}\bar{\alpha}, \tag{48}$$

$$\mu \bar{\alpha} = -(\beta \cdot n/\alpha)\bar{\alpha}, \tag{49}$$

$$\mu \bar{\pi} = -\frac{\sqrt{g}}{\alpha}\bar{g}_{AB}\delta^{AB} - (\beta \cdot n/\alpha)\bar{\pi}, \tag{50}$$

$$\mu \bar{\beta}_n = -(\beta \cdot n/\alpha)\bar{\beta}_n, \tag{51}$$

$$\mu \bar{\rho}_n = \frac{2}{\alpha}\bar{P}_{nn} - (\beta \cdot n/\alpha)\bar{\rho}_n. \tag{52}$$

This block has eigenvalues 0 and $-(\beta \cdot n/\alpha)$, each with multiplicity 4. There is only one eigenvector with eigenvalue 0, so the eigenvectors are not complete. The vector block is

$$\mu \bar{g}_{nA} = \frac{1}{\alpha}\bar{\beta}_A + \frac{2}{\sqrt{g}}\bar{P}_{nA}, \tag{53}$$

$$\mu \bar{P}_{nA} = 0, \tag{54}$$

$$\mu \bar{\beta}_A = -(\beta \cdot n/\alpha)\bar{\beta}_A, \tag{55}$$

$$\mu \bar{\rho}_A = \frac{2}{\alpha}\bar{P}_{nA} - (\beta \cdot n/\alpha)\bar{\rho}_A. \tag{56}$$

It has eigenvalues 0 and $-(\beta \cdot n/\alpha)$, each with multiplicity 2. There is only one eigenvector with eigenvalue 0, so the eigenvectors are not complete. Finally, the trace-free tensor block is

$$\mu \bar{g}^{tf}_{AB} = \frac{2}{\sqrt{g}}\bar{P}^{tf}_{AB}, \tag{57}$$

$$\mu \bar{P}^{tf}_{AB} = \frac{\sqrt{g}}{2}\bar{g}^{tf}_{AB}. \tag{58}$$

This block has eigenvalues ± 1 and a complete set of eigenvectors. Because the eigenvectors for the scalar and vector blocks are not complete, the system (25–32) is weakly hyperbolic.

5 Extending the ADM Formulation

In the previous section we modified the ADM Hamiltonian formulation of general relativity so that the lapse function α and shift vector β^a are treated as dynamical variables. Their canonical conjugates are denoted π and ρ_a, respectively. The undetermined multipliers for the constraints $\pi = 0$, $\rho_a = 0$ are Λ and Ω^a. These multipliers are freely specifiable functions of space and time. They determine the slicing and spatial coordinate conditions through the equations of motion $\dot{\alpha} = \Lambda$ and $\dot{\beta}^a = \Omega^a$.

Here is the key observation. The histories that extremize the action are unchanged if we replace the multipliers by $\Lambda \to \Lambda + \hat{\Lambda}$ and $\Omega^a \to \Omega^a + \hat{\Omega}^a$, where $\hat{\Lambda}$ and $\hat{\Omega}^a$ are quasilinear functions of the canonical variables. By quasilinear, I mean that the principal parts of $\hat{\Lambda}$ and $\hat{\Omega}^a$ are linear in the momenta (P^{ab}, π and ρ_a) and first spatial derivatives of the coordinates ($\partial_c g_{ab}$, $\partial_c \alpha$ and $\partial_c \beta^a$) with coefficients that depend on the coordinates. With these replacements the action becomes

$$S[g,P,\alpha,\pi,\beta,\rho,\Lambda,\Omega] = \int_{t_i}^{t_f} dt \int d^3x \Big\{ P^{ab}\dot{g}_{ab} + \pi\dot{\alpha} + \rho_a\dot{\beta}^a - \alpha\mathscr{H} - \beta^a\mathscr{M}_a - (\Lambda + \hat{\Lambda})\pi - (\Omega^a + \hat{\Omega}^a)\rho_a \Big\}, \tag{59}$$

and the Hamiltonian is

$$H = \int d^3x \left(\alpha\mathscr{H} + \beta^a \mathscr{M}_a + (\Lambda + \hat{\Lambda})\pi + (\Omega^a + \hat{\Omega}^a)\rho_a\right). \tag{60}$$

The solutions to the equations of motion are unaltered because the extra terms are proportional to the constraints $\pi = 0$, $\rho_a = 0$. In the Hamiltonian formulation we can dispense with the original multipliers Λ and Ω^a; that is, these quantities can be absorbed into the functions $\hat{\Lambda}$ and $\hat{\Omega}^a$.

With $\hat{\Lambda}$ and $\hat{\Omega}^a$ restricted to be quasilinear in the momenta and first spatial derivatives of the coordinates, the equations of motion become

$$\dot{g}_{ab} = \mathscr{L}_\beta g_{ab} + \frac{\alpha}{\sqrt{g}}(2P_{ab} - Pg_{ab}) + \frac{\partial\hat{\Lambda}}{\partial P^{ab}}\pi + \frac{\partial\hat{\Omega}^c}{\partial P^{ab}}\rho_c, \tag{61}$$

$$\begin{aligned}\dot{P}^{ab} &= \mathscr{L}_\beta P^{ab} + \frac{\alpha}{\sqrt{g}}(\delta^a_c\delta^b_d - g^{ab}g_{cd}/4)(PP^{cd} - 2P^{ce}P^d_e) \\ &\quad -\alpha\sqrt{g}G^{ab} + \sqrt{g}(D^aD^b\alpha - g^{ab}D_cD^c\alpha) \\ &\quad -\frac{\partial\hat{\Lambda}}{\partial g_{ab}}\pi - \frac{\partial\hat{\Omega}^c}{\partial g_{ab}}\rho_c + \partial_d\left(\frac{\partial\hat{\Lambda}}{\partial(\partial_d g_{ab})}\pi\right) + \partial_d\left(\frac{\partial\hat{\Omega}^c}{\partial(\partial_d g_{ab})}\rho_c\right),\end{aligned} \tag{62}$$

$$\dot{\alpha} = \Lambda + \hat{\Lambda} + \frac{\partial\hat{\Lambda}}{\partial\pi}\pi + \frac{\partial\hat{\Omega}^c}{\partial\pi}\rho_c, \tag{63}$$

$$\dot{\pi} = -\mathscr{H} - \frac{\partial\hat{\Lambda}}{\partial\alpha}\pi - \frac{\partial\hat{\Omega}^c}{\partial\alpha}\rho_c + \partial_d\left(\frac{\partial\hat{\Lambda}}{\partial(\partial_d\alpha)}\pi\right) + \partial_d\left(\frac{\partial\hat{\Omega}^c}{\partial(\partial_d\alpha)}\rho_c\right), \tag{64}$$

$$\dot{\beta}^a = \Omega^a + \hat{\Omega}^a + \frac{\partial\hat{\Lambda}}{\partial\rho_a}\pi + \frac{\partial\hat{\Omega}^c}{\partial\rho_a}\rho_c, \tag{65}$$

$$\dot{\rho}_a = -\mathscr{M}_a - \frac{\partial\hat{\Lambda}}{\partial\beta^a}\pi - \frac{\partial\hat{\Omega}^c}{\partial\beta^a}\rho_c + \partial_d\left(\frac{\partial\hat{\Lambda}}{\partial(\partial_d\beta^a)}\pi\right) + \partial_d\left(\frac{\partial\hat{\Omega}^c}{\partial(\partial_d\beta^a)}\rho_c\right), \tag{66}$$

$$\pi = 0, \tag{67}$$

$$\rho_a = 0, \tag{68}$$

Equations (61) and (62) are the usual ADM equations apart from terms proportional to the constraints, $\pi = 0$ and $\rho_a = 0$. The equations that govern the slicing and spatial coordinates are generalized by the presence of the functions $\hat{\Lambda}$ and $\hat{\Omega}^a$. Apart from terms that vanish with the constraints $\pi = 0$, $\rho_a = 0$, the evolution equation for the lapse becomes $\dot{\alpha} = \Lambda + \hat{\Lambda}$ and the evolution equation for the shift becomes $\dot{\beta}^a = \Omega^a + \hat{\Omega}^a$. The equations for $\dot{\pi}$ and $\dot{\rho}_a$ are modified, but once again we see that the complete set of constraints $\pi = 0$, $\rho_a = 0$, $\mathscr{H} = 0$, and $\mathscr{M}_a = 0$ is first class.

In principle we can choose $\hat{\Lambda}$ and $\hat{\Omega}^a$ to be any set of quasilinear functions of the canonical variables. In practice we might want $\hat{\Lambda}$ and $\hat{\Omega}^a$ to satisfy certain transformation properties. For example we can restrict $\hat{\Lambda}$ to be a scalar and $\hat{\Omega}^a$ to be a contravariant vector under spatial diffeomorphisms. This allows us to maintain a geometrical interpretation of the equations of motion. In particular this allows us to prepare and evolve identical geometrical data using different spatial coordinate systems.

Another property that can be imposed on the formalism is reparametrization invariance [13]. This is invariance under a change of coordinate labels t for the constant time slices. Consider the infinitesimal transformation $t \to t - \varepsilon(t)$. In the usual ADM system, the variables g_{ab} and P^{ab} transform as scalars under time reparametrization: $\delta g_{ab} = \varepsilon \dot{g}_{ab}$ and $\delta P^{ab} = \varepsilon \dot{P}^{ab}$. The time derivative of the metric, $\dot{g}_{ab}$, transforms as a covariant vector. In one dimension a covariant vector transforms in the same way as a scalar density of weight $+1$. It follows that the term $P^{ab}\dot{g}_{ab}$ that appears in the action is a weight $+1$ scalar density. For reparametrization invariance to hold, the integrand of the action should transform as a weight $+1$ scalar density since it is integrated over t. In particular the lapse function α and shift vector β^a, which multiply the scalars $\mathscr{H}$ and $\mathscr{M}_a$, must transform as scalar densities of weight $+1$.

Observe that the time derivatives $\dot{\alpha}$ and $\dot{\beta}^a$ are constructed from coordinate derivatives of scalar densities and, as a consequence, these quantities do not transform as tensors or tensor densities. We need to replace the coordinate derivatives (dots) with covariant derivatives. We can do this by choosing a background metric for the time direction. This should be viewed as part of the gauge fixing process. Now, the physical metric for the time manifold is α^2, so let $\tilde{\alpha}^2$ denote the background metric. The covariant derivative built from $\tilde{\alpha}^2$, acting on the densities α and β^a, is defined by

$$\mathring{\alpha} \equiv \dot{\alpha} - (\dot{\tilde{\alpha}}/\tilde{\alpha})\alpha, \tag{69}$$

$$\mathring{\beta}^a \equiv \dot{\beta}^a - (\dot{\tilde{\alpha}}/\tilde{\alpha})\beta^a. \tag{70}$$

The extra terms needed for reparametrization invariance can be built into the action or Hamiltonian by including a term $(\dot{\tilde{\alpha}}/\tilde{\alpha})\alpha$ in the function $\hat{\Lambda}$ and a term $(\dot{\tilde{\alpha}}/\tilde{\alpha})\beta^a$ in the function $\hat{\Omega}^a$.

With the appropriate terms included in $\hat{\Lambda}$ and $\hat{\Omega}^a$, the time derivatives of the lapse and shift appear in the action only in the combinations $\pi\mathring{\alpha}$ and $\rho_a\mathring{\beta}^a$. Since $\mathring{\alpha}$ and $\mathring{\beta}^a$ are covariant vector densities of weight $+1$, we see that π and ρ_a must transform as contravariant vectors with no density weight. In one dimension, contravariant vectors transform in the same way as a scalar density of weight -1. We will consider π and ρ_a to be scalar densities of weight -1 under time reparametrization. It follows that, apart from the terms $(\dot{\tilde{\alpha}}/\tilde{\alpha})\alpha$ and $(\dot{\tilde{\alpha}}/\tilde{\alpha})\beta^a$, the multipliers $\Lambda + \hat{\Lambda}$ and $\Omega^a + \hat{\Omega}^a$ should transform as scalar densities of weight $+2$.

We have now established the rules for adding terms to the functions $\hat{\Lambda}$ and $\hat{\Omega}^a$ such that the resulting formulation is invariant under time reparametrizations: these terms must be weight $+2$ densities built from the scalars g_{ab}, P^{ab}, the weight $+1$

densities α, β^a, and the weight -1 densities π, ρ_a. We can also insist that $\hat{\Lambda}$ and $\hat{\Omega}^a$ should be, respectively, a scalar and a contravariant vector under spatial diffeomorphisms. With these properties in mind, a fairly general form for $\hat{\Lambda}$ is

$$\hat{\Lambda} = (\dot{\tilde{\alpha}}/\tilde{\alpha})\alpha + \beta^a D_a \alpha - C_1 \frac{\alpha^2}{\sqrt{g}} P + C_4 \frac{\alpha^3}{\sqrt{g}} \pi. \tag{71}$$

The first term in required for reparametrization invariance. The second term will allow us to combine shift vector terms into a Lie derivative $\mathscr{L}_\beta$ acting on α. The terms multiplied by constants C_1 and C_4 are principal terms that will affect the hyperbolicity of the resulting equations. There are other principal terms that one can add, such as $\alpha^2 \beta^a \rho_a/\sqrt{g}$, but the form above will be general enough for present purposes. There are also lower order terms that one can add to $\hat{\Lambda}$.

For $\hat{\Omega}^a$ we must construct a spatial vector that [apart from the term $(\dot{\tilde{\alpha}}/\tilde{\alpha})\beta^a$] transforms as a weight $+2$ density under time reparametrizations. There are several obvious ways to construct a spatial vector from the canonical variables at hand. There are some possibilities that are not so obvious. Recall that the difference of two connections is a tensor. Thus, the combination $\Gamma^a_{bc} - \tilde{\Gamma}^a_{bc}$ is a spatial tensor if Γ^a_{bc} are the Christoffel symbols built from the physical metric g_{ab} and $\tilde{\Gamma}^a_{bc}$ are the Christoffel symbols built from a background metric $\tilde{g}_{ab}$. In setting up a numerical calculation, for example, on a logically Cartesian grid, we can take $\tilde{g}_{ab}$ to be the flat metric in Cartesian coordinates. Again, we view the introduction of the background structure $\tilde{g}_{ab}$ as part of the gauge fixing process.

The general form for $\hat{\Omega}^a$ that we will consider is

$$\begin{aligned}\hat{\Omega}^a &= (\dot{\tilde{\alpha}}/\tilde{\alpha})\beta^a + \beta^b \tilde{D}_b \beta^a + C_2 \alpha^2 (\Gamma^a_{bc} - \tilde{\Gamma}^a_{bc}) g^{bc} \\ &\quad + C_3 \alpha^2 (\Gamma^c_{cb} - \tilde{\Gamma}^c_{cb}) g^{ab} - C_5 \alpha D^a \alpha - C_6 \frac{\alpha^3}{\sqrt{g}} \rho^a.\end{aligned} \tag{72}$$

where $\tilde{D}_a$ is the covariant derivative compatible with $\tilde{g}_{ab}$. The first term is required for time reparametrization invariance. The second term will allow us to combine time derivatives and shift vector terms into the operator $\hat{\partial}_0$ in the principle parts of the equations for β^a and ρ_a. The remaining terms will modify the principal parts of the equations of motion and can affect the hyperbolicity of the system. There are other principal terms that we could add to $\hat{\Omega}^a$, such as $\alpha^2 \pi \beta^a/\sqrt{g}$ or $\alpha P^{ab} \beta_b/\sqrt{g}$. We can also add lower order terms.

With these expressions for $\hat{\Lambda}$ and $\hat{\Omega}^a$, we find the following equations of motion by varying the action (59):

$$\mathring{g}_{ab} = \mathscr{L}_\beta g_{ab} + \frac{\alpha}{\sqrt{g}}(2P_{ab} - P g_{ab}) - C_1 \frac{\alpha^2}{\sqrt{g}} \pi g_{ab}, \tag{73}$$

$$\begin{aligned}
\mathring{P}^{ab} &= \mathscr{L}_\beta P^{ab} + \frac{\alpha}{\sqrt{g}}(\delta^a_c\delta^b_d - g^{ab}g_{cd}/4)(PP^{cd} - 2P^{ce}P^d_e) - \alpha\sqrt{g}G^{ab} \\
&\quad + \sqrt{g}(D^aD^b\alpha - g^{ab}D_cD^c\alpha) + C_1\frac{\alpha^2}{\sqrt{g}}(P^{ab} - Pg^{ab}/2)\pi + C_4\frac{\alpha^3}{2\sqrt{g}}\pi^2 g^{ab} \\
&\quad + C_2 D^{(a}(\rho^{b)}\alpha^2) - C_5\alpha\rho^{(a}D^{b)}\alpha + \frac{1}{2}(C_3 - C_2)D_c(\alpha^2\rho^c)g^{ab} \\
&\quad + C_2\alpha^2\rho_e(\Gamma^e_{cd} - \tilde{\Gamma}^e_{cd})g^{ac}g^{bd} + C_3\alpha^2(\Gamma^d_{cd} - \tilde{\Gamma}^d_{cd})\rho^{(a}g^{b)c} \\
&\quad - C_6\frac{\alpha^3}{\sqrt{g}}(\rho^a\rho^b + \rho_c\rho^c g^{ab}/2), \qquad (74)
\end{aligned}$$

$$\mathring{\alpha} = \mathscr{L}_\beta\alpha + \Lambda - C_1\frac{\alpha^2}{\sqrt{g}}P + 2C_4\frac{\alpha^3}{\sqrt{g}}\pi, \qquad (75)$$

$$\begin{aligned}
\mathring{\pi} &= \mathscr{L}_\beta\pi - \mathscr{H} + 2C_1\frac{\alpha}{\sqrt{g}}P\pi - 3C_4\frac{\alpha^2}{\sqrt{g}}\pi^2 - 2C_2\alpha(\Gamma^a_{bc} - \tilde{\Gamma}^a_{bc})g^{bc}\rho_a \\
&\quad - 2C_3\alpha(\Gamma^b_{ab} - \tilde{\Gamma}^b_{ab})\rho^a - C_5\alpha D_a\rho^a + 3C_6\frac{\alpha^2}{\sqrt{g}}\rho_a\rho^a, \qquad (76)
\end{aligned}$$

$$\begin{aligned}
\mathring{\beta}^a &= \beta^b\tilde{D}_b\beta^a + \Omega^a + C_2\alpha^2(\Gamma^a_{bc} - \tilde{\Gamma}^a_{bc})g^{bc} + C_3\alpha^2(\Gamma^c_{bc} - \tilde{\Gamma}^c_{bc})g^{ab} \\
&\quad - C_5\alpha D^a\alpha - 2C_6\frac{\alpha^3}{\sqrt{g}}\rho^a, \qquad (77)
\end{aligned}$$

$$\mathring{\rho}_a = \tilde{D}_b(\beta^b\rho_a) - \rho_b\tilde{D}_a\beta^b - \mathscr{M}_a - \pi D_a\alpha, \qquad (78)$$

$$\pi = 0, \qquad (79)$$

$$\rho_a = 0. \qquad (80)$$

These equations are generally covariant under spatial diffeomorphisms and time reparametrizations. Equations (73–78) are generated through the Poisson brackets by the Hamiltonian (60). Note that for g_{ab} and P^{ab}, which are scalars under time reparametrization, the covariant time derivative (circle) is equivalent to a coordinate time derivative (dot).

6 Hyperbolicity of the Extended ADM Formulation

The principal parts of the extended ADM evolution equations (73–78) are:

$$\hat{\partial}_0 g_{ab} \cong 2g_{c(a}\partial_{b)}\beta^c + \frac{\alpha}{\sqrt{g}}(2P_{ab} - Pg_{ab}) - C_1\frac{\alpha^2}{\sqrt{g}}\pi g_{ab}, \qquad (81)$$

$$\begin{aligned}\hat{\partial}_0 P^{ab} &\cong \frac{\alpha\sqrt{g}}{2} g^{ac} g^{bd} g^{ef} (\partial_e \partial_f g_{cd} - 2\partial_e \partial_{(c} g_{d)f} + \partial_c \partial_d g_{ef}) \\ &\quad + \frac{\alpha\sqrt{g}}{2} g^{ab} g^{cd} g^{ef} (\partial_c \partial_e g_{df} - \partial_c \partial_d g_{ef}) + \sqrt{g}(g^{ac} g^{bd} - g^{ab} g^{cd}) \partial_c \partial_d \alpha \\ &\quad + \alpha^2 \left[C_2 g^{c(a} g^{b)d} + (C_3 - C_2) g^{ab} g^{cd}/2 \right] \partial_c \rho_d, \end{aligned} \tag{82}$$

$$\hat{\partial}_0 \alpha \cong -C_1 \frac{\alpha^2}{\sqrt{g}} P + 2C_4 \frac{\alpha^3}{\sqrt{g}} \pi, \tag{83}$$

$$\hat{\partial}_0 \pi \cong \sqrt{g} g^{ab} g^{cd} (\partial_a \partial_c g_{bd} - \partial_a \partial_b g_{cd}) - C_5 \alpha g^{ab} \partial_a \rho_b, \tag{84}$$

$$\begin{aligned}\hat{\partial}_0 \beta^a &\cong \alpha^2 \left[C_2 g^{ac} g^{bd} + (C_3 - C_2) g^{ab} g^{cd}/2 \right] \partial_b g_{cd} \\ &\quad - C_5 \alpha g^{ab} \partial_b \alpha - 2C_6 \frac{\alpha^3}{\sqrt{g}} \rho^a, \end{aligned} \tag{85}$$

$$\hat{\partial}_0 \rho_a \cong 2 g_{ac} \partial_b P^{bc}, \tag{86}$$

From here it is easy to construct the eigenvalue problem for the principal symbol. The symbol decomposes into scalar, vector, and trace-free tensor blocks under rotations orthogonal to the normal vector $n_a \equiv k_a/|k|$. For the scalar sector, we find

$$\mu \bar{g}_{nn} = \frac{2}{\alpha} \bar{\beta}_n + \frac{1}{\sqrt{g}} (\bar{P}_{nn} - \bar{P}_{AB} \delta^{AB}) - C_1 \frac{\alpha}{\sqrt{g}} \bar{\pi}, \tag{87}$$

$$\mu \bar{g}_{AB} \delta^{AB} = -\frac{2}{\sqrt{g}} \bar{P}_{nn} - 2C_1 \frac{\alpha}{\sqrt{g}} \bar{\pi}, \tag{88}$$

$$\mu \bar{P}_{nn} = \frac{\alpha}{2} (C_3 + C_2) \bar{\rho}_n, \tag{89}$$

$$\mu \bar{P}_{AB} \delta^{AB} = -\frac{1}{2} \sqrt{g} \bar{g}_{AB} \delta^{AB} - \frac{2}{\alpha} \sqrt{g} \bar{\alpha} + \alpha (C_3 - C_2) \bar{\rho}_n, \tag{90}$$

$$\mu \bar{\alpha} = -C_1 \frac{\alpha}{\sqrt{g}} (\bar{P}_{nn} + \bar{P}_{AB} \delta^{AB}) + 2C_4 \frac{\alpha^2}{\sqrt{g}} \bar{\pi}, \tag{91}$$

$$\mu \bar{\pi} = -\frac{\sqrt{g}}{\alpha} \bar{g}_{AB} \delta^{AB} - C_5 \bar{\rho}_n, \tag{92}$$

$$\mu \bar{\beta}_n = \frac{\alpha}{2} (C_3 + C_2) \bar{g}_{nn} + \frac{\alpha}{2} (C_3 - C_2) \bar{g}_{AB} \delta^{AB} - C_5 \bar{\alpha} - 2C_6 \frac{\alpha^2}{\sqrt{g}} \bar{\rho}_n, \tag{93}$$

$$\mu \bar{\rho}_n = \frac{2}{\alpha} \bar{P}_{nn}. \tag{94}$$

Again, the subscripts n and A denote contraction with n^a and e^a_A, respectively. The vector block is

$$\mu \bar{g}_{nA} = \frac{1}{\alpha}\bar{\beta}_A + \frac{2}{\sqrt{g}}\bar{P}_{nA}, \tag{95}$$

$$\mu \bar{P}_{nA} = \frac{\alpha}{2}C_2\bar{\rho}_A, \tag{96}$$

$$\mu \bar{\beta}_A = C_2\alpha\bar{g}_{nA} - 2C_6\frac{\alpha^2}{\sqrt{g}}\bar{\rho}_A, \tag{97}$$

$$\mu \bar{\rho}_A = \frac{2}{\alpha}\bar{P}_{nA}, \tag{98}$$

and the trace-free tensor block is unmodified from before:

$$\mu \bar{g}^{tf}_{AB} = \frac{2}{\sqrt{g}}\bar{P}^{tf}_{AB}, \tag{99}$$

$$\mu \bar{P}^{tf}_{AB} = \frac{\sqrt{g}}{2}\bar{g}^{tf}_{AB}. \tag{100}$$

The eigenvalues for the scalar block are $\pm\sqrt{2C_1}$ and $\pm\sqrt{C_2+C_3}$, each with multiplicity two. The eigenvalues for the vector block are $\pm\sqrt{C_2}$, each with multiplicity two. For the tensor block the eigenvalues are ± 1 and the eigenvectors are complete.

The eigenvalues are the characteristic speeds with respect to observers at rest in the spacelike hypersurfaces. Let us choose the C's such that the characteristics are ± 1, that is, along the physical light cone. Then we must have $C_1 = 1/2$, $C_2 = 1$, and $C_3 = 0$. With this choice a careful analysis of the scalar block shows that the eigenvectors are complete only if $C_4 = 1/8$, $C_5 = 1$, and $C_6 = 1/2$. For these values of the constants the eigenvectors for the vector block are complete as well. It can be shown that with these values for the C's the principal parts of the equations of motion are equivalent to the generalized harmonic formulation of relativity [9, 16, 20]. This will be discussed elsewhere [4].

The Hamiltonian system (73–80) is strongly hyperbolic with the choice of C's above. For this formulation the gauge conditions are

$$\mathring{\alpha} = \beta^a D_a\alpha + \Lambda - \frac{\alpha^2}{2}\frac{P}{\sqrt{g}} + \frac{\alpha^3}{4}\frac{\pi}{\sqrt{g}}, \tag{101}$$

$$\mathring{\beta}^a = \beta^b\tilde{D}_b\beta^a + \Omega^a + \alpha^2(\Gamma^a_{bc} - \tilde{\Gamma}^a_{bc})g^{bc} - \alpha D^a\alpha - \frac{\alpha^3}{\sqrt{g}}\rho^a \tag{102}$$

If we choose Λ and Ω^a to vanish, these gauge equations become

$$\dot{\alpha} - \beta^a\partial_a\alpha = \alpha(\dot{\tilde{\alpha}}/\tilde{\alpha}) - \alpha^2 K, \tag{103}$$

$$\dot{\beta}^a - \beta^b\partial_b\beta^a = \beta^a(\dot{\tilde{\alpha}}/\tilde{\alpha}) + \alpha^2(\Gamma^a_{bc} - \tilde{\Gamma}^a_{bc})g^{bc} - \alpha g^{ab}\partial_b\alpha, \tag{104}$$

to within terms that vanish with the constraints $\pi = 0$, $\rho_a = 0$. Here, $K \equiv P/(2\sqrt{g})$ is the trace of the extrinsic curvature. These gauge conditions agree, to within lower order (non-principal) terms, with the gauge conditions for the generalized harmonic formulation of gravity [16].

It is not difficult to find choices for the constants (the C's) that make the system (73–80) strongly hyperbolic. The trick is to find a strongly hyperbolic system with desirable coordinate and slicing conditions. Conditions similar to (103), supplemented with "gauge driver" equations, have been applied to the binary black hole problem with mixed results [16, 17, 20]. With the BSSN evolution system, the gauge conditions that work well for the binary black hole problem are "1 + log slicing,"

$$\dot{\alpha} - \beta^a \partial_a \alpha = -2\alpha K, \tag{105}$$

and "gamma-driver shift." The gamma-driver condition is usually written as a system of two first order (in time) equations for β^a and an auxiliary field $B^a = 4\dot{\beta}^a/3$. These equations, along with suitable initial conditions, can be integrated to yield a single first order equation for the shift [28]. Expressed as either two equations or one, the gamma-driver condition depends on the trace (in its lower indices) of the Christoffel symbols built from the conformal metric. In terms of the physical metric, we can write the single-equation form of the gamma-driver shift condition as

$$\dot{\beta}^a - \beta^b \partial_b \beta^a = \frac{3}{4}\sqrt{g}^{2/3}\left[\Gamma^a_{bc} g^{bc} + \frac{1}{3} g^{ab}\Gamma^c_{bc}\right] - \eta\beta^a, \tag{106}$$

where η is a constant parameter. The first term on the right-hand side (apart from the factor of $3/4$) is the trace of the conformal Christoffel symbols.

Let us see if we can find a set of functions $\hat{\Lambda}$ and $\hat{\Omega}^a$ that yield the gauge conditions above, and then ask whether the resulting system is strongly hyperbolic. In this example we dispense with any attempt to construct a formulation that is covariant under spatial diffeomorphisms or time reparametrizations. Comparing (63) and (65) with the 1 + log slicing condition (105) and the gamma-driver shift condition (106), we find

$$\hat{\Lambda} = \beta^a \partial_a \alpha - \alpha P/\sqrt{g} + F_1 \pi + F_2^a \rho_a, \tag{107}$$

$$\hat{\Omega}^a = \beta^b \partial_b \beta^a + \frac{3}{4}\sqrt{g}^{2/3}\left[\Gamma^a_{bc} g^{bc} + \frac{1}{3} g^{ab}\Gamma^c_{bc}\right] - \eta\beta^a + F_3^a \pi + F_4^{ab} \rho_b, \tag{108}$$

where the F's are functions of the coordinates g_{ab}, α, and β^a (and not their spatial derivatives). In this case the analysis of hyperbolicity is complicated by the fact that the scalar and vector blocks of the principal symbol are coupled by the terms F_2^a, F_3^a, and F_4^{ab}. The combined scalar/vector block has real eigenvalues, but completeness of the eigenvectors can be achieved only if F_4^{ab} depends on the normal direction $n_a \equiv k_a/|k|$. This is not acceptable; the Hamiltonian cannot depend on the propagation direction of a perturbative solution. We conclude that there is no Hamiltonian of the form (60) that yields strongly hyperbolic equations with 1 + log slicing (105) and gamma-driver shift (106).

7 Concluding Remarks

This paper outlines a procedure for constructing Hamiltonian formulations of Einstein's theory with dynamical gauge conditions and varying levels of hyperbolicity. One can use this procedure as a tool to help identify well posed formulations of the evolution equations that also maintain Hamiltonian and variational structures. The issues of dynamical gauge conditions and hyperbolicity cannot be separated. They are both dictated by the dependence of the multipliers $\hat{\Lambda}$ and $\hat{\Omega}^a$ on the canonical variables. There are many possibilities that one can explore for this dependence.

The Hamiltonian and variational formulations of general relativity have shaped our perspective and provided deep insights into the theory. In addition, there are a number of practical uses for a Hamiltonian/variational formulation. With an action principle we can pass between spacetime and space-plus-time formulations by adding or removing momentum variables. We can develop a fully first order multi-symplectic version of the theory. We can also develop new computational techniques such as variational and symplectic integrators [3, 18].

One important issue that has not been addressed here is the constraint evolution system. In numerical simulations it is important to control the growth of constraint violations. This might be accomplished in the present framework by including appropriate terms in $\hat{\Lambda}$ and $\hat{\Omega}^a$ to ensure that the constraints are damped. For example, a damping term $-C\pi$ (where C is a positive constant) can be added to the $\dot{\pi}$ equation by including a lower order term $C\alpha$ in $\hat{\Lambda}$. This issue will be explored in more detail elsewhere [4].

The formalism outlined here can be further extend by introducing dynamical equations for Λ and Ω^a. This is accomplished by introducing momentum variables conjugate to these multipliers. The new momentum variables are primary constraints and are accompanied by a new set of undetermined multipliers. Dynamical equations for Λ and Ω^a are introduced by allowing the new multipliers to depend on the canonical variables. In this way we can construct gauge driver conditions similar to the ones used with the generalized harmonic formulation [17, 20]. We can also allow for gauge conditions that are expressed as systems of PDEs, such as the two-equation version of the gamma-driver shift condition.

Acknowledgments I would like to thank Claudio Bunster for his guidance and support and, most of all, for his inspiration through the years. I also thank Lee Lindblom, Mark Scheel, and Manuel Tiglio for discussions that helped stimulate the ideas presented here. I especially thank Olivier Sarbach for his patient explanations of hyperbolicity. This work was supported by NSF Grant PHY-0600402.

References

1. Arnowitt, A., Deser, S., Misner, C.W.: The dynamics of general relativity. In: Witten, L. (ed.) Gravitation: An Introduction to Current Research, pp. 227–265. Wiley, New York (1962)
2. Baumgarte, T.W., Shapiro, S.L.: On the numerical integration of Einstein's field equations. Phys. Rev. **D59**, 024007 (1999)

3. Brown, J.D.: Midpoint rule as a variational-symplectic integrator: Hamiltonian systems. Phys. Rev. **D74**, 024001 (2006)
4. Brown, J.D.: in preparation
5. Brown, J.D., Lau, S.R., York, J.W.: Action and energy of the gravitational field. Ann. Phys. **297**, 175–218 (2002)
6. Calabrese, G., Pullin, J., Sarbach, O., Tiglio, M.: Convergence and stability in numerical relativity. Phys. Rev. **D66**, 041501 (2002)
7. DeWitt, B.S.: Quantum theory of gravity. 1. The canonical theory. Phys. Rev. **160** 1113–1148 (1967)
8. Dirac, P.A.M.: The theory of gravitation in Hamiltonian form. Proc. R. Soc. Lond. **A246**, 333–343 (1958)
9. Friedrich, H.: On the hyperbolicity of Einstein's and other gauge field equations. Commun. Math. Phys. **100**, 525–543 (1985)
10. Gundlach, C., Martin-Garcia, J.M.: Hyperbolicity of second-order in space systems of evolution equations. Class. Quant. Grav. **23**, S387–S404 (2006)
11. Gustafsson, B., Kreiss, H.-O., Oliger, J.: Time Dependent Problems and Difference Methods. Wiley, New York (1995)
12. Hanson, A., Regge, T., Teitelboim, C.: Constrained Hamiltonian Systems. Accademia Nazionale Dei Lincei, Rome (1976)
13. Henneaux, M., Teitelboim, C.: Quantization of Gauge Systems. Princeton University Press, Princeton (1992)
14. Hojman, S.A., Kuchar, K., Teitelboim, C.: Geometrodynamics Regained. Ann. Phys. **96**, 88–135 (1976)
15. Kreiss, H.-O., Winicour, J.: Problems which are well-posed in a generalized sense with applications to the Einstein equations. Class. Quant. Grav. **23**, S405–S420 (2006)
16. Lindblom, L., Scheel, M.A., Kidder, L.E., Owen, R., Rinne, O.:A new generalized harmonic evolution system. Class. Quant. Grav. **23**, S447–S462 (2006)
17. Lindblom, L., Matthews, K.D., Rinne, O., Scheel, M.A.: Gauge drivers for the generalized harmonic Einstein equations. Phys. Rev. **D77**, 084001 (2008)
18. Marsden, J.E., West, M.: Discrete mechanics and variational integrators. Acta Numerica **10**, 357–514 (2001)
19. Nagy, G., Ortiz, O.E., Reula, O.A.: Strongly hyperbolic second order Einstein's evolution equations. Phys. Rev. **D70**, 044012 (2004)
20. Pretorius, F.: Numerical relativity using a generalized harmonic decomposition. Class. Quant. Grav. **22**, 425–452 (2005)
21. Regge, T., Teitelboim, C.: Role of surface integrals in the Hamiltonian formulation of general relativity. Ann. Phys. **88**, 286 (1974)
22. Reula, O.A.: Hyperbolic methods for Einstein's equations. Living Rev. Rel. **1**, 3 (1998)
23. Reula, O.A.: Strongly hyperbolic systems in general relativity. gr-qc/0403007 (2004)
24. Shibata, M., Nakamura, T.: Evolution of three-dimensional gravitational waves: Harmonic slicing case. Phys. Rev. **D52**, 5428–5444 (1995)
25. Stewart, J.M.: The Cauchy problem and the initial boundary value problem in numerical relativity. Class. Quant. Grav. **15**, 2865–2889 (1998)
26. Teitelboim, C.: How commutators of constraints reflect the space–time structure. Ann. Phys. **79**, 542–557 (1973)
27. Teitelboim, C.: The Hamiltonian structure of space–time. In: Held, A. (ed.) General Relativity and Gravitation, vol. 1, pp. 195–225. Plenum, New York (1980)
28. van Meter, J.R., Baker, J.G., Koppitz, M., Choi, D.-I.: How to move a black hole without excision: Gauge conditions for the numerical evolution of a moving puncture. Phys. Rev. **D73**, 124011 (2006)

Black Hole Entropy and the Problem of Universality

S. Carlip

Abstract To derive black hole thermodynamics in any quantum theory of gravity, one must introduce constraints that ensure that a black hole is actually present. For a large class of black holes, the imposition of such "horizon constraints" allows the use of conformal field theory methods to compute the density of states, reproducing the correct Bekenstein–Hawking entropy in a nearly model-independent manner. This approach may explain the "universality" of black hole entropy, the fact that many inequivalent descriptions of quantum states all seem to give the same thermodynamic predictions. It also suggests an elegant picture of the relevant degrees of freedom, as Goldstone-boson-like excitations arising from symmetry breaking by a conformal anomaly induced by the horizon constraints.

1 The Problem of Universality

Nearly 35 years have passed since Bekenstein [1] and Hawking [2] first showed us that black holes are thermodynamic objects, with characteristic temperatures

$$T_H = \frac{\hbar\kappa}{2\pi c} \tag{1}$$

and entropies

$$S_{BH} = \frac{A}{4\hbar G}. \tag{2}$$

From the start, it was clear that a statistical mechanical description of these states would be rather peculiar: in contrast to the entropy of an ordinary thermodynamic system, black hole entropy is not extensive, depending on area rather than volume.

S. Carlip
Department of Physics, 1 Shields Ave., University of California at Davis, Davis, CA 95616, USA
e-mail: carlip@physics.ucdavis.edu

M. Henneaux, J. Zanelli (eds.), *Quantum Mechanics of Fundamental Systems: The Quest for Beauty and Simplicity*, DOI 10.1007/978-0-387-87499-9_7,

Moreover, by Wheeler's famous dictum, "a black hole has no hair": a classical black hole is determined completely by a few macroscopic characteristics, with no apparent room for additional microscopic states. Nevertheless, from the earliest days of black hole thermodynamics, the search for a microscopic understanding has been a vigorous area of research.

Until fairly recently, that search was largely unsuccessful. Some interesting ideas were suggested – entanglement entropy of quantum fields across the horizon [3], or the entropy of quantum fields near the horizon [4] – but these remained speculative. Today, in contrast, a great many physicists can tell you, often in great detail, exactly what microscopic degrees of freedom underlie black hole thermodynamics. The new problem is that they will offer you many *different* explanations. Depending on who you ask, black hole entropy may count

- Weakly coupled string and D-brane states [5,6]
- Horizonless "fuzzball" geometries [7]
- States in a dual conformal field theory "at infinity" [8,9]
- Spin network states crossing the horizon [10]
- Spin network states inside the horizon [11]
- Horizon states in a spin foam [12]
- "Heavy" degrees of freedom in induced gravity [13]
- Entanglement entropy [3] (maybe "holographic" [14, 15])
- No local states – it's inherently global [16]
- Nothing – it comes from quantum field theory in a fixed non-quantum background [2], which knows nothing of quantum gravity
- Maybe something else (points in a causal set in the horizon's domain of dependence [17]? Kolmogorov–Sinai entropy of strings spreading at the horizon [18]?)

There is, of course, nothing wrong with a healthy competition among candidates for the proper description of the quantum black hole. The relevant degrees of freedom are, after all, presumably quantum gravitational – the Bekenstein–Hawking entropy (2) involves both $\hbar$ and G – and we do not yet have an established quantum theory of gravity. But the fact that so many descriptions give exactly the same answer is a true puzzle.

To see this puzzle more clearly, consider one of the most successful approaches to black hole entropy, that of weakly coupled string theory. To count black hole microstates a la Strominger and Vafa [5], one should proceed as follows:

1. Start with an extremal, supersymmetric, charged black hole
2. Find the horizon area and express it as a function of the charges
3. "Tune down" the gravitational coupling to form a weakly coupled string/brane system
4. Count the states in this weakly coupled system, and express their number in terms of the charges
5. Argue that supersymmetry (or other properties [19]) guarantees that the number of states is the same at strong and weak coupling
6. Compare the results of steps 2 and 4 to determine the entropy as a function of the horizon area

The method is very effective, even away from extremality, and allows the computation not only of black hole entropy, but of Hawking radiation and even gray-body factors. But the fundamental relationship between entropy and area arises only indirectly, by way of the computation of charges, and this computation is different for each new type of black hole. One cannot use the results from, say, a three-charge black hole in five dimensions to conclude anything about a four-charge black hole in six dimensions, but must recalculate the entropy and horizon area for each new case. Weakly coupled string theory gives the Bekenstein–Hawking entropy, but it gives it one black hole at a time.

2 Conformal Field Theory and the Cardy Formula

The natural question, then, is whether some property of the classical black hole can explain this "universality" by determining the number of quantum states, independent of the details of their description. This is a lot to ask, and I know of only one case in which such a phenomenon occurs. Let us therefore take a brief detour to explore two-dimensional conformal field theory.

A conformal field theory is a field theory that is invariant under both diffeomorphisms (general covariance) and Weyl transformations ("local scale invariance" or "conformal invariance") [20]. In two dimensions, one can always choose complex coordinates; such a theory is then characterized by two symmetry generators $L[\xi]$ and $\bar{L}[\bar{\xi}]$, which generate holomorphic and antiholomorphic diffeomorphisms. The Poisson bracket algebra of these generators is given by the unique central extension of the algebra of two-dimensional diffeomorphisms, the Virasoro algebra:

$$\begin{aligned}
\{L[\xi],L[\eta]\} &= L[\eta\xi' - \xi\eta'] + \frac{c}{48\pi}\int dz\,(\eta'\xi'' - \xi'\eta'') \\
\{\bar{L}[\bar{\xi}],\bar{L}[\bar{\eta}]\} &= \bar{L}[\bar{\eta}\bar{\xi}' - \bar{\xi}\bar{\eta}'] + \frac{\bar{c}}{48\pi}\int d\bar{z}\,(\bar{\eta}'\bar{\xi}'' - \bar{\xi}'\bar{\eta}'') \\
\{L[\xi],\bar{L}[\bar{\eta}]\} &= 0,
\end{aligned} \tag{3}$$

where the central charges c and $\bar{c}$ (the "conformal anomalies") depend on the particular theory. The zero-mode generators $L_0 = L[\xi_0]$ and $\bar{L}_0 = \bar{L}[\bar{\xi}_0]$ are conserved charges, roughly analogous to energies; their eigenvalues are commonly referred to as "conformal weights" or "conformal dimensions."

In 1986, Cardy discovered a remarkable property of such theories [21,22]. Given *any* unitary two-dimensional conformal field theory for which the lowest eigenvalues Δ_0 of L_0 and $\bar{\Delta}_0$ of $\bar{L}_0$ are nonnegative, the asymptotic density of states at large eigenvalues Δ and $\bar{\Delta}$ takes the form

$$\ln\rho(\Delta,\bar{\Delta}) \sim 2\pi\sqrt{\frac{(c-24\Delta_0)\Delta}{6}} + 2\pi\sqrt{\frac{(\bar{c}-24\bar{\Delta}_0)\bar{\Delta}}{6}}, \tag{4}$$

with higher order corrections that are also determined by the symmetry [23–25]. The entropy is thus fixed by symmetry, independent of any details of the states being counted. Note that upon quantization, after making the usual substitutions $\{\bullet,\bullet\} \rightarrow [\bullet,\bullet]/i\hbar$ and $L_m \rightarrow L_m/\hbar$, a classical central charge c^{cl} contributes $c^{cl}/\hbar$ to the quantum central charge, and a classical conformal "charge" Δ^{cl} contributes $\Delta^{cl}/\hbar$ to the quantum conformal weight. The classical piece of a conformal field theory thus yields a term of order $1/\hbar$ in the entropy (4), reproducing the behavior of the Bekenstein–Hawking entropy (2).

At first sight, these results seem irrelevant to our problem. Black holes are not typically two dimensional, and neither are they conformally invariant. There is a sense, though, in which black holes are *nearly* two dimensional and *nearly* conformally invariant near their horizons. Consider, for example, a scalar field φ near a black hole horizon. If we write the metric in "tortoise" coordinates

$$ds^2 = N^2(dt^2 - dr_*{}^2) + ds_\perp{}^2 \quad \text{with } N \rightarrow 0 \text{ at the horizon,} \tag{5}$$

the Klein–Gordon operator becomes

$$(\Box - m^2)\varphi = \frac{1}{N^2}(\partial_t^2 - \partial_{r_*}^2)\varphi + \mathcal{O}(1), \tag{6}$$

and it is evident that the mass and transverse excitations become negligible as $N \rightarrow 0$. The field is thus effectively described by a two-dimensional conformal field theory [26]. A similar phenomenon occurs for other types of matter, and also, in a sense, for gravity: a generic black hole metric admits an approximate conformal Killing vector near the horizon [27].

Such an effective two-dimensional description has proven very useful in black hole thermodynamics. Building on old results of Chistensen and Fulling [28], Wilczek, Robinson, Iso, Morita, Umetsu, and others have recently shown that the Hawking radiation flux [29] and, indeed, the full thermal spectrum [30, 31] can be extracted from a two-dimensional conformal description, using methods that rely on the conformal anomalies $(c,\bar{c})$. Like most derivations of Hawking radiation, these arguments are based on quantum field theory in a fixed black hole background. But as Claudio showed long ago [32], essentially any effective two-dimensional description of gravity also involves a Virasoro algebra, typically with a nonvanishing central charge. We might therefore hope that the conformal description could also tell us about the statistical mechanics of the black hole states themselves.

3 $2+1$ Dimensions

There is one case in which a conformal field theory derivation of black hole entropy has been completely successful [33, 34]. The Bañados–Teitelboim–Zanelli black hole [35, 36] is a solution of the vacuum Einstein equations in three spacetime dimensions with a negative cosmological constant $\Lambda = -1/\ell^2$. Like all vacuum

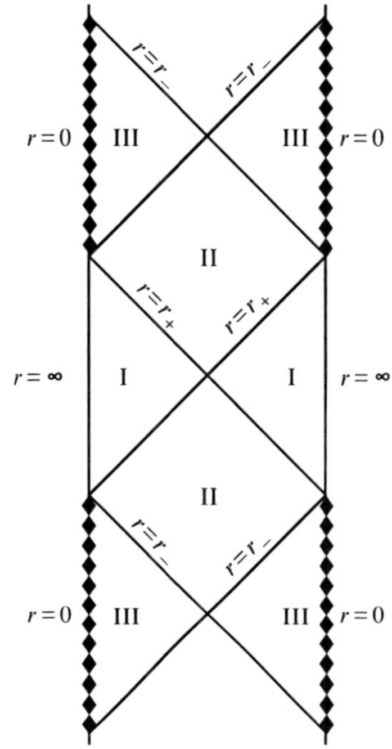

Fig. 1 The Carter–Penrose diagram for a nonextremal BTZ black hole

spacetimes in $2+1$ dimensions, the BTZ geometry has constant curvature, and can be in fact expressed as a quotient of anti-de Sitter space by a discrete group of isometries. Nevertheless, it is a real black hole:

- It has a genuine event horizon at $r=r_+$ and, if the angular momentum is nonzero, an inner Cauchy horizon at $r=r_-$, where $r_\pm$ are determined by the mass and angular momentum
- It occurs as the end point of the gravitational collapse of matter
- Its Carter–Penrose diagram, Fig. 1, is essentially the same as that of an ordinary Kerr-AdS black hole
- It exhibits standard black hole thermodynamics, with a temperature and entropy given by (1) and (2), where the horizon "area" is the circumference $A=2\pi r_+$

The conformal boundary of a $(2+1)$-dimensional asymptotically anti-de Sitter spacetime is a two-dimensional cylinder, so it is perhaps not surprising that the algebra of asymptotic symmetries of the BTZ black hole is a Virasoro algebra (3). It is rather more surprising that this algebra has a central extension, but as Brown and Henneaux showed in [37], the classical central charge, computed from the standard ADM constraint algebra, is nonzero:

$$c=\frac{3\ell}{2G}. \tag{7}$$

The appearance of this central charge can be traced back to the need for boundary terms in the canonical generators of diffeomorphisms, a phenomenon that we understand largely because of the pioneering work of Claudio and his collaborators [38]. Moreover, the classical conformal weights Δ and $\bar{\Delta}$ can be calculated in ordinary canonical general relativity, employing the same methods that are used to determine the ADM mass [37]. Indeed, for the BTZ black hole, the zero modes of the diffeomorphisms are linear combinations of time translations and rotations, and the corresponding conserved quantities are linear combinations of the ordinary ADM mass and angular momentum. A straightforward calculation gives

$$\Delta = \frac{1}{16G\ell}(r_+ + r_-)^2, \quad \bar{\Delta} = \frac{1}{16G\ell}(r_+ - r_-)^2, \tag{8}$$

and the Cardy formula (4) then yields an entropy

$$S = \log\rho \sim \frac{2\pi}{8G}(r_+ + r_-) + \frac{2\pi}{8G}(r_+ - r_-) = \frac{2\pi r_+}{4G}, \tag{9}$$

which may be recognized as precisely the Bekenstein–Hawking entropy.

This derivation is one of the first examples of Maldacena's celebrated AdS/CFT correspondence [8]: the entropy of an asymptotically anti-de Sitter spacetime is determined by the properties of a boundary conformal field theory. It is also a deeply mysterious result. Quantum gravity in three spacetime dimensions has no local degrees of freedom [39], and it is not at all clear where one can find enough degrees of freedom to account for the entropy (9). A review of current proposals can be found in [40]; I will return to this question, in a more general context, in Sect. 6.

The BTZ black hole demonstrates in principle that conformal field theory can be used to compute black hole entropy. Unfortunately, the generalization to higher dimensions is difficult. The derivation of [33, 34] depends crucially on the fact that the conformal boundary of $(2+1)$-dimensional asymptotically AdS space is a two-dimensional cylinder, which provides a setting for a two-dimensional conformal field theory. No higher-dimensional analog of the Cardy formula is known,[1] so one cannot, at least for now, use symmetries of a higher-dimensional boundary to constrain the density of states.

Moreover, the BTZ computations depend on a symmetry at infinity rather than at the horizon. In $2+1$ dimensions this may not matter, since there are no propagating degrees of freedom between the black hole and the conformal boundary, but in higher dimensions, it is less clear how to isolate black hole degrees of freedom. One might argue that a single black hole configuration should make the dominant contribution at infinity, but even this is now known to not always be true [41].

Despite these limitations, the BTZ results have proven surprisingly versatile. In particular, many near-extremal black holes – including most of the black holes whose entropy can be computed using weakly coupled string theory – have a near-horizon geometry of the form *BTZ* $\times$ *trivial*, allowing the application of the BTZ method in a more general setting [9]. For generic, nonextremal black holes, though, a more general extension is needed.

---

[1] Conformal field theory is qualitatively different in two and more than two dimensions: for $d > 2$, the symmetry group has a finite set of generators, but for $d = 2$ it has infinitely many [20].

4 Horizons and Constraints

While the conformal analysis of the BTZ black hole does not extend directly to higher dimensions, it does suggest some interesting directions. We should, perhaps, look for a hidden conformal symmetry, of the type discussed in Sect. 2, with a classical central charge; but we should look near the horizon.

To do so, we must first confront a fundamental conceptual issue. How, in a quantum theory of gravity, do we specify that a black hole is present? In a semiclassical approach, this is easy: we fix a background black hole metric and look at quantum fields and metric fluctuations in that background. In a full quantum theory of gravity, though, we cannot do that: the metric is a quantum operator whose components do not commute, and cannot be simultaneously specified. We must therefore look for a more limited set of constraints that are sufficient to guarantee the presence of the desired black hole while remaining quantum mechanically consistent. The simplest way to do this is to add conditions that ensure the presence of a horizon of some sort – say, an isolated horizon [42] – and study quantum gravity in the presence of these additional constraints. Physically, this amounts to asking questions about conditional probabilities: for instance, "What is the probability of detecting a Hawking radiation photon of energy E, *given* the presence of a horizon of area A?"

There are several ways to add such "horizon constraints," which are reviewed in [43]. One approach is to treat the horizon as a sort of boundary. At first sight, this seems a peculiar thing to do: a black hole horizon is certainly not a physical boundary for a freely falling observer. But a horizon *is* a hypersurface at which we can impose "boundary conditions" – namely, the conditions that it is, in fact, a horizon. As in the BTZ case, such restrictions require boundary terms in the generators of diffeomorphisms, whose presence affects their algebra. It can then be shown that in *any* spacetime of dimension greater than two, the subgroup of diffeomorphisms in the r–t plane becomes a Virasoro algebra with the right central charges and conformal weights to yield the Bekenstein–Hawking entropy [44–46].

Unfortunately, the diffeomorphisms whose algebra leads to this result are generated by vector fields that blow up at the horizon [47,48], and this divergence is poorly understood. Moreover, this method does not seem to work for the interesting case of the two-dimensional dilaton black hole. One can therefore look at a slightly different approach, in which the "horizon constraints" are literally imposed as constraints in canonical general relativity [49,50].

The basic steps of this approach can be summarized as follows:

1. Dimensionally reduce to the "r–t plane," which, as argued in Sect. 2, is the relevant setting for near-horizon conformal symmetry. Such a reduction is possible even in the absence of spherical or cylindrical symmetry, although it comes at the expense of an infinite-dimensional Kaluza–Klein gauge group [51]. The action then becomes

$$I = \frac{1}{2}\int d^2x\sqrt{g}\left[\varphi R + V[\varphi] - \frac{1}{2}W[\varphi]h_{IJ}F^I_{ab}F^{Jab}\right], \tag{10}$$

where the dilaton φ is the dimensionally reduced remnant of the transverse area and F^I is the field strength for the usual Kaluza–Klein gauge field A^I. The potentials V and W depend on the details of the higher-dimensional theory, and need not be further specified.

2. Continue to "Euclidean" signature, as Claudio has often advocated [52]. The metric can then be written in the form

$$ds^2 = N^2 f^2 dr^2 + f^2(dt + \alpha dr)^2. \tag{11}$$

For a black hole spacetime, the horizon shrinks to a point and time becomes an angular coordinate (see Fig. 2), with a period β determined by the geometry.[2] Rather than evolving in t, we borrow a trick from conformal field theory [20] and evolve radially, starting at a "stretched horizon" just outside $r = 0$.

3. Find the ordinary ADM-style constraints, which take the form

$$\begin{aligned}
\mathscr{H}_{\parallel} &= \dot{\varphi}\pi_\varphi - f\dot{\pi}_f = 0 \\
\mathscr{H}_{\perp} &= f\pi_f\pi_\varphi + f\left(\frac{\dot{\varphi}}{f}\right)^{\cdot} - \frac{1}{2}f^2\hat{V} = 0 \qquad \text{with } \hat{V} = V + \frac{h^{IJ}\pi_I\pi_J}{W} \\
\mathscr{H}_I &= \dot{\pi}_I - c^J{}_{IK}A^K\pi_J = 0.
\end{aligned} \tag{12}$$

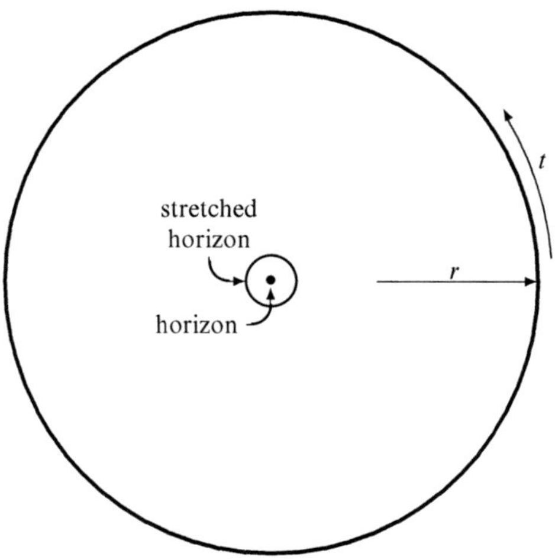

Fig. 2 A "Euclidean" black hole spacetime

[2] Claudio and his collaborators were among the first to study such black holes in two-dimensional dilaton gravity [53].

These can be combined to form Virasoro generators

$$L[\xi] = \frac{1}{2}\int dt\,\xi\,(\mathscr{H}_{\parallel} + i\mathscr{H}_{\perp})$$
$$\bar{L}[\bar{\xi}] = \frac{1}{2}\int dt\,\bar{\xi}\,(\mathscr{H}_{\parallel} - i\mathscr{H}_{\perp})\,, \tag{13}$$

which satisfy the algebra (3) with vanishing central charge.

4. Determine the geometrical quantities that characterize the black hole:

$$\text{the expansion} \qquad s = \varphi\vartheta = f\pi_f - i\dot{\varphi} \tag{14}$$
$$\text{the surface gravity} \qquad \hat{\kappa} = \pi_\varphi - i\dot{f}/f + f^2\frac{d\omega}{d\varphi}.$$

The surface gravity is not unique – in standard general relativity, it depends on the normalization of the Killing vector at the horizon [42], which here appears as a conformal factor ω that will be determined later.

5. Impose horizon constraints to ensure that our initial surface is a stretched horizon. As Claudio noted in [52], the actual horizon is determined by the conditions $s = \bar{s} = 0$. A stretched horizon with surface gravity $\hat{\kappa}_H$ is naturally specified by the slightly loosened conditions

$$K = s - a(\hat{\kappa} - \hat{\kappa}_H) = 0$$
$$\bar{K} = \bar{s} - a(\bar{\hat{\kappa}} - \bar{\hat{\kappa}}_H) = 0, \tag{15}$$

where the constant a will be determined below.

6. Note that the horizon constraints K and $\bar{K}$ do not commute with the Virasoro generators (13), which are therefore not symmetries of the constrained system. Cure this problem by using the Bergmann–Komar formulation of Dirac brackets [54]. Let $\{K_i\}$ be a set of constraints for which the inverse Δ_{ij} of $\{K_i, K_j\}$ exists (in Dirac's language, a set of second class constraints). Then for any observable $\mathscr{O}$, the new observable

$$\mathscr{O}^* = \mathscr{O} - \sum_{i,j}\int du dv \{\mathscr{O}, K_i(u)\}\Delta_{ij}(u,v)K_j(v) \tag{16}$$

will have vanishing Poisson brackets with the K_i. Since $\mathscr{O}^*$ differs from $\mathscr{O}$ only by a multiple of the constraints K_i, the two are physically equivalent. The Poisson bracket $\{\mathscr{O}_1^*, \mathscr{O}_2^*\}$ can be shown to be equal to the Dirac bracket of $\mathscr{O}_1$ and $\mathscr{O}_2$.

7. Work out the Poisson algebra of the modified Virasoro generators $L^*[\xi]$ and $\bar{L}^*[\bar{\xi}]$. The conformal factor ω in (14) and the constant a in (15) are both fixed by the requirement that these brackets be "nice," and in particular that they reduce to an ordinary Virasoro algebra at the horizon. Choosing modes

$$\xi_n = \frac{\beta}{2\pi}e^{2\pi i n t/\beta} \tag{17}$$

for the vector fields used to smear the Virasoro generators, we obtain central charges and conformal weights

$$c = \bar{c} = \frac{3\varphi_H}{4G}, \qquad \Delta = \bar{\Delta} = \frac{\varphi_H}{16G}\left(\frac{\kappa_H \beta}{2\pi}\right)^2. \tag{18}$$

8. Use the Cardy formula (4) to obtain an entropy

$$S = \frac{2\pi\varphi_H}{4G}\left(\frac{\kappa_H \beta}{2\pi}\right). \tag{19}$$

This is *almost* the correct Bekenstein–Hawking entropy; it differs from the correct expression by a factor of 2π. I believe this factor has a simple physical explanation: entropy should count the black hole degrees of freedom at a fixed time, but because of our choice of radial evolution, we have computed the entropy at the horizon for *all* times, effectively integrating over a circle of circumference 2π.

5 Universality Again

Now, however, let us recall our original motivation, which was to understand the "universality" of black hole entropy. If the conformal field theory/horizon constraint picture is to explain this universality, it must be the case that the symmetry of the preceding section is secretly present in all of the other derivations of black hole entropy. This is certainly not obvious, but there are a few hopeful signs.

Let us first compare the horizon constraint method to the conformal approach to the BTZ black hole described in Sect. 3. We can start by comparing the central charges and conformal weights:

	BTZ	**Horizon CFT**	
modes	$\xi_n \sim e^{in(t\pm\ell\phi)/\ell}$	$\xi_n \sim e^{in\kappa_H t}$	
c	$\dfrac{3\ell}{2G}$	$2\pi\cdot\dfrac{3\varphi_H}{4\pi G}$	(20)
$\Delta,\ \bar{\Delta}$	$\dfrac{(r_+ \pm r_-)^2}{16G\ell}$	$2\pi\cdot\dfrac{\varphi_H}{32\pi G}\left(\dfrac{\kappa_H\beta}{2\pi}\right)^2$	

While the entropies agree, it appears that the central charges and conformal weights do not. In fact, though, these disagreements can be traced to two simple sources [50]: the periodicities of the modes do not match, and the BTZ results are based on a coordinate system that is not corotating at the horizon, as one would desire for dimensional reduction. Once these differences are accounted for, the central charges and conformal weights agree precisely. As noted in Sect. 3, the BTZ approach applies also to most of the black holes that can be exactly analyzed with weakly coupled string theory, so this agreement is a significant step.

For loop quantum gravity, the connection is less clear. There is, however, an interesting coincidence that may point toward something deeper. The horizon states of a spin network described in [10] are characterized by a constrained $SL(2,\mathbb{R})$ Chern–Simons theory with coupling constant $k = iA/8\pi\gamma G$, where γ is the Immirzi parameter. Any three-dimensional Chern–Simons theory has an associated two-dimensional conformal field theories, a Wess–Zumino–Witten model that appears, for example, in the description of boundary states [55]. In the present case, this conformal field theory is Liouville theory, and its central charge is approximately $6k$. If we choose Ashtekar's original self-dual formulation of loop quantum gravity [56], for which $\gamma = i$, this central charge agrees precisely with the value obtained by the horizon constraint approach.

The central charge (18) also matches that of the "horizon as boundary" approach of [45], and the conformal weights can be obtained as a Komar integral, as suggested in a slightly different context by Emparan and Mateos [57]. I believe it should also be possible to relate this method to the Euclidean path integral approach to black hole entropy; work on this question is in progress.

6 What are the States?

If near-horizon conformal symmetry really provides a universal explanation for black hole statistical mechanics, it had better *not* give us a unique description of the relevant microstates. The problem, after all, is that many different microscopic descriptions seem to yield the same macroscopic thermal properties; picking out one "right" description would miss the point. Nevertheless, it is possible that the derivation of Sect. 4 might give a useful *effective* description of the microscopic degrees of freedom.

Consider the standard Dirac treatment of constraints in quantum mechanics. A set of classical (first class) constraints $L[\xi] = \bar{L}[\bar{\xi}] = 0$ translates to a quantum restriction on the space of states:

$$L[\xi]|\text{phys}\rangle = \bar{L}[\bar{\xi}]|\text{phys}\rangle = 0. \tag{21}$$

But in the presence of a central charge, such a restriction is inconsistent with the Virasoro algebra (3). This is not new, of course, and it is well known how to fix the problem [20]; for example, one can require that only the positive frequency pieces of $L[\xi]$ and $\bar{L}[\bar{\xi}]$ annihilate physical states. The net result, though, is that some states that would have been unphysical in the absence of a central charge must now be considered physical. Equivalently [58], the presence of boundaries or constraints can remove gauge degeneracies among otherwise physically equivalent states, turning "would-be gauge transformations" into new dynamical degrees of freedom.

This phenomenon may have first been observed by Claudio. In an underappreciated passage in [32], he points out that the presence of a central charge in dilaton gravity is quantum mechanically consistent, but results in the appearance of a new degree of freedom. In the present context, we are imposing the constraints (15) only at the horizon, so it is only there that a central charge appears, but the new horizon degree of freedom is essentially the same as Claudio's.

As Kaloper and Terning have observed [59], this process is also somewhat reminiscent of the Goldstone mechanism, in which a spontaneously broken symmetry gives rise to massless excitations in the direction of the "broken" generators. Here, of course, the broken symmetry is a gauge symmetry, and the corresponding degrees of freedom are therefore new. But as in the Goldstone mechanism, the pattern of symmetry breaking may give us a universal effective description of the degrees of freedom, while not touching on their "real" structure in terms of the fundamental underlying quantum gravitational states.

For asymptotically anti-de Sitter spacetimes in three dimensions, an explicit description of the symmetry breaking and the corresponding degrees of freedom at infinity is possible [60]. The resulting effective field theory is a Liouville theory. This two-dimensional conformal field theory has the correct central charge and conformal weights to yield the Bekenstein–Hawking entropy via the Cardy formula, but there is still a debate as to whether it really contains enough degrees of freedom [36]. A similar induced action can be found in five-dimensional asymptotically anti-de Sitter gravity [61], although the problem of counting states has not been solved. One might hope for a more general result in arbitrary dimension, perhaps focusing on the horizon rather than infinity; Claudio is responsible for an interesting effort in this direction [52].

One avenue for further research may be to look more carefully at the path integral measure, which is in some sense a count of the number of states, in the presence of a Virasoro algebra with a nonzero central charge. It is known that when second class constraints $\{C_i\}$ are present, the path integral acquires a Fadeev–Popov-like determinant $\det|\{C_i, C_j\}|^{1/2}$ [62]. For a Virasoro algebra, this is

$$D = \det|\{L_m, L_n\}|^{1/2}. \tag{22}$$

A naive evaluation of this expression, using the algebra (3), *almost* gives the Cardy formula: one finds $D \sim \exp\{2\pi\sqrt{6\Delta/c}\}$, which differs from (4) by a flip from $c/6$ to $6/c$. This is a bit too simple, though, since the Virasoro algebra contains an $\mathrm{SL}(2,\mathbb{R})$ subgroup, generated by $\{L_0, L_{\pm 1}\}$, whose algebra remains first class. This adds a large degeneracy, increasing the density of states; we really need to evaluate a determinant of the form

$$D = \det\left|-\frac{c}{12}\frac{d^3}{dx^3} + \frac{d}{dx}L + L\frac{d}{dx}\right|^{1/2} \quad \text{with } L = L_0 + L_1 e^{2ix} + L_{-1}e^{-2ix} \tag{23}$$

and trace over appropriate $\mathrm{SL}(2,\mathbb{R})$ states. It is not yet clear whether this approach is still too naive; it may be that we need more detailed information about the $\mathrm{SL}(2,\mathbb{R})$ representations than can be obtained from a constraint analysis alone.

7 What Next?

While I have given some evidence for the proposal that black hole thermodynamics is effectively determined by near-horizon conformal symmetry, the hypothesis remains very far from being proven. I can think of two main directions to proceed.

First, we should try to connect the near-horizon symmetry more closely to other derivations of black hole entropy. In loop quantum gravity, does the numerical coincidence discussed in Sect. 5 have any deeper significance? Is there a way of using the associated Liouville theory to count states? In the "fuzzball" approach to black holes in string theory [7], no single configuration is expected to have a near-horizon conformal symmetry (or, indeed, a horizon); can a sum over configurations recover such a symmetry? In induced gravity [13], a connection to conformal field theory is already known [63]; can it be tied to the near-horizon symmetry discussed here? Can the determinant (23) be evaluated, and will it give the correct density of states? Does the spin foam method of [12], which relies on a treatment of the horizon as an effective boundary, contain a hidden conformal symmetry?

Second, we should keep in mind that there is more to black hole thermodynamics than the Bekenstein–Hawking entropy. As I noted in Sect. 2, the intensity and spectrum of Hawking radiation can be obtained from an effective two-dimensional theory near the horizon, using conformal field theory methods applied to matter rather than gravity [29–31]. It is natural to hope that these matter degrees of freedom can be coupled to the near-horizon gravitational degrees of freedom to obtain a dynamical description of Hawking radiation. In $2+1$ dimensions, Emparan and Sachs have shown that something of this sort may be possible [64]: a classical scalar field can be coupled to the conformal boundary degrees of freedom of the BTZ black hole, and conformal methods then yield the correct description of Hawking radiation. If this result could be generalized to arbitrary dimensions, with a coupling at the horizon, it would provide very strong evidence for the conformal description of black hole thermodynamics.

Acknowledgments This work was supported in part by Department of Energy grant DE-FG02-91ER40674.

References

1. J. D. Bekenstein, Phys. Rev. D7 (1973) 2333.
2. S. W. Hawking, Nature 248 (1974) 30.
3. L. Bombelli, R. K. Koul, J. Lee, and R. Sorkin, Phys. Rev. D34 (1986) 373.
4. G. 't Hooft, Nucl. Phys. B 256 (1985) 727.
5. A. Strominger and C. Vafa, Phys. Lett. B379 (1996) 99, hep-th/9601029.
6. A. W. Peet, Class. Quant. Grav. 15 (1998) 3291, hep-th/9712253.
7. S. D. Mathur, Class. Quant. Grav. 23 (2006) R115, hep-th/0510180.
8. O. Aharony, S. S. Gubser, J. M. Maldacena, H. Ooguri, and Y. Oz, Phys. Rept. 323 (2000) 183, hep-th/9905111.
9. K. Skenderis, Lect. Notes Phys. 541 (2000) 325, hep-th/9901050.

10. A. Ashtekar, J. Baez, A. Corichi, and K. Krasnov, Phys. Rev. Lett. 80 (1998) 904, gr-qc/9710007.
11. E. R. Livine and D. R. Terno, Nucl. Phys. B 741 (2006) 131, gr-qc/0508085.
12. J. Manuel Garcia-Islas, "BTZ black hole entropy: a spin foam model description," Class. Quant. Grav. 25 (2008) 245001, arXiv:0804.2082 [gr-qc].
13. V. P. Frolov and D. V. Fursaev, Phys. Rev. D56 (1997) 2212, hep-th/9703178.
14. S. Ryu and T. Takayanagi, Phys. Rev. Lett. 96 (2006) 181602, hep-th/0603001.
15. D. V. Fursaev, JHEP 0609 (2006) 018, hep-th/0606184.
16. S. W. Hawking and C. J. Hunter, Phys. Rev. D59 (1999) 044025, hep-th/9808085.
17. D. Rideout and S. Zohren, Class. Quant. Grav. 23 (2006) 6195, gr-qc/0606065.
18. K. Ropotenko, "Kolmogorov–Sinai and Bekenstein–Hawking entropies," Class. Quant. Grav. 25 (2008) 195005, arXiv:0711.3131 [gr-qc].
19. K. Goldstein, N. Iizuka, R. P. Jena, and S. P. Trivedi, Phys. Rev. D72 (2005) 124021, hep-th/0507096.
20. P. Di Francesco, P. Mathieu, and D. Sénéchal, *Conformal Field Theory* (Springer, Berlin, 1997).
21. J. A. Cardy, Nucl. Phys. B 270 (1986) 186.
22. H. W. J. Blöte, J. A. Cardy, and M. P. Nightingale, Phys. Rev. Lett. 56 (1986) 742.
23. S. Carlip, Class. Quant. Grav. 17 (2000) 4175, gr-qc/0005017.
24. R. Dijkgraaf, J. M. Maldacena, G. W. Moore, and E. P. Verlinde, "A black hole Farey tail," (unpublished) hep-th/0005003.
25. D. Birmingham, I. Sachs, and S. Sen, Int. J. Mod. Phys. D10 (2001) 833, hep-th/0102155.
26. D. Birmingham, K. S. Gupta, and S. Sen, Phys. Lett. B 505 (2001) 191, hep-th/0102051.
27. A. J. M. Medved, D. Martin, and M. Visser, Phys. Rev. D 70 (2004) 024009, gr-qc/0403026.
28. S. M. Christensen and S. A. Fulling, Phys. Rev. D15 (1977) 2088.
29. S. P. Robinson and F. Wilczek, Phys. Rev. Lett. 95 (2005) 011303, gr-qc/0502074.
30. S. Iso, T. Morita, and H. Umetsu, Phys. Rev. D75 (2007) 124004, hep-th/0701272.
31. S. Iso, T. Morita, and H. Umetsu, Phys. Rev. D77 (2008) 045007, arXiv:0710.0456 [hep-th].
32. C. Teitelboim, Phys. Lett. B126 (1983) 41.
33. A. Strominger, JHEP 02 (1998) 009, hep-th/9712251.
34. D. Birmingham, I. Sachs, and S. Sen, Phys. Lett. B424 (1998) 275, hep-th/9801019.
35. M. Bañados, C. Teitelboim, and J. Zanelli, Phys. Rev. Lett. 69 (1992) 1849, hep-th/9204099.
36. S. Carlip, Class. Quant. Grav. 12 (1995) 2853, gr-qc/9506079.
37. J. D. Brown and M. Henneaux, Commun. Math. Phys. 104 (1986) 207.
38. T. Regge and C. Teitelboim, Ann. Phys. 88 (1974) 286.
39. S. Carlip, Living Rev. Rel. 8 (2005) 1, gr-qc/0409039.
40. S. Carlip, Class. Quant. Grav. 22 (2005) R85, gr-qc/0503022.
41. F. Denef and G. W. Moore, Gen. Rel. Grav. 39 (2007) 1539, arXiv:0705.2564 [hep-th].
42. A. Ashtekar, C. Beetle, and S. Fairhurst, Class. Quant. Grav. 16 (1999) L1, gr-qc/9812065.
43. S. Carlip, "Horizon constraints and black hole entropy," to appear in *The Kerr spacetime: Rotating black holes in general relativity*, edited by S. Scott, M. Visser, and D. Wiltshire (Cambridge University Press, Cambridge), gr-qc/0508071.
44. S. Carlip, Phys. Rev. Lett. 82 (1999) 2828, hep-th/9812013.
45. S. Carlip, Class. Quant. Grav. 16 (1999) 3327, gr-qc/9906126.
46. M. Cvitan, S. Pallua, and P. Prester, Phys. Rev. D 70 (2004) 084043, hep-th/0406186.
47. O. Dreyer, A. Ghosh, and J. Wisniewski, Class. Quant. Grav. 18 (2001) 1929, hep-th/0101117.
48. J. Koga, Phys. Rev. D 64 (2001) 124012, gr-qc/0107096.
49. S. Carlip, Class. Quant. Grav. 22 (2005) 1303, hep-th/0408123.
50. S. Carlip, Phys. Rev. Lett. 99 (2007) 021301, gr-qc/0702107.
51. J. H. Yoon, Phys. Lett. B 451 (1999) 296, gr-qc/0003059.
52. C. Teitelboim, Phys. Rev. D 53 (1996) 2870, hep-th/9510180.
53. J. D. Brown, M. Henneaux, and C. Teitelboim, Phys. Rev. D33 (1986) 319.
54. P. G. Bergmann and A. B. Komar, Phys. Rev. Lett. 4 (1960) 432.
55. E. Witten, Commun. Math. Phys. 121 (1989) 351.
56. A. Ashtekar, Phys. Rev. Lett. 57 (1986) 2244.
57. R. Emparan and D. Mateos, Class. Quant. Grav. 22 (2005) 3575, hep-th/0506110.

58. S. Carlip, Nucl. Phys. B 362 (1991) 111.
59. N. Kaloper and J. Terning, personal communication.
60. S. Carlip, Class. Quant. Grav. 22 (2005) 3055, gr-qc/0501033.
61. R. Aros, M. Romo, and N. Zamorano, Phys. Rev. D75 (2007) 067501, hep-th/0612028.
62. M. Henneaux and C. Teitelboim, *Quantization of Gauge Systems* (Princeton University Press, Princeton, NJ, 1992), Chapter 16.
63. V. P. Frolov, D. Fursaev, and A. Zelnikov, JHEP 0303 (2003) 038, hep-th/0302207.
64. R. Emparan and I. Sachs, Phys. Rev. Lett. 81 (1998) 2408, hep-th/9806122.

Sources for Chern–Simons Theories

José D. Edelstein and Jorge Zanelli

Abstract The coupling between Chern–Simons Theories and matter sources defined by branes of different dimensionalities is examined. It is shown that the standard coupling to membranes, such as the one found in supergravity or in string theory, does not operate in the same way for CS theories; the only p-branes that naturally couple seem to be those with $p = 2n$; these p-branes break the gauge symmetry (and supersymmetry) in a controlled and sensible manner.

1 Introduction

Chern–Simons (CS) theories have a number of appealing attributes that make them interesting candidates for the description of natural phenomena. In spite of their promise, they also present a number of puzzling features that set them in a different class from other gauge theories, such as those that have been successfully used for the description of the Standard Model. In the case of higher (than three) dimensional CS theories, aiming at describing gravitational physics, several important difficulties emerge. Prominent among these, stands the problem of how to couple them to different forms of matter such as, for instance, branes of different dimensionalities.

These extended objects, whose existence is familiar in the context of string theory, play a natural role in CS gravitational theories based on supersymmetric extensions of both the Poincaré and the anti de Sitter (AdS) groups. In the particular case of 11D, while M-theory is well-known to possess two kinds of branes – the electric M2- and the magnetic M5-branes – the very same objects naively appear in a CS theory based on the M-algebra [1]. It is not clear, however, whether these objects actually belong to the spectrum of CS theory, and how they couple to the remaining fields. We shall address both problems, on general grounds, in the present article.

J.D. Edelstein
Department of Particle Physics and IGFAE, University of Santiago de Compostela, Spain
e-mail: jedels-at-usc-dot-es

M. Henneaux, J. Zanelli (eds.), *Quantum Mechanics of Fundamental Systems: The Quest for Beauty and Simplicity*, DOI 10.1007/978-0-387-87499-9_8,

Interactions with sources provide a handle to probe the perturbative structure of quantum theories, but that requires a well defined expansion of the interacting theory as a power series in a weak coupling parameter. In this manner, currents generated by point charges – and more generally, by charged extended objects such as strings or higher dimensional branes – are standard mechanisms that allow extracting predictions from the effective low energy limit in string/supergravity theories, which are in principle experimentally testable.

On the other hand, the perturbative expansion as a power series in the coupling constant seems to be of little use in a CS system: CS theories are highly nonlinear and self-interacting in a way that it is not possible to distinguish between the "free" and the "interacting" parts of the action without making a severe mutilation of the system. CS Lagrangians have no adjustable coupling constants (dimensionful or otherwise). This a priori appealing feature has a downside: the separation between background and perturbations is not clear-cut either.

A further complication is that since CS actions do not involve a metric, there is no notion of energy, and hence no energy scale is naturally defined in them. An energy scale can be introduced only at the cost of breaking gauge symmetry. In this sense, CS systems can be viewed as the analog of noble gases in chemistry, because they would not interact or bind to any other form of matter. It may seem as if they are inert, subtle beautiful structures to be admired, unrelated to the physical reality of the world. Here we will argue that this is not quite true: there is no obstruction to the coupling between membranes whose worldvolumes are odd-dimensional and non-Abelian CS systems.

On the road to understand such couplings, one faces the issue of uncovering the BPS spectrum of supersymmetric higher dimensional CS theories. Besides the expected presence of BPS branes preserving one half or one quarter of the original supersymmetries, it is interesting to seek for possible states preserving all but one supercharges that may be understood as constituents of the former. These so-called BPS preon states were proposed in [2], and they were recently alleged to exist in the $osp(32|1)$ CS theory (and presumably in other related theories) [3]. The existence of preons has also been recently ruled out as solutions of the presumed low-energy limit of M-theory, the Cremmer–Julia–Scherk (CJS) supergravity [4]. Here we show that the arguments presented in [2] are appropriate as well for CS theories based on extensions of the AdS algebra, but they need a subtle improvement for the case of Poincaré based theories such as the CS supergravity for the M-algebra.

Claudio Bunster was a promoter of the idea that the world history of a point particle, as well as that of the entire Universe, can be viewed as similar objects, to be treated quantum mechanically in a similar way [5]. Bunster was also a pioneer in considering currents with support on branes as sources coupled to p-form gauge potentials. In [6, 7], he showed that it is impossible to minimally couple a non-Abelian connection to a p-brane for $p > 0$. It is therefore a suitable form of tribute to celebrate his 60th birthday, to discuss a context in which this obstruction can be circumvented.

2 Remarks on the BPS Spectrum of CS Supergravity

A key feature of higher dimensional supergravity theories (and, more generally, of string/M-theory) is their natural coupling to certain branes. The dimensionality of these objects is strongly restricted by the tensorial properties of the field content of the theory. In 11D supergravity, for instance, the three-form field couples naturally to an electric 2-brane (M2) or to a magnetic 5-brane (M5). Moreover, these branes are BPS states, their mass being quantum mechanically equal to their charges.

The BPS spectrum of higher dimensional CS theories is not yet well understood. There are some scattered examples in the literature but no general results or exhaustive studies have been undertaken so far. In the present section we will illustrate some of the difficulties that this problem embodies. We will focus on the case of the CS supergravity for the M-algebra, though our results are quite general. We will argue that it may be necessary to reconsider the way sources and couplings come into place in these theories. A concrete proposal is then presented in the forthcoming section.

2.1 CS Supergravity for the M-Algebra

One of the nicest features of CS supergravities is that, being gauge theories, their dynamical variables are connections living in a given Lie algebra. Their fiber bundle structure seems an auspicious starting point towards a quantization program. However, this is an intricate problem, mostly due to the existence of highly nontrivial vacua with radically different dynamical content and the lack of a perturbative expansion around many of them [8, 9].

It is possible to write down a CS supergravity theory with the symmetry dictated by the so-called M-algebra explicitly realized off-shell [1]. Soon after the discovery of M-theory [10, 11], it was suggested that CS supergravity might provide a covariant non-perturbative formulation of quantum M-theory [12] based upon $osp(32|1)$, the minimal supersymmetric extension of the AdS group in 11D. The observation that this theory violates parity conservation (a symmetry that, for consistency, should be present in M-theory [13, 14]), prompted the suggestion of a CS theory based on $osp(32|1) \times osp(32|1)$ [15]. These CS theories have a number of nice features that include the presence of a central extension whose tensorial character matches that of an extended object like the M5-brane. The M2-brane, instead, enters the game in a less natural way. A CS theory based on the M-algebra puts both basic constituents of M-theory in a more democratic ground.

The M-algebra includes, apart from the Poincaré generators $\mathbf{J}_{ab}$ and $\mathbf{P}_a$, a Majorana supercharge $\mathbf{Q}_\alpha$ and two additional bosonic generators, $\mathbf{Z}_{ab}$ and $\mathbf{Z}_{abcde}$ that close the supersymmetry algebra [16],

$$\begin{aligned}\{\mathbf{Q}_\alpha, \mathbf{Q}_\beta\} &= (C\Gamma^a)_{\alpha\beta}\, \mathbf{P}_a + (C\Gamma^{ab})_{\alpha\beta}\, \mathbf{Z}_{ab} + (C\Gamma^{abcde})_{\alpha\beta} \mathbf{Z}_{abcde} \\ &\equiv \mathbf{P}_{\alpha\beta}\,,\end{aligned} \tag{1}$$

where the charge conjugation matrix C is antisymmetric. The "central charges" $\mathbf{Z}_{ab}$ and $\mathbf{Z}_{abcde}$ are tensors under Lorentz rotations but are otherwise Abelian generators. It must be stressed that the M-algebra is not the same as $osp(32|1)$, nor a subalgebra of the latter, and not even a contraction of it. The M-algebra can be obtained through an expansion of $osp(32|1)$, which corresponds to an analytic continuation of the Maurer–Cartan form [17–19]. This mechanism was used to obtain the actions for the corresponding algebras in [20].

The field content of the theory is thus given by a connection in the M-algebra,

$$\mathscr{A} = \frac{1}{2}\omega^{ab}\,\mathbf{J}_{ab} + e^{a}\,\mathbf{P}_{a} + \frac{1}{\sqrt{2}}\psi^{\alpha}\,\mathbf{Q}_{\alpha} + b^{ab}_{[2]}\,\mathbf{Z}_{ab} + b^{abcde}_{[5]}\,\mathbf{Z}_{abcde}\,, \tag{2}$$

where e and ω describe the metric and affine features of the spacetime geometry (including torsion); ψ is the gravitino, and $b_{[2]}$ and $b_{[5]}$ are Abelian gauge fields in antisymmetric tensor representations of the Lorentz group. The field strength $\mathscr{F} = d\mathscr{A} + \mathscr{A}\wedge\mathscr{A}$ reads

$$\mathscr{F} = \frac{1}{2}R^{ab}\,\mathbf{J}_{ab} + \tilde{T}^{a}\,\mathbf{P}_{a} + \frac{1}{\sqrt{2}}\mathscr{D}\psi^{\alpha}\,\mathbf{Q}_{\alpha} + \tilde{F}^{ab}_{[2]}\,\mathbf{Z}_{ab} + \tilde{F}^{abcde}_{[5]}\,\mathbf{Z}_{abcde}\,, \tag{3}$$

where $R^{ab} = d\omega^{ab} + \omega^{a}_{\ c}\wedge\omega^{cb}$ is the curvature 2-form, and

$$\begin{aligned} \tilde{T}^{a} &= \mathscr{D}e^{a} - \frac{1}{4}\bar{\psi}\Gamma^{a}\psi\,, \\ \tilde{F}^{a_1\cdots a_k}_{[k]} &= \mathscr{D}b^{a_1\cdots a_k}_{[k]} - \frac{1}{4}\bar{\psi}\Gamma^{a_1\cdots a_k}\psi\,. \end{aligned} \tag{4}$$

It is important to specify at this point the expression for the covariant derivative acting on the gravitino, $\mathscr{D}\psi^{\alpha} = d\psi^{\alpha} + \frac{1}{4}\omega^{\alpha}{}_{\beta}\,\psi^{\beta}$.

The CS form[1] is the Lagrangian of the theory, constructed through the standard requirement that $d\mathscr{C}_{11} = \langle\mathscr{F}\wedge\cdots\wedge\mathscr{F}\rangle$, where the bracket $\langle\cdots\rangle$ stands for a multilinear form of the M-algebra generators $\mathbf{G}_{A}$ whose only non-vanishing bosonic components are

$$\begin{aligned} \langle\mathbf{J}_{a_1a_2}\,\mathbf{J}_{a_3a_4}\,\mathbf{J}_{a_5a_6}\,\mathbf{J}_{a_7a_8}\,\mathbf{J}_{a_9a_{10}}\,\mathbf{P}_{a_{11}}\rangle &= \frac{16}{3}\,\varepsilon_{a_1a_2a_3a_4a_5a_6a_7a_8a_9a_{10}a_{11}}\,, \\ \langle\mathbf{J}_{a_1a_2}\,\mathbf{J}_{a_3a_4}\,\mathbf{J}_{a_5a_6}\,\mathbf{J}_{a_7a_8}\,\mathbf{J}^{a_9a_{10}}\,\mathbf{Z}_{abcde}\rangle &= -\frac{4\alpha}{9}\,\varepsilon_{a_1a_2a_3a_4a_5a_6a_7a_8abc}\,\delta^{a_9a_{10}}_{de}\,, \\ \langle\mathbf{J}_{a_1a_2}\,\mathbf{J}_{a_3a_4}\,\mathbf{J}_{a_5a_6}\,\mathbf{J}^{a_7a_8}\,\mathbf{J}^{a_9a_{10}}\,\mathbf{Z}^{ab}\rangle &= \frac{16(1-\alpha)}{3}\left[\delta^{a_7\ldots a_{10}ab}_{a_1\ldots a_6} - \delta^{a_9a_{10}ab}_{a_1\ldots a_4}\,\delta^{a_7a_8}_{a_5a_6}\right]. \end{aligned} \tag{5}$$

[1] In general, for a $2n-1$ theory, the Chern–Simons form is given as $d\mathscr{C}_{2n-1} = \langle\mathscr{F}\wedge\cdots\wedge\mathscr{F}\rangle$ (n times), $\langle\cdots\rangle$ being an invariant tensor for the corresponding Lie algebra.

As a consequence of this, the equations of motion are quintic polynomials in the curvature,

$$\langle \mathscr{F} \wedge \cdots \wedge \mathscr{F}\, \mathbf{G}_A \rangle = 0 \,. \tag{6}$$

2.2 Preons in the CS M-Theory

The spectrum of BPS states in M-theory goes beyond the M2- and M5-brane. By algebraic reasoning, one would expect to have so-called preons. These are states that preserve the supersymmetric invariance due to all but one real supercharge component. For instance, in 11D, this amounts to 31 real supercharges. It has been argued that CS supergravity possesses BPS preons in its spectrum [3]. That proof, however, is somewhat biased by the assumption that the relevant Lie algebra is $osp(32|1)$. In this section we will closely follow the compelling algebraic reasoning presented in [3] in support of a proof of existence of BPS states in CS M-theory.

Assume that we are interested in a BPS preon solution preserving 31 real supercharges. This means that there are 31 (real components of the generalized) Killing spinors, $\varepsilon_J{}^\alpha$, $J = 1, \ldots 31$. They are defined in terms of the differential operator that generates the supersymmetry transformation of the gravitino, $\delta\psi^\alpha = \mathscr{D}\varepsilon_J{}^\alpha = d\varepsilon_J{}^\alpha + \Omega_\alpha{}^\beta \varepsilon_J{}^\beta$. There is a single bosonic spinor λ_α, orthogonal to $\varepsilon_J{}^\alpha$, $\varepsilon_J{}^\alpha \lambda_\alpha = 0$, that characterizes the expected preonic state. It is natural, then, to call it $|\lambda\rangle$, which schematically satisfies $\varepsilon_J{}^\alpha \mathbf{Q}_\alpha |\lambda\rangle = 0$, for all J. Thus, $\mathbf{Q}_\alpha |\lambda\rangle = \lambda_\alpha |\lambda\rangle$, and therefore, $\mathbf{P}_{\alpha\beta} |\lambda\rangle = \lambda_\alpha \lambda_\beta |\lambda\rangle$. It is useful to complete the basis of spinors both with indices up, $\{\varepsilon_J{}^\alpha, w^\alpha\}$, and down, $\{u^J{}_\alpha, \lambda_\alpha\}$. It is always possible to choose both bases in such a way that they are orthogonal $w^\alpha \lambda_\alpha = 1$, $\varepsilon_J{}^\alpha u^I{}_\alpha = \delta^I_J$, $w^\alpha u^J{}_\alpha = \varepsilon_J{}^\alpha \lambda_\alpha = 0$.

The $\varepsilon_J{}^\alpha$ are Killing spinors, $\mathscr{D}\varepsilon_J{}^\alpha = 0$, and since $d(\varepsilon_J{}^\alpha \lambda_\alpha) = 0$, it turns out that $\mathscr{D}\lambda_\alpha$ is proportional to λ_α. Let us thus define the one-form A as $\mathscr{D}\lambda_\alpha = A\,\lambda_\alpha$. The application of two consecutive covariant derivatives yields $\mathscr{D}\mathscr{D}\lambda_\alpha = \mathscr{R}_\alpha{}^\beta \lambda_\beta$, where $\mathscr{R}_\alpha{}^\beta$ is the generalized curvature two-form

$$\mathscr{R}_\alpha{}^\beta = d\,\Omega_\alpha{}^\beta - \Omega_\alpha{}^\gamma \wedge \Omega_\gamma{}^\beta \,. \tag{7}$$

On the other hand, by applying the exterior derivative to the remaining orthogonality relations, the expression for the covariant derivatives of the remaining spinors $\mathscr{D}u^I{}_\alpha = B^I \lambda_\alpha$, where B^I is a collection of 31 1-forms, and $\mathscr{D}w^\alpha = -A\,w^\alpha - B^I \varepsilon_I{}^\alpha$, are easily obtained. Performing now the same trick on $w^\alpha \mathscr{D}\lambda_\alpha$ and $\varepsilon_J{}^\alpha \mathscr{D}\lambda_\alpha$, one gets $\mathscr{D}\mathscr{D}\lambda_\alpha = dA\;\lambda_\alpha$. It is easy to play the same trick with $u^I{}_\alpha$, with the result $\mathscr{D}\mathscr{D}u^I{}_\alpha = \mathscr{R}_\alpha{}^\beta u^I{}_\beta = \nabla B^I \lambda_\alpha$, where $\nabla B^I = dB^I + A \wedge B^I$. All in all, we can decompose both Ω and $\mathscr{R}$ in the spinorial basis as [3]

$$\begin{aligned}
\Omega_\alpha{}^\beta &= A\,\lambda_\alpha\, w^\beta + B^I \lambda_\alpha\, \varepsilon_I{}^\beta - d\lambda_\alpha\, w^\beta - du^I{}_\alpha\, \varepsilon_I{}^\beta \,, \\
\mathscr{R}_\alpha{}^\beta &= dA\,\lambda_\alpha\, w^\beta + \nabla B^I \lambda_\alpha\, \varepsilon_I{}^\beta \,.
\end{aligned} \tag{8}$$

Now, let us complete the argument used in [3] for the case of the smallest AdS superalgebra $osp(32|1)$. The bosonic part of the connection is a one-form in the subalgebra $sp(32)$

$$\Omega^{osp(32|1)} = \frac{1}{2}\, e_a \Gamma^a + \frac{1}{4}\, \omega_{ab} \Gamma^{ab} + \frac{1}{240}\, b_{abcde} \Gamma^{abcde} \, . \tag{9}$$

Thus, $\Omega^{\alpha\beta} = \Omega^{\beta\alpha}$, that is, $\Omega^{[\alpha\beta]} = 0$. Both Ω and $\mathscr{R}$ belong to $sp(32)$. Thus, they are traceless $\Omega_\alpha{}^\alpha = \mathscr{R}_\alpha{}^\alpha = 0$. This means $A = 0$, then $\mathscr{R}_\alpha{}^\beta = dB^I \lambda_\alpha \varepsilon_I{}^\beta$. This implies that the generalized supercovariant curvature is nilpotent [3]

$$\mathscr{R}_\alpha{}^\gamma \wedge \mathscr{R}_\gamma{}^\beta = 0 \, , \tag{10}$$

due to the orthogonality $\varepsilon_I{}^\alpha \lambda_\alpha = 0$. Now, since the gauge connection in the $osp(32|1)$ CS supergravity, $\mathscr{A}$ precisely matches Ω, then $\mathscr{F} = \mathscr{R}$. The equation of motion (that formally looks like (6)) is, then, always satisfied for a BPS preonic configuration. Still, it is necessary to check the actual integrability of (10), as there might be, for instance, topological obstructions.

Now, coming back to the case of the M-algebra, it is important to stress that $\Omega^{M-algebra} = \frac{1}{4}\, \omega_{ab} \Gamma^{ab}$, which is not $\mathscr{A}$. Then, $\mathscr{R} = \frac{1}{2} R^{ab} \mathbf{J}_{ab} \neq \mathscr{F}$. The nilpotency of $\mathscr{R}$, thus, does not guarantee a priori the solution of the CS equations of motion. This is a generic feature of all theories built as extensions of Poincaré CS supergravity (see, for example, [21]). Notice that this is qualitatively different to the behavior of AdS-based Lie algebras. The connection between both kinds of theories, though, is well understood [20]. However, recalling that the nonvanishing components of the invariant tensor for all these theories look like those displayed earlier in (5), we can conclude that at least four factors in (6) admit the replacement $\mathscr{F} \to \mathscr{R}$, and this guarantees that the preonic configuration – if no topological obstruction arises and it actually solves (10) – always satisfies the CS supergravity equations.

2.3 Difficulties with the Standard M-Brane Construction in CS Theory

In standard supergravity one typically "deduces" from a given SUSY algebra that there are BPS states. Applying the same analysis to a CS theory in a straightforward manner meets with severe difficulties that cast doubt on the viability of the strategy. In order to fix ideas, let us begin by recalling how it is that standard supergravity couples to a membrane. The exercise will suggest why a similar strategy would not work for a CS supergravity. Let us present the argument through an example.

Consider a flat M2-brane extended in the x^1–x^2 plane. It should be associated with a non-zero value of Z_{12} (the very presence of the M2-brane breaks the Lorentz group from $SO(1,10)$ to $SO(1,2) \times SO(8)$). Let us choose the Majorana representation in which $C = \Gamma^0$ $[(\Gamma^0)^2 = 1]$. In that case, for a static membrane,

$$\{\mathbf{Q}_\alpha, \mathbf{Q}_\beta\} = \delta_{\alpha\beta} P_0 + (\Gamma^{012})_{\alpha\beta} Z_{12} . \quad (11)$$

In 11D, the Majorana spinors $\mathbf{Q}_\alpha$ are real. So, the left hand side is manifestly positive definite. The sign of Z_{12} can be flipped by replacing a membrane by an anti-membrane. Instead, $P_0 \geq 0$. Thus, as a consequence of the positive definite bracket, using Witten–Olive's construction [22], it turns out that

$$P_0 \geq |Z_{12}| . \quad (12)$$

A BPS M2-brane is expected to saturate the bound, $P_0 = Z_{12}$,

$$\{\mathbf{Q}_\alpha, \mathbf{Q}_\beta\} = P_0 \left[1 \mp \Gamma^{012}\right]_{\alpha\beta} . \quad (13)$$

Spinors ε satisfying $\Gamma^{012}\varepsilon = \pm\varepsilon$ are eigenspinors of $\{\mathbf{Q}_\alpha, \mathbf{Q}_\beta\}$ with zero eigenvalue. These are the spinors corresponding to the $1/2$ unbroken supersymmetries. A similar argument holds for the M5-brane. This argument is *naively independent of the dynamics*, i.e., whether it is given by a CS Lagrangian or that of Cremmer–Julia–Scherk. We will come back to this point shortly.

In standard CJS supergravity, $\delta\psi = (d + \Omega^{CJS})\,\varepsilon$, with the connection given by the 32×32 matrix valued 1-form

$$\Omega^{CJS} = \frac{1}{4}\,\omega_{ab}\,\Gamma^{ab} + \frac{i}{18}\,e^a\,F_{ab_1b_2b_3}\,\Gamma^{b_1b_2b_3} + \frac{i}{144}\,e_a\,F_{b_1b_2b_3b_4}\,\Gamma^{ab_1b_2b_3b_4} . \quad (14)$$

An M2-brane has non-vanishing F_{012r}, r being the transverse radial direction. Imposing $\delta\psi = 0$ leads to three different equations, $\delta\psi_m = \delta\psi_r = \delta\psi_s = 0$, where $m = 0,1,2$, r amounts for the radial direction and s runs over transverse indices. The second equation just provides a differential equation that dictates the radial dependence $\varepsilon(r)$. Now, since the spinor ε obeys a chirality prescription, we see that,

$$\begin{aligned}
\delta\psi_m &= \frac{1}{4}\,\omega_{m\,ab}\Gamma^{ab}\varepsilon - \frac{1}{12}\,\varepsilon_{mnp}\,\Gamma^{npr}\,F_{012r}\,\varepsilon = 0 , \\
\delta\psi_s &= \frac{1}{4}\,\omega_{s\,ab}\Gamma^{ab}\varepsilon + \frac{1}{12}\,\Gamma^{012r}{}_s\,F_{012r}\,\varepsilon = 0 ,
\end{aligned} \quad (15)$$

and we see that, provided the only non-vanishing components of the spin connection are ω_m^{mr} and ω_s^{sr}, which is the case in standard supergravity for a natural D-brane ansatz, previous equations would only lead to non-trivial solutions provided, precisely, the chirality condition $\Gamma^{012}\varepsilon = \pm\varepsilon$ is imposed on the spinor. The whole picture is self-consistent.

Instead, in CS M-theory supergravity, the supersymmetry transformation of the gravitino is dramatically simpler, $\delta\psi_\mu = D_\mu\varepsilon = 0$, the only difference having to do with the fact that now ω can have a contorsion contribution, $\omega = \omega^{(0)} + \kappa$. The naive expectation is that κ should play the role of the $A_{[3]}$ form. For example, $A_{[3]} = e^a \wedge e^b \wedge \kappa_{ab}$ [23]. However, whatever is the case, the above equation would lead schematically to

$$\left(a_1 \Gamma^{ab} + a_2 \Gamma^{cd}\right) \varepsilon = 0\,, \tag{16}$$

and this could never reduce to $\Gamma^{012}\varepsilon = \pm\varepsilon$, the projection that, according to our simple algebraic argument, is necessary for the M2-brane. At best, the resulting condition, if consistent with the M2-brane projection, would lead to a $1/4$ supersymmetric configuration that does not correspond to the M2-brane.

This is not the end of the story. For Chern–Simons theories one cannot rely on the naive analysis performed using the M-algebra. We know that the canonical structure of these theories is intricate. We should first check whether we are working in a degenerate sector or in a generic one, and use Dirac's formalism thoroughly to determine the exact form of the supersymmetry algebra on the constraint surface. This was partially done for 5D CS supergravity in [24]. A necessary step to put our conclusions on a firm ground involves a generalization of this analysis to the 11D case, which is not an easy job. It is still intriguing to figure out how the inconsistency between the Γ matrix structure of the algebra and the supersymmetry transformation laws shall be solved. For this to happen, the actual Γ matrix structure should change after the Dirac analysis. This would lead to the very interesting scenario in which the starting point might be quite a rather strange looking CS theory whose constrained algebra looks like the M-algebra. This seems very hard to implement.

In what follows, we present an alternative route to couple a CS theory to a brane, taking as a model the electromagnetic coupling to a the worldline of a point-charge (0-brane). Starting from the observation that the electromagnetic coupling is also the integral of a CS form, the coupling is generalized to higher-dimensional branes and to non-Abelian connections. The resulting structure may not be the most general form, but it has the advantage that it exploits the geometric features of the CS forms to bring about the interactions (no metric required, topological origin, quantized charges, etc.).

3 CS Actions as Brane Couplings

A Chern–Simons action is a functional for a Lie-algebra valued one-form $\mathscr{A}$, defined in a topological space of dimension $D = 2n+1$,

$$I_{2n+1}[\mathscr{A}] = \frac{\kappa}{n+1} \int_{\Gamma^{2n+1}} \sum_{k=0}^{n} c_k \left\langle \mathscr{A}^{2k+1} \wedge (d\mathscr{A})^{n-k} \right\rangle, \tag{17}$$

where $\langle \cdots \rangle$ denotes the symmetrized trace in some representation of the Lie algebra, $\mathscr{A}^p$ means $\mathscr{A} \wedge \cdots \wedge \mathscr{A}$ (p times), $(d\mathscr{A})^q$ should be analogously understood as $d\mathscr{A} \wedge \cdots \wedge d\mathscr{A}$ (q times), c_k are specific coefficients ($c_0 = 1$), and κ is a constant, known as the level of the theory. The fundamental difference between CS theories and the vast majority of physical actions is the absence of a metric structure and of dimensionful parameters in the former. This makes the theory simultaneously scale invariant, covariant under general coordinate transformations, and background independent.

The simplest example of a CS action is the familiar minimal coupling between an electric point charge and the electromagnetic potential,

$$I_{Int} = \int_{M^D} j^{\mu} \mathscr{A}_{\mu} d^D x \,. \tag{18}$$

Since the current density j^{μ} has support on the worldline of the charge, (18) can also be written as an integral over a $(0+1)$-dimensional manifold, which corresponds to the case $n = 0$ in (17),

$$I_{0+1}[\mathscr{A}] = \kappa \int_{\Gamma^1} \langle \mathscr{A} \rangle \,. \tag{19}$$

Here the manifold Γ^1 is the worldline of the charge, a one-dimensional submanifold embedded in the higher-dimensional space M^D, which is identified as the spacetime. The D-dimensional embedding spacetime may have a metric which induces a natural metric on the worldline, but this metric is not necessary to construct the action.

In [6, 7], Bunster analyzed the generalization of (18) to describe the coupling between a $(p-1)$-brane to a gauge potential, with an interacion of the form

$$I_{Int} = \int_{M^D} j^{\mu_1 \mu_2 \cdots \mu_p} A_{\mu_1 \mu_2 \cdots \mu_p} d^D x \,. \tag{20}$$

He showed that this form of minimal coupling can only be defined (for any $p > 1$) if the connection is Abelian, i.e., it transforms as $A \to A + d\Lambda$, where Λ is a (real valued) $(p-1)$-form. The extension to non-Abelian connections was shown to be inconsistent due to the noncommutativity of the Hamiltonian at different times. As we show below, this obstruction does not arise if the branes couple to CS forms, Abelian or otherwise.

3.1 0+1 CS Theories

As emphasized in [25], the same expression (18) can also be interpreted as the action, in Hamiltonian form, for an arbitrary mechanical system of finitely many degrees of freedom [26]. Therefore all mechanical systems are also examples of CS theories. Moreover, the Bohr–Sommerfeld rules of quantum mechanics, as well as Dirac's quantization rule for electric–magnetic charges, can be seen as consequences of the topological origin of CS theories, the Chern classes. So, CS theories are far from exceptional, they seem to be rather commonplace in physics.

For most Lie groups of physical interest (unitary, orthonormal), the generators are traceless and therefore $\langle \mathscr{A} \rangle = 0$. The only important exception is the $U(1)$ group, and therefore one should look at the coupling between an electric charge $e = \kappa$ and the electromagnetic field,

$$I = e \int_{\Gamma^1} \mathscr{A}_\mu(z)\, dz^\mu \,, \tag{21}$$

where z^μ are the embedding coordinates of the worldline Γ^1, giving the position of the charge in M^D. The interesting point is that the electromagnetic interaction is a model that captures the essential features of the coupling between higher-dimensional CS theories and branes.

The point charge is described by a delta function with support at the position of the charge on the spatial section $x^0 =$ constant. The interaction term is

$$I = \int_{M^D} j_0^{(D-1)} \wedge \mathscr{A} \,, \tag{22}$$

where

$$j_0^{(D-1)} = \kappa \delta^{(D-1)}(x-z) d\Omega^{D-1} \,. \tag{23}$$

Here $d\Omega^{D-1}$ is the volume form of the spatial section in the rest frame of the charge. The action (21) by itself can be varied with respect to the embedding coordinates z^μ which, in the mechanical interpretation, are the enlarged phase space coordinates, $z^\mu \leftrightarrow (p^i, q_i, t)$ [25]. This means, in particular, that the embedding space must be odd-dimensional. This underscores the fact that a CS theory in a spacetime of dimension $D = (2n+1)$ can be naturally coupled to a $0+1$ Chern–Simons action defined on a one-dimensional worldline. This idea may be easily generalized to include CS actions for all lower (odd-) dimensional worldvolumes generated by $2p$-branes, with $p < n$, as we show next.

3.2 (2n+1)-Dimensional Abelian CS Theories and 2p Branes

Comparing (17) with the expression for the coupling between a point charge and the electromagnetic potential (19), it is clear that a $(2p+1)$ CS action, can be viewed as the coupling between the connection and a $2p$-brane [25, 27, 28]. One is then led to consider the general coupling between an Abelian connection $\mathscr{A}$ and external sources with support on the $(2p+1)$-worldvolumes of all possible $2p$-branes that can be embedded in M^{2n+1},

$$I_{2n+1}[\mathscr{A}] = \int_{\Gamma^{2n+1}} \sum_{p=0}^{n} j_{2p}^{(2n-2p)} \wedge \mathscr{C}_{2p+1} = \sum_{p=0}^{n} \kappa_p \int_{\Gamma^{2p+1}} \mathscr{C}_{2p+1} \,. \tag{24}$$

Here the levels κ_p are independent dimensionless coupling constants that can be identified with the "electromagnetic" charges.[2] For simplicity we set $\kappa_n = 1$ here. The simplest rendering of this form is $p = 0$, $n = 1$: a point charge acting as the source of a $2+1$ Abelian CS connection. The action reads

$$I[\mathscr{A}] = \int_{\Gamma^{2+1}} \left[\frac{1}{2} \mathscr{A} \wedge d\mathscr{A} + j_0^{(2)} \wedge \mathscr{A} \right] . \tag{25}$$

Assuming Γ^{2+1} to be compact and without boundary, the action can be varied with respect to $\mathscr{A}$. The field equation reads, not surprisingly,

$$\mathscr{F} = j_0^{(2)} , \tag{26}$$

where $j_0^{(2)}$ is the 2-form charge density describing a point charge at rest,

$$j_0^{(2)} = \kappa_0 \delta^{(2)}(\mathbf{z}) dx \wedge dy . \tag{27}$$

The field $\mathscr{A}$ is given by

$$\mathscr{A} = \frac{\kappa_0}{2\pi} d\phi , \tag{28}$$

as shown by direct integration of the field equation $d\mathscr{A} = \kappa_0 \delta(x,y) dx \wedge dy$ on a disc, and using Stokes' theorem on a manifold that is topologically $\mathbb{R}^2 - \{0\}$. This source produces a magnetic field ($\mathscr{F}_{0i} = 0$) concentrated along the worldline of the charge, like an infinitely thin solenoid (with the only peculiarity that the solenoid is infinitely long in the time direction) [29,30]. So, this configuration is the electromagnetic field produced by a magnetic point source (monopole).

Similarly, a $(2n+1)$-CS form couples to the worldvolume of a charged $2p$-brane through the interaction

$$I[\mathscr{A}] = \int_{\Gamma^{2n+1}} j_{2p}^{(2n-2p)} \wedge \mathscr{C}_{2p+1} , \tag{29}$$

where $j_{2p}^{(2n-2p)}$ stands for a $(2n-2p)$ form with support on the worldvolume of the $2p$ brane embedded in the $(2n+1)$-dimensional spacetime, and the field equations read

$$\mathscr{F}^n = \sum_p j_{2p}^{(2n-2p)} \wedge \mathscr{F}^p . \tag{30}$$

where $\mathscr{F}^k = \mathscr{F} \wedge \cdots \wedge \mathscr{F}$ (k times).

[2] Local gauge invariance of the Chern–Simons form guarantees that $I_{2n+1}[\mathscr{A}]$ is gauge invariant provided the currents $j_{2p}^{(2n-2p)}$ are closed (conserved), $dj_{2p}^{(2n-2p)} = 0$. Under quite general arguments, analogous to Dirac's for the quantization of the electric/magnetic charges, it can be shown that these charges must also be quantized.

3.3 Coupling of Non-Abelian CS Actions to 2p-Branes

The above construction can be extended to non-Abelian connections simply allowing $\mathscr{C}_{2p+1}$ to be a $(2p+1)$ CS form for the same non-Abelian connection[3] $\mathscr{A}$. However, the generalization meets an important new constraint: the invariant tensor $\tau_{a_1 a_2 \cdots a_{n+1}}$ of a given Lie algebra, with generators $\mathbf{G}_a$, $a = 1, 2, \ldots r$,

$$\tau_{a_1 a_2 \cdots a_{n+1}} := \langle \mathbf{G}_{a_1} \mathbf{G}_{a_2} \cdots \mathbf{G}_{a_{n+1}} \rangle \,, \tag{31}$$

required for the CS action in $2n+1$ dimensions (see, for example, (5)), may not be defined for all values of n. It is an open problem how many invariant tensors of a given rank there exist for a given Lie algebra. This puts a severe restriction on the kinds of allowed couplings between a non-Abelian connection $\mathscr{A}$ and a $2p$-brane. Generically, one could write (29) as in the previous case, but since there is no guarantee that a given Lie algebra admits an invariant tensor of a certain rank, many CS forms $\mathscr{C}_{2p+1}$ may vanish identically.

An alternative possibility is that the $2p$-brane couples to a $(2p+1)$-CS form constructed with the invariant tensor for a *subalgebra* of the Lie algebra defining the local symmetries of the theory in the embedding space M^{2n+1}. In fact, the presence of the brane generically produces a topological defect that partially breaks the original gauge symmetry. The surviving symmetry forms a subalgebra that admits an invariant tensor that can be used to construct a CS form on the worldvolume of the brane/defect. This was observed to occur in the presence of a codimension 2 topological defect [31, 32]. There, the defect breaks the gauge symmetry $SO(D-1,2)$ down to $SO(D-2,1)$, giving rise to a gravitational action in $D-1$ dimensions out of a topological invariant in $D+1$ dimensions.

Another interesting feature of this mechanism of symmetry breaking by a $2p$-brane is this: suppose one couples a connection for the AdS algebra in $2n+1$ dimensions; the worldvolume of the brane,[4] is a manifold of dimension $2p+1$, and the maximal symmetry of the tangent space is $SO(2p,2)$. Since the number of components of a spinor representation goes as $2^{[D/2]}$, for every reduction by two in the dimension of the brane, there is a reduction by half in the number of components of the possible Killing spinors admitted by the configuration. This means that one can expect to generate $1/2$, $1/4$ (in general $1/2^k$) BPS states in this manner.

One alternative to the breaking would occur if the fermions in the space with a defect are combined in complex representations. For instance, starting from 11D and an $osp(32|1)$ real spinor with 32 components, the presence of a codimension two defect would break the spacetime symmetry down to $so(8,2) \times so(2)$ admitting a complex spinor with 16 components. The supersymmetric extension of the AdS

[3] Note that although $\mathscr{A}$ may be a non-Abelian connection, $\mathscr{C}_{2p+1}$ is in the center of the algebra and hence commuting. In this way, the obstruction presented in [6, 7] can be circumvented.

[4] These arguments may be extended to the case of spacelike worldvolumes. The fate of supersymmetry is unclear in this case, though.

group in 9D is $SU(8,8|1)$. Topological defects will host Killing spinors living in representations of the latter group and this generically implies the breaking of a fraction of the original supersymmetries. For particular values of the parameters, however, it might happen that the defect preserves all the supercharges (see the example below).

An interesting case that deserves further discussion is that of a membrane coupled to an 11D CS theory for the M-algebra. It is not hard to be tempted to identify such an object as the M2-brane. Notice that its coupling to the gauge connection is given through a 3D CS action based at most on the maximal supersymmetric extension of $so(2,1) \times so(8)$. This term is reminiscent of the action for multiple M2-branes recently unveiled by Bagger and Lambert [33]. No doubt that there are important differences, such as the presence of extra scalar fields (and their supersymmetric partners) and the fact that the CS Lagrangian in the BL theory is based upon a 3-algebra. Our approach attempts to address how these branes couple to the 11D fields while BL theory aims at describing the dynamics of multiple M2-branes on their own. They are not on equal footing. However, we find striking that our independent proposal for the introduction of M2-branes in a CS theory based on the M-algebra possesses these similarities and consider that this is a worth exploring avenue for further research.

3.3.1 Example: Topological Defects

Branes are, in a broad sense, topological obstructions, like boundaries and defects. They restrict the continuous differentiable propagation on the manifold, which have topological consequences for the allowed orbits and for the spectrum of the differential operators. CS systems are particularly sensitive to the topological structure of the spacetime on which they are defined, and therefore the coupling between a connection dynamically governed by a CS action and a brane is necessarily nontrivial. Instead of developing a general theory for this problem, we illustrate this with an example. The discussion will remain at an introductory level and the reader is encouraged to look for the relevant sources as they become available.

Consider a $(2n+1)$-dimensional AdS spacetime where a point has been removed from the spatial section. The evolution in time of the missing point is a removed one-dimensional worldline. The resulting topology is not that of AdS and it allows for nontrivial winding numbers for S^{D-2} spheres mapped onto the spacetime. In principle, there could be an angular defect concentrated on the removed line, which measures the strength of the singularity. The defect is produced by an identification in the angular directions whereby the solid angle Ω_{D-2} of the S^{D-2} sphere is shrunk to $(1-\alpha)\Omega_{D-2}$. The metric produced by this angular defect can be written as

$$ds^2 = -(1+\rho^2)dT^2 + (1+\rho^2)^{-1}d\rho^2 + \alpha^2\rho^2 d\Omega_{D-2}^2 \, . \tag{32}$$

It is straightforward to check that this metric has a naked curvature singularity in the angular components of the Riemann tensor,

$$R^{\alpha\beta}{}_{\theta\phi} = \left[-1 + \frac{\sqrt{1+M}}{r^2}\right] \delta^{\alpha\beta}{}_{\theta\phi} \, , \qquad (33)$$

while the remaining components are those of a constant curvature, $R^{0r}{}_{0r} = -1$, $R^{0\alpha}{}_{0\beta} = R^{r\alpha}{}_{r\beta} = -\delta^{\alpha}_{\ \beta}$. This looks like the standard $(2n+1)$ AdS CS black hole (in $r(\rho)$ radial coordinate) [34]. However, for $0 < \alpha < 1$, the (dimensionless) mass parameter corresponds to a "negative mass black hole," $-1 < M(\alpha) < 0$, which is just a naked singularity [35].

The resulting space does not admit Killing spinors and the singularity cannot correspond to a BPS configuration, except for $D = 3$ in the limit when the angular defect becomes maximal ($\alpha \to 1$). This special case, $M = 0$, corresponds to a massless 2+1 black hole and the space now admits half of the Killing spinors of the AdS spacetime. In this case, the defect can legitimately be called a BPS 0-brane[5] [35].

The massless 2+1 black hole can also be generated through a particular identification by a Killing vector in AdS_3 space and this mechanism can be repeated in higher dimensions: starting with AdS_{2n+1}, an identification with a Killing vector that has a fixed point generates a topological defect at the fixed point and breaks the symmetry from $SO(2n,2)$ down to $SO(2n-1,1) \times R$. The AdS space has maximal supersymmetry with 2^n component spinors; the topological defect can have at most 2^{n-1} local supersymmetries, that is $1/2$ BPS. Additional breakings generated by further identifications with Killing spinors, would reduce the supersymmetry to $1/4$, $1/8$, etc. [35, 37]. This topological symmetry breaking was recently exploited in [31, 32] to generate an effective Einstein–Hilbert action in four dimensions from a topological defect in a six-dimensional topological field theory.

4 Summary and Gambling on Future Directions

We have presented a proposal for the coupling of sources in CS theories. When the spacetime dimensions is $D \leq 3$, this amounts to the standard minimal gauge coupling. However, for higher dimensional CS theories this produces new interaction terms with a number of consequences on which this article just offers a first glance. The coupling is entirely given in terms of the connection of the original theory and does not require (the otherwise problematic) non-Abelian p-forms to couple directly with extended objects such as branes. We argue that this suggests the need to revisit the exploration of the BPS spectrum of CS theories, a certainly difficult subject, in a way that possibly circumvents naive obstructions to the existence of expected objects such as the M2-brane in a CS theory based on the M-algebra.

[5] For $D > 3$ the massless black holes also have a curvature singularity and are not BPS, as shown in [36].

We have explored the existence of preons in CS theories. Following the algebraic reasoning introduced in [3], we have shown that it applies to CS theories based on extensions of the AdS algebra, but needs some improvement for the case of Poincaré-based theories such as the CS supergravity for the M-algebra. We should emphasize that even if the integrability conditions of the local Killing spinor conditions are consistent with the equations of motion, it is still necessary to scrutinize the actual integrability of the equations, as there might be, for example, topological obstructions.

Our proposal for the introduction of sources implies a novel mechanism of symmetry breaking through the presence of defects in CS theories. Since the couplings are given in terms of CS forms with support in lower dimensional submanifolds, they will be written generally in terms of subalgebras of the original algebra.

The quantization of CS theories is an important problem. The case of $0+1$ is just the old Bohr–Sommerfeld quantization [25]. The $2+1$ case is well-understood: the path integral is given in terms of knot invariants [38, 39]. The corresponding quantization for higher dimensions remains an open problem.

The emergence of a dimensionful physical scale in the theory may arise through condensates of the form $\langle \mathscr{A} \bullet \mathscr{A} \rangle \neq 0$, where the bracket is an invariant tensor of the reduced (physical) symmetry. For instance, in a CS theory based on the AdS group broken down to the Lorentz group, $\langle \mathscr{A} \bullet \mathscr{A} \rangle$ reduces to $e^a_\mu e^b_\nu \eta_{ab} = \ell^2 g_{\mu\nu}$. The remnant Lorentz symmetry suggests the possibility that the induced dynamics may be governed by the Einstein–Hilbert action. This is indeed the case in the scheme studied in [31, 32], and might be a promising avenue to explore the (still open) connection between CS theories and ordinary supergravity.

Acknowledgments We are pleased to thank Mokhtar Hassaïne, Olivera Mišković and Steve Willison, for many interesting discussions and comments. We are also thankful to Claudio Bunster for so many years of collaboration and friendship. This work is supported in part by MEC and FEDER (grant FPA2005-00188), by Xunta de Galicia (Consellería de Educación and grant PGIDIT06PXIB206185PR), by the European Commission (grant MRTN-CT-2004-005104), and by the Spanish Consolider-Ingenio 2010 Programme CPAN (CSD2007-00042). JDE is a *Ramón y Cajal* Fellow. The Centro de Estudios Científicos (CECS) is funded by the Chilean Government through the Millennium Science Initiative and the Centers of Excellence Base Financing Program of Conicyt. CECS is also supported by a group of private companies which at present includes Antofagasta Minerals, Arauco, Empresas CMPC, Indura, Naviera Ultragas and Telefónica del Sur. This work was partially supported by Fondecyt grants #1061291 and #7070117.

References

1. Hassaïne, M., Troncoso R., Zanelli, J.: Eleven-dimensional supergravity as a gauge theory for the M-algebra. Phys. Lett. B **596**, 132 (2004)
2. Bandos, I.A., de Azcárraga, J.A., Izquierdo, J.M., Lukierski, J.: BPS states in M-theory and twistorial constituents. Phys. Rev. Lett. **86**, 4451 (2001)
3. Bandos, I.A., de Azcárraga, J.A., Izquierdo, J.M., Picon, M., Varela, O.: BPS preons, generalized holonomies and 11D supergravities. Phys. Rev. D **69**, 105010 (2004)

4. Gran, U., Gutowski, J., Papadopoulos, G., Roest, D.: N = 31, D = 11. JHEP **0702**, 043 (2007)
5. Teitelboim, C.: Quantum mechanics of the gravitational field. Phys. Rev. D **25**, 3159 (1982)
6. Teitelboim, C.: Gauge invariance for extended objects. Phys. Lett. B **167**, 63 (1986)
7. Henneaux, M., Teitelboim, C.: p-Form Electrodynamics. Found. Phys. **16**, 593 (1986)
8. Zanelli, J.: Lecture notes on Chern–Simons (super-)gravities. Second edition (February 2008) (unpublished). arXiv:hep-th/0502193
9. Edelstein, J.D., Zanelli, J.: (Super-)gravities of a different sort. J. Phys. Conf. Ser. **33**, 83 (2006)
10. Hull, C.M., Townsend, P.K.: Unity of superstring dualities. Nucl. Phys. B **438**, 109 (1995)
11. Witten, E.: String theory dynamics in various dimensions. Nucl. Phys. B **443**, 85 (1995)
12. Troncoso, R., Zanelli, J.: New gauge supergravity in seven and eleven dimensions. Phys. Rev. D **58**, 101703 (1998)
13. Horava, P., Witten, E.: Heterotic and type I string dynamics from eleven dimensions. Nucl. Phys. B **460**, 506 (1996)
14. Witten, E.: On flux quantization in M-theory and the effective action. J. Geom. Phys. **22**, 1 (1997)
15. Horava, P.: M-theory as a holographic field theory. Phys. Rev. D **59**, 046004 (1999)
16. Townsend, P.K.: P-brane democracy (unpublished). arXiv:hep-th/9507048
17. Hatsuda, M., Sakaguchi, M.: Wess–Zumino term for the AdS superstring and generalized Inonu–Wigner contraction. Prog. Theor. Phys. **109**, 853 (2003)
18. de Azcárraga, J.A., Izquierdo, J.M., Picón, M., Varela, O.: Generating Lie and gauge free differential (super)algebras by expanding Maurer–Cartan forms and Chern–Simons supergravity. Nucl. Phys. B **662**, 185 (2003)
19. de Azcárraga, J.A., Izquierdo, J.M., Picón, M., Varela, O.: Expansions of algebras and superalgebras and some applications. Int. J. Theor. Phys. **46**, 2738 (2007)
20. Edelstein, J.D., Hassaïne, M., Troncoso R., Zanelli, J.: Lie-algebra expansions, Chern–Simons theories and the Einstein–Hilbert Lagrangian. Phys. Lett. B **640**, 278 (2006)
21. Hassaïne, M., Romo, M.: Local supersymmetric extensions of the Poincare and AdS invariant gravity. JHEP **0806**, 018 (2008)
22. Witten, E., Olive, D.I.: Supersymmetry algebras that include topological charges. Phys. Lett. B **78**, 97 (1978)
23. Bañados, M.: The linear spectrum of OSp(32|1) Chern–Simons supergravity in eleven dimensions. Phys. Rev. Lett. **88**, 031301 (2002)
24. Mišković, O., Troncoso, R., Zanelli, J.: Canonical sectors of 5d Chern–Simons theories. Phys. Lett. B **615**, 277 (2005)
25. Zanelli, J.: Uses of Chern–Simons actions. In: Edelstein, J.D., Grandi, N., Núñez, C., Schvellinger, M. (eds.) Ten years of AdS/CFT, pp. 115–129. American Institute of Physics, New York (2008)
26. Dunne, G.V., Jackiw, R., Trugenberger, C.A.: Topological (Chern–Simons) quantum mechanics. Phys. Rev. D **41**, 661 (1990)
27. Mora, P., Nishino, H.: Fundamental extended objects for Chern–Simons supergravity. Phys. Lett. B **482**, 222 (2000)
28. Mora, P.: Chern–Simons supersymmetric branes. Nucl. Phys. B **594**, 229 (2001)
29. Jackiw R., Templeton, S.: How superrenormalizable interactions cure their infrared divergences. Phys. Rev. D **23**, 2291 (1981)
30. Schonfeld, J.F.: A mass term for three-dimensional gauge fields. Nucl. Phys. B **185**, 157 (1981)
31. Anabalón, A., Willison, S., Zanelli, J.: General relativity from a gauged WZW term. Phys. Rev. D **75** 024009 (2007)
32. Anabalón, A., Willison, S., Zanelli, J.: The Universe as a topological defect. Phys. Rev. D **77** 044019 (2008)
33. Bagger J., Lambert, N.: Gauge symmetry and supersymmetry of multiple M2-branes. Phys. Rev. D **77**, 065008 (2008)
34. Bañados, M.,Teitelboim, C., Zanelli, J.: Dimensionally continued black holes. Phys. Rev. D **49** 975 (1994)
35. Mišković, O., Zanelli, J.: Couplings between Chern-Simons gravities and 2p-branes. To appear (2009) (unpublished)

36. Aros, R., Martínez, C., Troncoso, R., Zanelli, J.: Supersymmetry of gravitational ground states. JHEP **0205**, 020 (2002)
37. Edelstein, J.D., Zanelli, J.: In progress
38. Witten, E.: Quantum field theory and the Jones polynomial. Commun. Math. Phys. **121**, 351 (1989)
39. Bar-Natan, D., Witten, E.: Perturbative expansion of Chern–Simons theory with noncompact gauge group. Commun. Math. Phys. **141**, 423 (1991)

The Emergence of Fermions and the E_{11} Content

François Englert and Laurent Houart

Abstract Claudio's warm and endearing personality adds to our admiration for his achievements in physics a sense of friendliness. His constant interest in fundamental questions motivated the following presentation of our attempt to understand the nature of fermions. This problem is an essential element of the quantum world and might be related to the quest for quantum gravity. We shall review how space–time fermions can emerge out of bosons in string theory and how this fact affects the extended Kac–Moody approach to the M-theory project.

1 Introduction

Despite the impressive theoretical developments of superstring theory, the quantization of gravity remains elusive. The difficulties encountered in coping with the non-perturbative level may well hide non-technical issues. A crucial point is whether the assumed quantum theoretical framework can cope with the quantum nature of space–time, in particular when confronted to the existence of black hole and the cosmological horizons. In this essay, we inquire into the fundamental nature of fermions, which constitute an essential element of the quantum world. We shall review how in string theory space–time fermions can be constructed out of bosons and we shall discuss how this fact affects the extended Kac–Moody approach to the M-theory project for quantum gravity.

In Sect. 2 we unveil the fermionic subspaces of the bosonic closed strings compactified on sublattices of a $E_8 \times \widetilde{SO}(16)$ weight lattice, where $\widetilde{SO}(16)$ is the universal covering group of $SO(16)$ [4]. All modular invariant fermionic closed

F. Englert
Service de Physique Théorique and The International Solvay Institutes, Université Libre de Bruxelles, Campus Plaine C.P.225, 1050 Bruxelles, Belgium
e-mail: fenglert@ulb.ac.be

M. Henneaux, J. Zanelli (eds.), *Quantum Mechanics of Fundamental Systems: The Quest for Beauty and Simplicity*, DOI 10.1007/978-0-387-87499-9_9,

strings, supersymmetric or not, are obtained from the parent bosonic strings by a universal truncation performed on both left and right sectors, or on the right sector for the heterotic strings. Supersymmetry arises when the sublattice of the $\widetilde{SO}(16)$ weight lattice is taken to be the E_8 root lattice so that the bosonic gauge group in each sector is $\mathscr{G} = E_8 \times E_8$. We found that not only the closed string spectra of the fermionic string, but also the charges, the chiralities and the tensions of all the fermionic D-branes are encoded in the bosonic strings [4]. In addition, the universal truncation applied to the unique tadpole-free unoriented bosonic string with Chan–Paton group $SO(2^{13})$ yields all tadpole- and anomaly-free open unoriented fermionic strings [4, 19].

In Sect. 3 we review the attempt to formulate the M-theory project in terms of the very-extended Kac–Moody algebra $E_{11} \equiv E_8^{+++}$ [41] (or the overextended $E_{10} \equiv E_8^{++}$ [29]). Along this line of thought, the inclusion of the bosonic string suggest the introduction of the algebra D_{24}^{+++} [32] (or D_{24}^{++}). However it is easily shown that the D_{24}^{+++} fields cannot accommodate the degrees of freedom needed to generate the fermionic subspaces of the bosonic string. More generally we argue that, without extending the D_{24}^{+++} algebra, one cannot encode genuine bosonic string degrees of freedom and, similarly, that E_{11} alone does not encode genuine superstring degrees of freedom.

2 Fermions and Supersymmetry from the Bosonic String

It is well-known that ten-dimensional fermionic strings can be analyzed in terms of bosonic operators, a consequence of the boson-fermion equivalence in two dimensions. The approach taken here is different. We wish to show that the Hilbert space of all the perturbative fermionic closed strings, and of all their tadpole- and anomaly-free open descendants, are subspaces of the 26-dimensional closed bosonic string theory, and of its tadpole-free open descendant, compactified on suitable 16-dimensional manifolds [3, 4, 18, 19].

2.1 The Fermionic Subspaces of the Closed Bosonic String

To accommodate space–time fermions in the left and/or the right sector of the 26-dimensional closed bosonic string one must meet three requirements:

1. A continuum of bosonic zero modes must be removed. This can be achieved by compactifying $d = 24 - s$ transverse dimensions on a d-dimensional torus. This leaves $s + 2$ non-compact dimensions with transverse group $SO_{trans}(s)$.

2. Compactification must generate an internal group $SO_{int}(s)$ admitting spinor representations.[1] This can be achieved by toroidal compactification on the weight lattice of a simply laced Lie group $\mathcal{G}$ of rank d containing a subgroup $SO_{int}(s)$. The latter is then mapped onto $SO_{trans}(s)$ in such a way that the diagonal algebra $SO_{diag}(s) = \mathrm{diag}[SO_{trans}(s) \times SO_{int}(s)]$ becomes identified with a new transverse algebra. In this way, the spinor representations of $SO_{int}(s)$ describe fermionic states because a rotation in space induces a half-angle rotation on these states. This mechanism is distinct from the two-dimensional world-sheet equivalence of bosons and fermions. It is reminiscent of a similar mechanism at work in monopole theory: there, the diagonal subalgebra of space–time rotations and isospin rotations can generate space–time fermions from a bosonic field condensate in spinor representations of the isospin group [25–27].
3. The consistency of the above procedure relies on the possibility of extending the diagonal algebra $SO_{diag}(s)$ to the new full Lorentz algebra $SO_{diag}(s+1,1)$, a highly non trivial constraint. To break the original Lorentz group $SO(25,1)$ in favor of the new one, a truncation consistent with conformal invariance must be performed on the physical spectrum of the bosonic string. Actually, the states described by 12 compactified bosonic fields must be projected out, except for momentum zero-modes of unit length [3, 18]. The removal of 12 bosonic fields accounts for the difference between the bosonic and fermionic light cone gauge central charges. Namely, in units where the central charge of a boson is 1, this difference counts 11 for the superghosts and $(1/2).2$ for time-like and longitudinal Majorana fermions. The zero-modes of length $\ell = 1$ kept in the 12 truncated dimensions contribute a constant $\ell^2/2$ to the mass.[2] They account for the removal by truncation of the oscillator zero-point energies in these dimensions, namely for $-(-1/24).12 = +1/2$. Moreover, the need to generate an internal group $SO_{int}(s)$ via toroidal compactification requires $s/2$ compactified bosons which can account for s transverse Majorana fermions (we hereafter take s to be even, in which case $s/2$ is the rank of the internal group). Therefore, one must ensure that the total number $d = 24 - s$ of compactified dimensions is at least $12 + s/2$. In other words,

$$s \leq 8, \tag{1}$$

and the highest available space–time dimension accommodating fermions is therefore $s+2 = 10$ [3,18]. Here, we restrict our discussion to the case $s+2 = 10$.

To realize this program we choose a compactification of the closed string at an enhanced symmetry point with gauge group $\mathcal{G}_L \times \mathcal{G}_R$ where $\mathcal{G}_L = \mathcal{G}_R = \mathcal{G}$ (or $\mathcal{G}_R = \mathcal{G}$

[1] Throughout this paper we shall denote by $SO(s)$ all the groups locally isomorphic to the rotational group of order $s(s-1)/2$. When specifically referring to its universal covering group, we shall use the notation $\widetilde{SO}(s)$. Also we shall keep the same notation for groups and their Lie algebra.

[2] We choose units in which the string length squared $\alpha' = 1/2$.

for the heterotic string) and $\mathscr{G} = E_8 \times SO(16)$ (or $E_8 \times \widetilde{SO}(16)/Z_2 = E_8 \times E_8$). Recall that in terms of the left and right compactified momenta, the mass spectrum is

$$\begin{aligned} \frac{m_L^2}{8} &= \frac{\mathbf{p}_L{}^2}{2} + N_L - 1, \\ \frac{m_R^2}{8} &= \frac{\mathbf{p}_R{}^2}{2} + N_R - 1, \end{aligned} \tag{2}$$

and

$$m^2 = \frac{m_L^2}{2} + \frac{m_R^2}{2} \quad ; \quad m_L^2 = m_R^2. \tag{3}$$

In (2) N_L and N_R are the oscillator numbers in 26-dimensions and the zero-modes $\mathbf{p}_L$, $\mathbf{p}_R$ span a $2d$-dimensional even self-dual Lorentzian lattice with negative (resp. positive) signature for left (resp. right) momenta. This ensures modular invariance of the closed string spectrum [33]. The massless vectors $\alpha^{\mu}_{-1,R}\ \alpha^{i}_{-1,L}|0_L, 0_R\rangle$ and $\alpha^{\mu}_{-1,L}\ \alpha^{i}_{-1,R}|0_L, 0_R\rangle$, where the indices μ and i refer respectively to non-compact and compact dimensions, generate for generic toroidal compactification a local symmetry $[U_L(1)]^d \times [U_R(1)]^d$. But more massless vectors arise when $\mathbf{p}_L$ and $\mathbf{p}_R$ are roots of simply laced groups $\mathscr{G}_L$ and $\mathscr{G}_R$ of rank d (with root length $\sqrt{2}$). The gauge symmetry is then enlarged to $\mathscr{G}_L \times \mathscr{G}_R$.

2.1.1 The Group $\mathscr{G} = E_8 \times \widetilde{SO}(16)$

The compactification lattices in both sectors (or in the right sector only for the heterotic strings) are taken to be sublattices of the $\mathscr{G} = E_8 \times \widetilde{SO}(16)$ weight lattice. These sublattices must preserve modular invariance, which means that the left and right compactified momenta $\mathbf{p}_L$, $\mathbf{p}_R$ must span a $2d$-dimensional even self-dual Lorentzian lattice. All closed fermionic strings follow then from the properties of the $\widetilde{SO}(16)$ weight lattice and the subsequent truncation.

The weight lattice $\Lambda_{\widetilde{SO}(2n)}$ split into four sublattices.

$$\Lambda_{\widetilde{SO}(2n)} = \left\{ \begin{array}{ll} (o)_{2n} : \mathbf{p}_o + \mathbf{p} & (s)_{2n} : \mathbf{p}_s + \mathbf{p} \\ (v)_{2n} : \mathbf{p}_v + \mathbf{p} & (c)_{2n} : \mathbf{p}_c + \mathbf{p} \\ & \\ \mathbf{p}_o = (\underbrace{0,0,0,\ldots 0}_{n}) & \mathbf{p}_s = (\underbrace{\tfrac{1}{2},\tfrac{1}{2},\tfrac{1}{2},\ldots \tfrac{1}{2}}_{n}) \\ & \\ \mathbf{p}_v = (\underbrace{1,0,0,\ldots 0}_{n}) & \mathbf{p}_c = (\underbrace{\tfrac{-1}{2},\tfrac{1}{2},\tfrac{1}{2},\ldots \tfrac{1}{2}}_{n}) \end{array} \right\} \tag{4}$$

and the E_8 weight (and root) lattice is

$$\Lambda_{E_8} = (o)_{16} + (s)_{16}. \tag{5}$$

Here $\mathbf{p}$ is a vector of the root lattice Λ_R of $SO(16)$.

The partition functions o, v, s, c corresponding to the lattices $(o), (v), (s), (c)$ are

$$P_{j_{2n}} = \sum_{\mathbf{p} \in \Lambda_R; N^{(c)}} \exp\left\{2\pi i\tau\left[\frac{(\mathbf{p}+\mathbf{p}_j)^2}{2} + N^{(c)} - \frac{n}{24}\right]\right\} \quad j = o, v, s, c. \tag{6}$$

Here $N^{(c)}$ is the oscillator number in the compact dimensions. Note that the additive group of the four sublattices of the weight lattice of $SO(2n)$ is isomorphic to the center of the covering group $\widetilde{SO}(2n)$, that is Z_4 for n odd and $Z_2 \times Z_2$ for n even.

2.1.2 The Modular Invariant Truncation

These lattices combine to form four modular invariant partition functions which after the truncation generate the four non-heterotic fermionic strings in ten dimensions [4]. In what follows, we shall only write down the $SO(16)$ characters in the integrand of amplitudes: the E_8 characters in $E_8 \times \widetilde{SO}(16)$ will be entirely truncated and will play no role and we do not display the contribution of the eight light-cone gauge non-compact dimensions. For the four bosonic ancestors, we get

$$OB_b = \bar{o}_{16}\, o_{16} + \bar{v}_{16}\, v_{16} + \bar{s}_{16}\, s_{16} + \bar{c}_{16}\, c_{16} \tag{7}$$

$$OA_b = \bar{o}_{16}\, o_{16} + \bar{v}_{16}\, v_{16} + \bar{s}_{16}\, c_{16} + \bar{c}_{16}\, s_{16} \tag{8}$$

$$IIB_b = \bar{o}_{16}\, o_{16} + \bar{s}_{16}\, o_{16} + \bar{o}_{16}\, s_{16} + \bar{s}_{16}\, s_{16} \tag{9}$$

$$IIA_b = \bar{o}_{16}\, o_{16} + \bar{c}_{16}\, o_{16} + \bar{o}_{16}\, s_{15} + \bar{c}_{16}\, s_{16} \tag{10}$$

where the bar superscript labels the left sector partition functions.

The universal truncation from $E_8 \times \widetilde{SO}(16)$ to $SO_{int}(8)$ is defined by decomposing $SO(16)$ into $SO'(8) \times SO(8)_{int}$ and erasing the E_8 and $SO'(8)$ lattices except, in accordance with item 3 of the above discussion, for unit vectors of $SO'(8)$. In this way the internal momenta $\mathbf{p}[SO(8)]$ are related to $\mathbf{p}[\mathscr{G}]$ by

$$\boxed{\frac{\mathbf{p}^2[\mathscr{G}]}{2} = \frac{\mathbf{p}^2[SO(8)]}{2} + \frac{1}{2}} \tag{11}$$

The unit vectors are identified as follows. The decomposition of an $SO(16)$ lattice in terms of $SO'(8) \times SO(8)$ lattices yields

$$\begin{aligned}
(o)_{16} &= [(o)_{8'} \oplus (o)_8] + [(v)_{8'} \oplus (v)_8], \\
(v)_{16} &= [(v)_{8'} \oplus (o)_8] + [(o)_{8'} \oplus (v)_8], \\
(s)_{16} &= [(s)_{8'} \oplus (s)_8] + [(c)_{8'} \oplus (c)_8], \\
(c)_{16} &= [(s)_{8'} \oplus (c)_8] + [(c)_{8'} \oplus (s)_8].
\end{aligned} \tag{12}$$

The vectors of norm one in $SO'(8)$ are the 4-vectors $\mathbf{p}'_v$, $\mathbf{p}'_s$ and $\mathbf{p}'_c$ defined in (4). We choose one vector $\mathbf{p}'_v$ and one vector $\mathbf{p}'_s$. (One might equivalently have chosen $\mathbf{p}'_c$ instead of $\mathbf{p}'_s$.) In this way we get from (12)

$$o_{16} \to v_8, \qquad v_{16} \to o_8, \tag{13}$$

$$s_{16} \to -s_8, \qquad c_{16} \to -c_8. \tag{14}$$

It follows from the closure of the Lorentz algebra that states belonging to v_8 or o_8 are bosons while those belonging to the spinor partition functions s_8 and c_8 are space–time fermions. The shift of sign in the fermionic amplitudes, which is consistent with the decomposition of s_{16} and c_{16} into $SO'(8) \times SO(8)_{int}$, is required by the spin-statistic theorem and is needed to preserve modular invariance in the truncation.

The four ten dimensional fermionic string partition functions are

$$OB_b \to \bar{o}_8\, o_8 + \bar{v}_8\, v_8 + \bar{s}_8\, s_8 + \bar{c}_8\, c_8 \equiv OB \tag{15}$$

$$OA_b \to \bar{o}_8\, o_8 + \bar{v}_8\, v_8 + \bar{s}_8\, c_8 + \bar{c}_8\, s_8 \equiv OA \tag{16}$$

$$IIB_b \to \bar{v}_8\, v_8 - \bar{s}_8\, v_8 - \bar{v}_8\, s_8 + \bar{s}_8\, s_8 \equiv IIB \tag{17}$$

$$IIA_b \to \bar{v}_8\, v_8 - \bar{c}_8\, v_8 - \bar{v}_8\, s_8 + \bar{c}_8\, s_8 \equiv IIA \tag{18}$$

Note that the partition functions of supersymmetric strings IIA and IIB arise from $E_8 \times E_8$ sublattices of the $E_8 \times \widetilde{SO}(16)$ weight lattice.

The same procedure can be used to obtain all the heterotic strings, supersymmetric or not, by selecting the modular partition functions which are truncated in the right channel only. It will later be extended to D-branes and open descendants.

2.1.3 The Configuration Space Torus Geometry

The four modular invariant theories can be formulated in terms of the actions

$$S = \frac{-1}{2\pi} \int d\sigma d\tau \left[\{ g_{ab} \partial_\alpha X^a \partial^\alpha X^b + b_{ab} \varepsilon^{\alpha\beta} \partial_\alpha X^a \partial_\beta X^b \} + \eta_{\mu\nu} \partial_\alpha X^\mu \partial^\alpha X^\nu \right], \tag{19}$$

with g_{ab} a constant metric and b_{ab} a constant antisymmetric tensor in the compact directions $(a, b = 1, \ldots, 16)$, $\eta_{\mu\nu} = (-1; +1, \ldots)$ for $\mu, \nu = 1, \ldots, 10$ and $0 \leq \sigma \leq \pi$. The fields X^a are periodic with period 2π. In this formalism the left and right momenta are given by

$$\mathbf{p}_R = \left[\frac{1}{2} m_b + n^a (b_{ab} + g_{ab}) \right] \mathbf{e}^b,$$

$$\mathbf{p}_L = \left[\frac{1}{2} m_b + n^a (b_{ab} - g_{ab}) \right] \mathbf{e}^b, \tag{20}$$

where $\{\mathbf{e}^a\}$ is the dual of the basis $\{\mathbf{e}_a\}$ defining the configuration space torus[3]

$$\mathbf{x} \equiv \mathbf{x} + 2\pi n^a \mathbf{e}_a \qquad n^a \in \mathscr{Z}, \tag{21}$$

and the lattice metric is given by

$$g_{ab} = \mathbf{e}_a . \mathbf{e}_b . \tag{22}$$

Explicit forms of the g_{ab} and b_{ab} tensors for the four models are given in [4].

The D9-branes pertaining to the four different bosonic theories compactified on the $E_8 \times \widetilde{SO}(16)$ lattices provide an easy way to construct their configuration space tori. We shall find that these tori are linked to each other through global properties of the universal covering group $\widetilde{SO}(16)$.

The tree channel amplitudes $\mathscr{A}_{tree}$ of the D9-branes are obtained from the torus partition functions (7)–(10) by imposing Dirichlet boundary conditions on the compact space. For open strings the latter do not depend on b_{ab} and are given by

$$\partial_\tau X^a = 0, \tag{23}$$

where τ is the worldsheet time coordinate and σ the space one. Using the worldsheet duality which interchanges the roles of τ and σ, these equations yield the following relation between the left and right momenta:

$$\mathbf{p}_L - \mathbf{p}_R = 0, \tag{24}$$

as well as a match between left and right oscillators in the tree channel. The conditions (24) determine the closed strings which propagate in the annulus amplitude. Imposing them on the four tori amounts to keep *all* characters which appear diagonally in (7)–(10). Up to a normalization, the annulus amplitudes, written as closed string tree amplitudes, are

$$\begin{aligned}
\mathscr{A}_{tree}(OB_b) &\sim (o_{16} + v_{16} + s_{16} + c_{16}), \\
\mathscr{A}_{tree}(OA_b) &\sim (o_{16} + v_{16}), \\
\mathscr{A}_{tree}(IIB_b) &\sim (o_{16} + s_{16}), \\
\mathscr{A}_{tree}(IIA_b) &\sim \ o_{16}.
\end{aligned} \tag{25}$$

We express suitably normalized $\mathscr{A}_{tree}$ as a loop amplitude $\mathscr{A}$ for a single open string (i.e., without Chan–Paton multiplicity) by performing a change of variable and the S-transformation on the modular parameter ($\tau \to -1/\tau$). The result is given in Table 1.

[3] In the previous sections momenta compactification was defined in both left and right channels. Both compactifications are obtained in the action formalism from the compactification on the configuration torus and the (quantized) b-field.

Table 1 Characters of the bosonic D9-branes in the tree and loop channels

	$\mathscr{A}_{tree}$	$\mathscr{A}$
OB_b	$\sim(o_{16}+v_{16}+s_{16}+c_{16})$	o_{16}
OA_b	$\sim(o_{16}+v_{16})$	$o_{16}+v_{16}$
IIB_b	$\sim(o_{16}+s_{16})$	$o_{16}+s_{16}$
IIA_b	$\sim o_{16}$	$o_{16}+v_{16}+s_{16}+c_{16}$

The configuration space tori of the four bosonic theories (7)–(10) are defined by lattices with basis vectors $\{2\pi\mathbf{e}_a\}$ according to (21). We note that the Dirichlet condition (24) reduces (20) to $\mathbf{p}_L = \mathbf{p}_R = (1/2)m_a\mathbf{e}^a$ (independent of b_{ab}). Using the general expression for lattice partition functions (6), we then read off for each model the dual of its $SO(16)$ weight sublattice from the four tree amplitudes in Table 1. We then deduce the $\{\mathbf{e}_a\}$ from the duality between the root lattice $(o)_{16}$ and the weight lattice $(o)_{16}+(v)_{16}+(c)_{16}+(s)_{16}$, and from the self-duality of $(o)_{16}+(v)_{16}$ and $(o)_{16}+(s)_{16}$. We get

$$\mathbf{e}_a = (1/2)\mathbf{w}_a, \tag{26}$$

where the $\mathbf{w}_a$ are weight vectors forming a basis of a sublattice $(r)_{16}$ of the weight lattice of $SO(16)$. The sublattice $(r)_{16}$ for each theory is

$$\begin{aligned}
(OB_b) &: (r)_{16} = (o)_{16},\\
(OA_b) &: (r)_{16} = (o)_{16}+(v)_{16},\\
(IIB_b) &: (r)_{16} = (o)_{16}+(s)_{16},\\
(IIA_b) &: (r)_{16} = (o)_{16}+(v)_{16}+(s)_{16}+(c)_{16}.
\end{aligned} \tag{27}$$

These tori can be visualized by the projection depicted in Fig. 1.

The tori $\widetilde{t}$ of the four bosonic theories are, as group spaces, the maximal toroids $\widetilde{\mathscr{T}}/Z_c$ of the locally isomorphic groups $E_8\times\widetilde{SO}(16)/Z_c$ where Z_c is a subgroup of the center $Z_2\times Z_2$ of the universal covering group $\widetilde{SO}(16)$. We write

$$\begin{aligned}
\widetilde{t}(OB_b) &= \widetilde{\mathscr{T}},\\
\widetilde{t}(OA_b) &= \widetilde{\mathscr{T}}/Z_2^d,\\
\widetilde{t}(IIB_b) &= \widetilde{\mathscr{T}}/Z_2^+ \;\text{ or }\; \widetilde{\mathscr{T}}/Z_2^-,\\
\widetilde{t}(IIA_b) &= \widetilde{\mathscr{T}}/(Z_2\times Z_2),
\end{aligned} \tag{28}$$

where $Z_2^d = diag(Z_2\times Z_2)$ and the superscripts $\pm$ label the two isomorphic IIB_b theories obtained by interchanging $(s)_{16}$ and $(c)_{16}$.

There is thus a unified picture for the four theories related to the global properties of the $SO(16)$ group [4]. The OB_b theory built upon $\widetilde{\mathscr{T}}$ plays in some sense the role of the "mother theory" of the others. One may view the different maximal toroids (28) as resulting from the identification of center elements of $\widetilde{SO}(16)$, which are

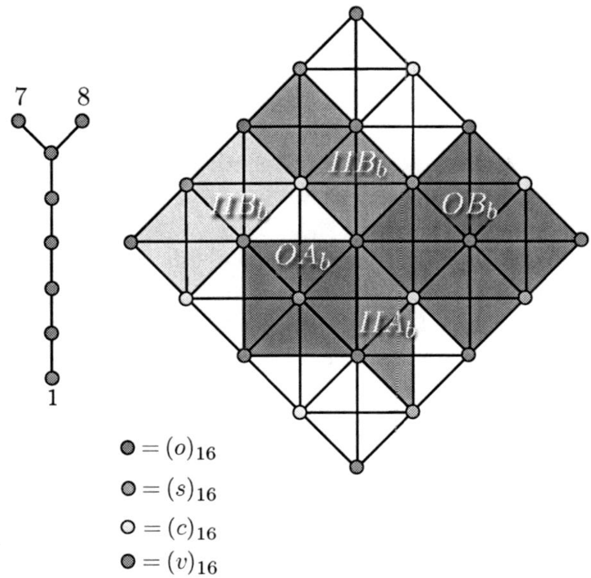

Fig. 1 Projected weight lattice of $\widetilde{SO}(16)$ in the 7–8 plane of the $SO(16)$ Dynkin diagram depicted in the figure. We see from (21), (26) that the volumes ξ_i of the unit cells, exhibited in *shaded areas*, must be multiplied by $(2\pi)^8 \times 2^{-8}$ to yield the $SO(16)$ compactification space torus volume of the four bosonic theories (in units where $\alpha' = 1/2$). The two IIB_b theories defined by the two *rectangles* are isomorphic and differ by the interchange of s_{16} and c_{16}

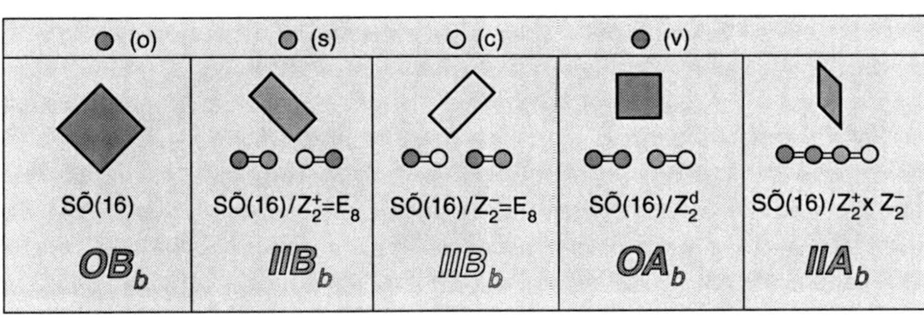

Fig. 2 Identification of center elements of $\widetilde{SO}(16)$ in the four closed string bosonic theories

represented by weight lattice points, with its unit element. These identifications give rise to the smaller shaded cells of Fig. 2. In this way, the unit cell of the IIB_b theory is obtained from the OB_b one by identifying the $(o)_{16}$ and $(s)_{16}$ lattice points (or alternatively the $(o)_{16}$ and the $(c)_{16}$ lattice points) and therefore also the $(v)_{16}$ and $(c)_{16}$ lattice (or the $(v)_{16}$ and the $(s)_{16}$ lattice), as seen in Fig. 2. It is therefore equal to the unit cell of the E_8 lattice.[4] The unit cell of the OA_b theory is obtained by identification of (o) and (v), and of $(s)_{16}$ and $(c)_{16}$.

[4] The latter however does not contain $(v)_{16}$ and $(c)_{16}$ lattice points.

2.2 The Fermionic D-Branes

Table 1 lists the partition functions of a single "elementary" D9-brane for the four $SO(16)$ bosonic strings. In Table 2 below, using the universal truncation, we list the corresponding fermionic D9-branes and their loop partition function $\mathscr{A}_{trunc}$. We now generalize the analysis to encompass several D-branes [4].

First, we remark that the relative position of the different D9-branes in the eight compact dimensions of the $SO(16)$ torus is not arbitrary. Group symmetry requires that the partition function of an open string with end points on D9-branes be a linear combination of the four $SO(16)$ characters. The vector $\mathbf{d}$ separating the two points where two distinct D9-branes meet the $SO(16)$ torus determines the partition function of the string starting at one point and ending at the other after winding any number of times around the torus. The smallest eigenvalue of the string Hamiltonian is $(1/2)\,\mathbf{d}.\mathbf{d}/\pi^2$. Therefore $\mathbf{d}/\pi$ must be a weight and the D9-branes can only be separated in the compact space (rescaled by a factor π^{-8}) by a weight vector. Consider for instance two branes in the OB_b theory, one located at $(o)_{16}$ and the other located at $(v)_{16}$. The partition function of a string beginning and ending on the same brane is o_{16}, while the partition function of an open string stretching between them is v_{16}. For the other theories, the partition function of a string beginning and ending on the same brane will then contain, in addition to o_{16}, the characters corresponding to the strings stretched between $(o)_{16}$ and all points identified with $(o)_{16}$. This can be checked by comparing the identifications indicated in Fig. 2 with the partition functions $\mathscr{A}$ listed in Table 1.

If one chooses the location of one elementary brane as the origin of the weight lattice, the other D9-branes can then only meet the $SO(16)$ torus (rescaled by π^{-8}) at a weight lattice point. The number of distinct elementary branes is, for each of the four bosonic theories, equal to the number of distinct weight lattice points in the unit cell. For the mother theory OB_b there are four possible elementary D9-branes. We label them by their positions in the unit cell, namely by $(o)_{16}, (v)_{16}, (s)_{16}$ and $(c)_{16}$. Note that these weight lattice points represent the center elements of the $\widetilde{SO}(16)$. For the other theories the unit cells are smaller and there are fewer possibilities. The unit cell of the IIB_b theory allows only for two distinct branes $(o)_{16} = (s)_{16}$, $(c)_{16} = (v)_{16}$ (or those obtained by the interchange of $(s)_{16}$ and $(c)_{16}$, as seen in Fig. 2). Similarly for the OA_b theory, we have the two branes $(o)_{16} = (v)_{16}$ and $(s)_{16} = (c)_{16}$, and finally for the "smallest" theory IIA_b, we have only one elementary brane $(o)_{16} = (v)_{16} = (s)_{16} = (c)_{16}$. Finally to describe several D9-branes meeting at the same point of the $SO(16)$ torus, one uses the appropriate Chan–Paton factors.

2.2.1 Charge Conjugation

Charge conjugation of the truncated fermionic strings is encoded in their bosonic parents. A brane sitting at $(v)_{16}$ can always be joined by an open string to a brane sitting at $(o)_{16}$. The partition function of such a string is given by the character v_{16} and therefore the two branes can exchange closed strings with tree amplitude

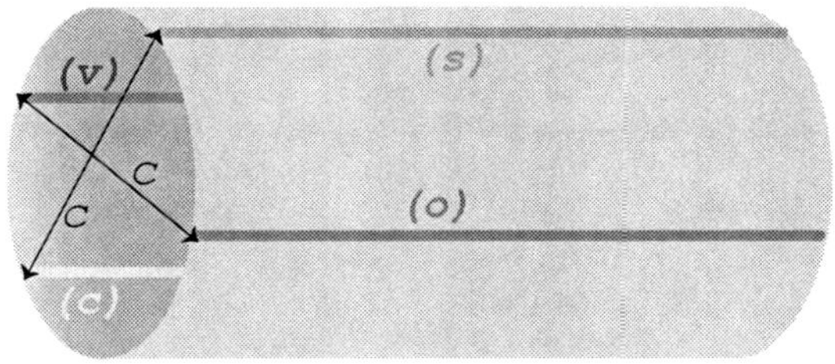

Fig. 3 Charge conjugation of the fermionic D9-branes from the position of their bosonic ancestors on the $\widetilde{SO}(16)$ torus. (Subscripts in the labeling of the lattice points is omitted.) The charge conjugate branes are linked by the lines C

Table 2 Fermionic D9-branes charges encoded in their bosonic ancestors

	$\mathscr{A}$	$\mathscr{A}^{trunc}$	Fermionic D9-branes	Stability
$OB_b \to OB$	o_{16}	v_8	$D_1^+ + D_2^+ + D_1^- + D_2^-$	Stable
$OA_b \to OA$	$o_{16} + v_{16}$	$v_8 + o_8$	$D_1^0 + D_2^0$	Unstable
$IIB_b \to IIB$	$o_{16} + s_{16}$	$v_8 - s_8$	$D^+ + D^-$	Stable
$IIA_b \to IIA$	$o_{16} + v_{16} + s_{16} + c_{16}$	$v_8 + o_8 - s_8 - c_8$	D^0	Unstable

$\mathscr{A}_{tree} = o_{16} + v_{16} - s_{16} - c_{16}$ as follows from the S-transformation of the characters. Namely the closed string exchange describing the interaction between these two branes has opposite sign for the $(s)_{16}$ and $(c)_{16}$ contribution as compared to the closed string exchange between D9-branes located at the same point. This shift of sign persists in the truncation to the fermionic theories where the above tree amplitude becomes $o_8 + v_8 + s_8 + c_8$. This shift of sign thus describes the RR-charge conjugation between fermionic D9-branes. It is encoded in the bosonic string as a shift by the lattice vector (v) (see Fig. 3). In particular, when $(o)_{16}$ and $(v)_{16}$ are identified, all branes of the fermionic offsprings are neutral. These are always unstable branes, as the truncation of v_{16} is o_8 and contains a tachyon. Charged branes are always stable.

The distinct fermionic D9-branes and their charge conjugates can thus be directly read off from Fig. 1. They are included in Table 2, where the charge is indicated by a superscript $+, -$ or 0, and additional quantum numbers by a subscript.

2.2.2 Chirality

We now consider the truncation of bosonic D-branes to lower dimensional fermionic Dp-branes ($p < 9$). This is a non-trivial problem for the following reason. In fermionic string theories, a T-duality interchanges type *IIA* with *IIB*, and type *OA* with *OB* while transmuting D9-branes to D8-branes without changing their corresponding $\mathscr{A}^{trunc}$ amplitudes given in Table 2. More generally, while the amplitudes of Dp-branes for p odd are essentially the same as the D9-brane amplitude,

Table 3 Fermionic Dp-brane partition functions ($p \leq 9$) and their bosonic ancestors

	$\mathscr{A}_p$, p odd	$\mathscr{A}_{p+8}$, p even	$\mathscr{A}_p^{trunc}$, p odd	$\mathscr{A}_p^{trunc}$, p even
$OB_b \to OB$	o_{16}	$o_{16}+v_{16}$	v_8	o_8+v_8
$OA_b \to OA$	$o_{16}+v_{16}$	o_{16}	o_8+v_8	v_8
$IIB_b \to IIB$	$o_{16}+s_{16}$	$o_{16}+v_{16}+s_{16}+c_{16}$	v_8-s_8	$o_8+v_8-s_8-c_8$
$IIA_b \to IIA$	$o_{16}+v_{16}+s_{16}+c_{16}$	$o_{16}+s_{16}$	$o_8+v_8-s_8-c_8$	v_8-s_8

those of the p even would require, at the bosonic level, the interchange of IIA_b with IIB_b and OA_b with OB_b for the universal truncation to yield the correct chirality.

This problem is beautifully solved [4] by noticing that the truncation of bosonic Dp-branes with p even would violate Lorentz invariance for the fermionic strings except if this A, B interchange could be performed at the bosonic parent level. And such interchange is indeed a symmetry of the compactified bosonic string! One has to perform an "odd" E-duality which interchanges on the $SO(16)$ torus Dirichlet with Neumann boundary conditions and *simultaneously* performs the required A, B interchange. One thus obtains in this way the loop amplitudes of the bosonic parents which lead from the universal truncation to the correct fermionic amplitudes consistent with Lorentz invariance. This is summarized in Table 3.

2.2.3 Tensions

We recall that the tension $T_{Dp}^{bosonic}$ of a Dp-brane in the 26-dimensional uncompactified theory is [36]

$$T_{Dp}^{bosonic} = \frac{\sqrt{\pi}}{2^4 \kappa_{26}} (2\pi\alpha'^{1/2})^{11-p}, \tag{29}$$

where $\kappa_{26}^2 = 8\pi G_{26}$ and G_{26} is the Newtonian constant in 26 dimensions. The tensions of the Dirichlet D9-branes of the four compactified theories are obtained from (29) by expressing κ_{26} in term of the ten-dimensional coupling constant κ_{10}. Recalling that $\kappa_{26} = \sqrt{V}\kappa_{10}$ where V is the volume of the configuration space torus, one finds from Fig. 1,

$$T_{OB_b} = \frac{\sqrt{\pi}}{\sqrt{2}\kappa_{10}} (2\pi\alpha'^{1/2})^{-6}, \tag{30}$$

$$T_{OA_b} = T_{IIB_b} = \frac{\sqrt{\pi}}{\kappa_{10}} (2\pi\alpha'^{1/2})^{-6}, \tag{31}$$

$$T_{IIA_b} = \frac{\sqrt{2}\sqrt{\pi}}{\kappa_{10}} (2\pi\alpha'^{1/2})^{-6}. \tag{32}$$

We now perform the universal truncation on the loop amplitudes $\mathscr{A}$ listed in Table 3 to compute the tensions of the fermionic branes. Tensions are conserved in the truncation, as proven in [19]. The tensions of the different bosonic D9-branes given in (30)–(32) are thus equal, when measured with the same gravitational constant κ_{10}, to the tensions of the corresponding fermionic D9-branes [30, 37]. This is indeed a correct prediction.

2.3 Tadpole-Free and Anomaly-Free Fermionic Open Strings

The open descendants of the closed bosonic theories will be determined by imposing the *tadpole condition* [1] on the bosonic string, namely by imposing that divergences due to massless tadpoles cancel in the vacuum amplitudes. We will show that the bosonic OB_b, IIB_b and OA_b theories admit tadpole-free open bosonic descendants and that those descendants give after truncation the three open fermionic string theories which are anomaly- or tadpole-free [4, 19]. Compactification of the bosonic string plays of course a crucial role, as the following analysis for the uncompactified unoriented bosonic string would recover the unique consistent Chan–Paton group $SO(2^{13})$.

A first step in obtaining the open descendant corresponding to the four bosonic string theories characterized by the tori amplitudes $\mathscr{T}$ (7)–(10) is the construction of the Klein bottle amplitudes $\mathscr{K}$ (Fig. 4). These are obtained from the amplitudes $\mathscr{T}/2+\mathscr{K}$, which are the torus closed string partition functions $\mathscr{T}$ with the projection operator $(1+\Omega)/2$ inserted, where Ω interchanges the left and right sectors: $\Omega|L,R>=|R,L>$. This can be done for OB_b, IIB_b and OA_b but not for IIA_b, because Ω in that case is not a symmetry of the theory. The IIA_b theory does not admit any open descendant. The projection on Ω eigenstates amounts to impose the condition

$$\mathbf{p_R} = \mathbf{p_L}, \tag{33}$$

on the closed string momenta equations (20). Acting with $\Omega/2$ on the three different tori (7)–(9), one finds the three Klein bottle amplitudes[5]

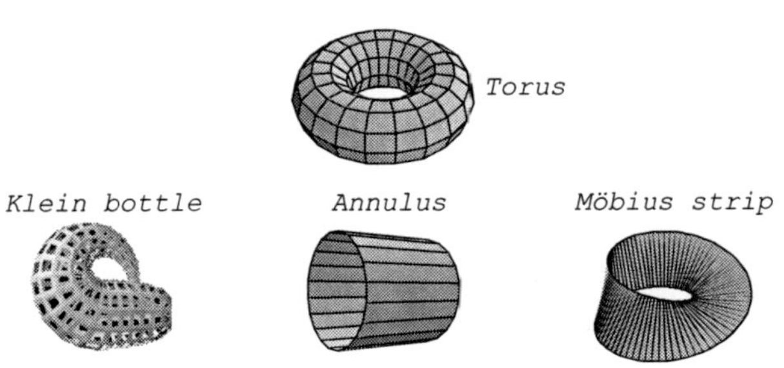

Fig. 4 World-sheet contributions to unoriented open strings

[5] Recall that we display only the $SO(16)$ contribution to the amplitudes.

$$\mathscr{K}(OB_b) = \frac{1}{2}(o_{16} + v_{16} + s_{16} + c_{16}), \tag{34}$$

$$\mathscr{K}(IIB_b) = \frac{1}{2}(o_{16} + s_{16}), \tag{35}$$

$$\mathscr{K}(OA_b) = \frac{1}{2}(o_{16} + v_{16}). \tag{36}$$

The two remaining amplitudes with vanishing Euler characteristic, the annulus $\mathscr{A}$ and the Möbius strip $\mathscr{M}$, determine the open string partition function (Fig. 4). The annulus amplitudes of D9-branes with generic Chan–Paton multiplicities are

$$\mathscr{A}(OB_b) = \frac{1}{2}(n_o^2 + n_v^2 + n_s^2 + n_c^2)\,o_{16} + (n_o n_v + n_s n_c)\,v_{16}$$
$$+(n_o n_s + n_v n_c)\,s_{16} + (n_o n_c + n_v n_s)c_{16}, \tag{37}$$

$$\mathscr{A}(OA_b) = \frac{1}{2}(n_o^2 + n_s^2)\,(o_{16} + v_{16}) + n_o n_s\,(s_{16} + c_{16}), \tag{38}$$

$$\mathscr{A}(IIB_b) = \frac{1}{2}(n_o^2 + n_v^2)\,(o_{16} + s_{16}) + n_o n_v\,(v_{16} + c_{16}). \tag{39}$$

To get the Möbius amplitudes $\mathscr{M}$ and to implement the tadpole condition we express the Klein bottle and annulus amplitudes (34)–(36) and (37)–(39) as closed string tree channel amplitudes using the S-transformation of the characters. From the resulting amplitudes $\mathscr{K}_{tree}$ and $\mathscr{A}_{tree}$ one obtains the Möbius amplitudes $\mathscr{M}_{tree}$ by requiring that each term in the power series expansion of the total tree channel amplitude $\mathscr{K}_{tree} + \mathscr{A}_{tree} + \mathscr{M}_{tree}$ be a perfect square. One gets

$$\mathscr{M}_{tree}(OB_b) = \varepsilon_1\,(n_o + n_v + n_s + n_c)\,\hat{o}_{16}, \tag{40}$$

$$\mathscr{M}_{tree}(IIB_b) = \varepsilon_2\,(n_o + n_v)\,\hat{o}_{16} + \varepsilon_3\,(n_o - n_v)\,\hat{s}_{16}, \tag{41}$$

$$\mathscr{M}_{tree}(OA_b) = \varepsilon_4\,(n_o + n_s)\,\hat{o}_{16} + \varepsilon_5\,(n_o - n_s)\,\hat{v}_{16}, \tag{42}$$

where $\varepsilon_i = \pm 1$ will be determined by tadpole conditions. The "hat" notation in the amplitudes (40)–(42) means that the overall phase present in the characters r_{16} is dropped. This phase arises because the modulus over which $\mathscr{M}$ is integrated (and which is not displayed here) is not purely imaginary but is shifted by 1/2, inducing in the partition functions i_{16} an alternate shift of sign in its power series expansion as well as a global phase. This one half shift is needed to preserve the group invariance of the amplitudes. A detailed discussion of the shift in general cases can be found in [1].

We now impose the tadpole conditions on the three theories, namely we impose the cancellation of the divergences due to the massless mode exchanges in the total amplitudes $\mathscr{K}_{tree} + \mathscr{A}_{tree} + \mathscr{M}_{tree}$. One determines in this way the Chan–Paton of the tadpole-free compactified unoriented bosonic open strings. The universal truncation preserves these factors and one gets in this way the correct tadpole-free and anomaly-free open fermionic strings [1], as indicated in Table 4.

Table 4 Chan–Paton group of tadpole- and anomaly-free fermionic strings

	Chan–Paton group
$OB_b \to OB \to B$	$[SO(32-n) \times SO(n)]^2$
$OA_b \to OA \to A$	$SO(32-n) \times SO(n)$
$IIB_b \to IIB \to I$	$SO(32)$
$IIA_b \to IIA$	–

2.4 The Fermionic Content of the Bosonic String: Summary

- From the torus compactification of the bosonic string on the weight lattice of $E_8 \times \widetilde{SO}(16)$ and the universal truncation to $SO(8)$ keeping $\mathbf{p}'_v, \mathbf{p}'_s \in SO'(8)$ one recovers all the closed fermionic string spectra. Those resulting from the compactification on the $E_8 \times E_8$ sublattice are supersymmetric.
- Spectra, charges, chiralities, tensions of all the fermionic D-branes are obtained from their bosonic parents by the same universal truncation.
- Tadpole and anomaly cancellation of unoriented open fermionic strings follow from the tadpole free unoriented bosonic string by the same universal truncation.

3 The Generalized Kac–Moody Approach

The five consistent superstring theories appear to be related by U-dualities and a conjectured non-perturbative formulation encompassing all of them has been labeled M-theory. Attempts to understand its symmetries has led to an approach to the M-theory project based on generalized Kac–Moody algebras. We shall analyze to what extend the connection between bosonic and fermionic strings found in Sect. 2 survives in this Kac–Moody approach. This will shed some light on its significance.

3.1 $E_{11} \equiv E_8^{+++}$ and 11-Dimensional Supergravity

Among the consistent superstring theories, type *IIA* and type *IIB* are maximally supersymmetric, i.e., they are characterized by 32 supercharges. We will focus on such maximally supersymmetric phase of M-theory, whose classical limit is assumed to be 11-dimensional supergravity and whose dimensional reduction to ten dimensions yields the low energy effective action of type *IIA* superstring. The bosonic action of 11-dimensional supergravity is given by

$$\mathscr{S}^{(11)} = \frac{1}{16\pi G_{11}} \int d^{11}x \sqrt{-g^{(11)}} \left(R^{(11)} - \frac{1}{2.4!} F_{\mu\nu\sigma\tau} F^{\mu\nu\sigma\tau} + CS\text{-term} \right). \quad (43)$$

Scalars in the dimensional reduction of the action (43) to three space–time dimensions realize non-linearly the maximal non-compact form of the Lie group E_8 as a coset $E_8/SO(16)$ where $SO(16)$ is its maximal compact subgroup. Here, the symmetry of the $(2+1)$ dimensionally reduced action has been enlarged from the $GL(8)$ deformation group of the compact torus T^8 to the simple Lie group E_8. This symmetry enhancement stems from the detailed structure of the action (43).

Coset symmetries were first found in the dimensional reduction of 11-dimensional supergravity [7] to four space–time dimensions [6] but appeared also in other theories. They have been the subject of much study, and some classic examples are given in [5,8,23,28,29,40]. In fact, all simple maximally non-compact Lie group $\mathscr{G}$ can be generated from the reduction down to three dimensions from actions of gravity coupled to suitably chosen matter fields [9].

It has been suggested that these actions, or possibly some unknown extensions of them, possess a much larger symmetry than the one revealed by their dimensional reduction to three space–time dimensions in which all fields, except $(2+1)$-dimensional gravity itself, are scalars. Such hidden symmetries would be, for each simple Lie group $\mathscr{G}$, the Lorentzian [24] "overextended" $\mathscr{G}^{++}$ [12] or "very-extended" $\mathscr{G}^{+++}$ [20, 32] infinite Kac–Moody algebras generated respectively by adding two or three nodes to the Dynkin diagram defining $\mathscr{G}$. One first adds the affine node, then a second node connected to it by a single line to get the $\mathscr{G}^{++}$ Dynkin diagram and then similarly a third one connected to the second to generate $\mathscr{G}^{+++}$. In particular, the E_8 invariance of the dimensional reduction to three dimensions of 11-dimensional supergravity would be enlarged to $E_8^{++} \equiv E_{10}$ [29] or to $E_8^{+++} \equiv E_{11}$, as first proposed in [41]. The extension of the Dynkin diagram of E_8 to E_{11} is depicted in Fig. 5. The horizontal line in Fig. 5 form the Dynkin diagram of the A_{10} subalgebra of E_{11}. It is labeled *the gravity line*, as the nodes 4–10 of Fig. 5 arise from the reduction of gravitational part of the action (43).

To explore the possible significance of these huge symmetries a Lagrangian formulation [13] *explicitly* invariant under E_{10} has been proposed. It was constructed as a reparametrization invariant σ-model of fields depending on one parameter t, identified as a time parameter, living on the coset space E_{10}/K_{10}^{+}. Here K_{10}^{+} is the subalgebra of E_{10} invariant under the Chevalley involution. The σ-model contains an infinite number of fields and is built in a recursive way by a level expansion of

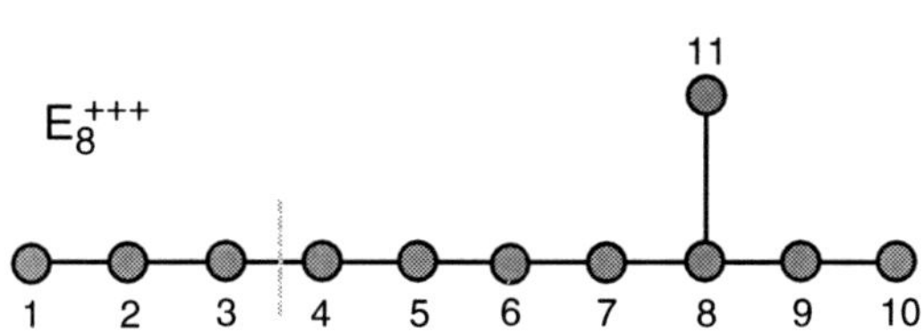

Fig. 5 Dynkin diagram of $E_{11} \equiv E_8^{+++}$

E_{10} with respect to its subalgebra A_9 [13, 34] whose Dynkin diagram is the "gravity line" defined in Fig. 5, with the node 1 deleted.[6] The level of an irreducible representation of A_9 occurring in the decomposition of the adjoint representation of E_{10} counts the number of times the simple root α_{11} not pertaining to the gravity line appears in the decomposition. The σ-model, limited to the roots up to level 3 and height 29, reveals a perfect match with the bosonic equations of motion of 11-dimensional supergravity in the vicinity of the space-like singularity of the cosmological billiards [10, 11, 14], where fields depend only on time. It was conjectured that space derivatives are hidden in some higher level fields of the σ-model [13]. We shall label this σ-model S^{cosmo}.

An alternate E_{10} σ-model parametrized by a space variable x^1 can be formulated on a coset space E_{10}/K_{10}^-, where K_{10}^- is invariant under a "temporal" involution ensuring the Lorentz invariance $SO(1,9)$ at each level in the A_9 decomposition of E_{10}. This σ-model provides a natural framework for studying static solutions [16, 17]. It yields all the basic BPS solutions of 11D supergravity, namely the KK-wave, the M2 brane, the M5 brane and the KK6-monopole, smeared in all space dimensions but one, as well as their exotic counterparts. We shall label the action of this σ-model S^{brane}. The algebras K_{10}^+ and K_{10}^- are both subalgebras of the algebra K_{11}^- invariant under the temporal involution defined on E_{11}, which selects the Lorentz group $SO(1,10) = K_{11}^- \cap A_{10}$ in the A_{10} decomposition of E_{11} [16, 21].

The underlying algebraic structure in this approach is thus E_{11} and the infinite number of covariant fields are the parameters of the coset E_{11}/K_{11}^- which can be recursively determined by the level decomposition with respect to A_{10}. In this section, we adopt this algebraic description of the field content of Kac–Moody algebras.

The first three levels contain the space–time degrees of freedom of 11-dimensional gravity along with their duals as depicted in Table 5. These levels are labeled classical.

In this E_{11} algebraic approach the crucial problem is to elucidate the role of the huge number of fields beyond the classical levels (level ≥ 3, height >29) and to find their significance. Addressing this problem could bring an answer to two fundamental questions of this approach. First, is space–time itself encoded in the algebra and is then E_{11} a symmetry of the uncompactified theory? Second, does E_{11}

Table 5 Low level E_{11} fields

Level	Field	Supergravity content
Level 0	$g_{\mu\nu}$	Gravity
Level 1	$A_{\mu\nu\lambda}$	3-Form potential
Level 2	$A_{\mu\nu\lambda\rho\sigma\tau}$	6-Form dual potential
Level 3	$A_{\mu\nu\lambda\rho\sigma\tau\upsilon\zeta\|\zeta}$	Dual graviton
Level ≥ 3	Non "classical"	?

[6] Level expansions of G^{+++} algebras in terms of a subalgebra A_{D-1} have been considered in [31, 42].

describes the degrees of freedom of 11-dimensional maximal supergravity or more, for instance string degrees of freedom in some tensionless limit?

It is fair to say that up to now there is no clear and totally satisfying answer to these questions. But recent progress to be discussed below points toward an answer to the second question.

It has been conjectured that the fields corresponding to the real roots of $E_9 \subset E_{11}$ are dual fields and are not new degrees of freedom [38]. It was indeed shown that these fields express non-closing Hodge-like dualities relating between themselves the usual degrees of freedom of maximal 11-dimensional supergravity. Explicitly, from the E_{11} fields parametrizing the coset E_{11}/K_{11}^{-}, the subset of real roots of $E_9 \subset E_{11}$ generate, using these non-closing dualities realized as E_9 Weyl reflections, an infinite U-duality E_9 multiplet of BPS static solutions of 11-dimensional supergravity [22].

In another development [2, 39], it was shown that another class of E_{11} fields contain all those needed to describe all the maximal gauged supergravity in $D \leq 11$ dimensions. Namely the $D-1$ forms and the D-forms content, present in the E_{11} algebraic description interpreted in D dimensions, matches the embedding tensor description [15] of all the gauged maximal supergravities. Hence the E_{11} algebraic approach appears to contain the algebraic structure of all maximal non-abelian supergravities (with 32 supercharges). However again, although these transcend 11D ungauged maximal supergravity they do not contain new degrees of freedom.

There is still an infinite number of other fields characterized by A_{10} representations with mixed Young tableaux. Their significance is hitherto unclear.

We now turn to the generalized Kac–Moody approach to the bosonic string. This will give some new information on the field content in the algebraic approach.

3.2 D_{24}^{+++} and the Bosonic String

The low-energy effective action of the $D = 26$ bosonic string (without tachyon) contains gravity, the NS-NS three form field strength and the dilaton. It is given by

$$S = \int d^{26}x \sqrt{-g} \left(R - \frac{1}{2} \partial_\mu \phi \partial^\mu \phi - \frac{1}{2.3!} e^{-\frac{1}{3}\phi} H_{\mu\nu\sigma} H^{\mu\nu\sigma} \right), \tag{44}$$

and $H = db$.

Upon dimensional reduction to three space–time dimensions, one would have expected to have a $GL(23) \times U(1)$ symmetry (the U(1) coming from the dilaton). Again, there is an enhancement to a simple Lie algebra, namely D_{24} in its split form. The symmetry is non-linearly realized and the scalar lives in the coset $SO(24,24)/SO(24) \times SO(24)$. The corresponding Dynkin diagram of D_{24} is the part of the diagram Fig. 6 on the right of the dashed line.

Having this symmetry in three dimensions, the discussion of the preceding section suggests that the "very extended" Kac–Moody algebra D_{24}^{+++} could encode a

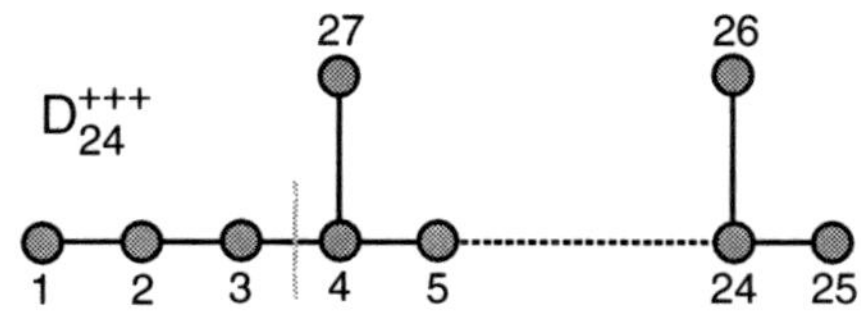

Fig. 6 Dynkin diagram of D_{24}^{+++}

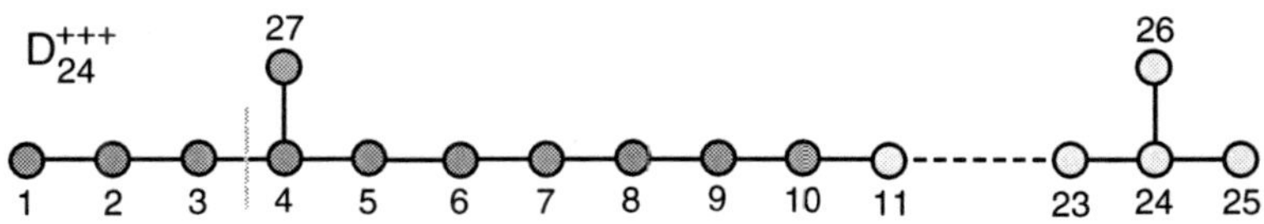

Fig. 7 The decomposition of D_{24}^{+++} into $A_9 \times D_{16}$ with levels defined by the nodes α_{27} and α_{10}

Table 6 Low level representations in the decomposition of D_{24}^{+++} into $A_9 \times D_{16}$. Their Dynkin labels are p_r and p_i, their dimensions d_r and d_i. r^2 is the root norm and μ the outer multiplicity

l_1, l_2	p_i	p_r	r^2	d_r	d_i	μ
0, 0	000000000	000000000000000	0	1	1	2
0, 0	100000001	000000000000000	2	99	1	1
0, 0	000000000	010000000000000	2	1	496	1
1, 0	000000001	100000000000000	2	10	32	1
0, 2	100000100	000000000000000	2	1155	1	1
2, 0	000000010	000000000000000	2	45	1	1
0, 1	000100000	000000000000000	2	210	1	1
1, 1	001000000	100000000000000	2	120	32	1

symmetry of the bosonic string [32]. The physical fields of this algebraic approach would then live in a coset D_{24}^{+++}/K_{24}^{-} where K_{24}^{-} is the maximal subalgebra invariant under the temporal involution. The corresponding Dynkin diagram is depicted in Fig. 6.

To make contact with the analysis of Sect. 2 and uncover a possible relation through truncation with the fermionic strings in ten space–time dimensions we consider the decomposition of D_{24}^{+++} into $A_9 \times D_{16}$ where the diagram of A_9 is the gravity line of a ten dimensional space–time and D_{16} a symmetry arising from a torus compactification of 16 dimensions. In this decomposition the unbroken subalgebra K_{24}^{-} decomposes into $SO(1,9) \times SO(16) \times SO(16)$. Physical fields appear in the double level decomposition with respect to the nodes α_{27} and α_{10} in Fig. 7.

The level l_1 (resp. l_2) counts the number of times the root α_{10} (resp. α_{27}) appears in the decomposition of the adjoint representation of D_{24}^{+++} into irreducible representations of A_9 and $SO(16,16)$. The first levels are listed in Table 6 obtained from the SimpLie program [35].

Thus, the internal space of the the physical fields is the coset $SO(16,16)/SO(16) \times SO(16)$. This is exactly the moduli space of modular invariant compactifications of the closed bosonic string on a 16-dimensional torus. At a generic point, one has $U(1)_L^{16} \times U(1)_R^{16}$ where L (resp. R) stands for left (resp. right). We thus expect to find 32 abelian gauge fields and these indeed appear in Table 6 at the level $(1,0)$ in the fundamental representation of $SO(16,16)$. As explained in Sect. 2, the compactifications needed for uncovering by truncation the fermionic strings are the special points of enhanced symmetry in the coset where the torus is identified to one of the four maximal toroids of $[\widetilde{SO}(16)/Z_i] \times E_8$, where Z_i is an element of the center $Z_2 \times Z_2$ of the universal covering $\widetilde{SO}(16)$. For $Z_i = Z_2$, (resp. $Z_i = Z_2 \times Z_2$) this yields the gauge group $(E_8 \times E_8)_L \times (E_8 \times E_8)_R$, which yields after truncation to the maximal supersymmetric type IIB (resp. IIA) string theory.

One may first ask if the non-abelian extension of the gauged $U(1)_L^{16} \times U(1)_R^{16}$ to $(E_8 \times E_8)_L \times (E_8 \times E_8)_R$, which appear at these enhanced symmetry points are encoded in D_{24}^{+++}, as do the non-abelian gauging of maximal supergravities in E_8^{+++}. The answer is no, as spinor representations of D_{16} cannot appear in the adjoint representation of D_{24}^{+++}. This means that one would have to extend the algebra of D_{24}^{+++} to include fields not contained in the adjoint representation of the generators if one wishes to recover the information encoded in the torus compactification at these enhanced symmetry points.

The problem is not limited to enhanced symmetry points involving spinor representations of the group. The stringy nature of the massless degrees of freedom enlarging the abelian gauging to a non abelian one at enhanced symmetry points has no counterpart in the non-abelian gauging of maximal supergravities which appear in E_8^{+++} and which are studied in [2, 39]. These indeed do not introduce new degrees of freedom. One might then expect that the D_{24}^{+++} fields do not comprise the genuine string degrees of freedom of the bosonic string. Similarly, despite the fact that E_8^{+++} does contain spinor representations of orthogonal groups, the E_8^{+++} fields are not expected to comprise genuine superstring degrees of freedom such as the massless fields resulting from torus compactifications at enhanced symmetry points. In that case, if the the full set of string degrees of freedom are to be included in some M-theory project, its algebraic description would transcend the description by the E_8^{+++} fields.

Acknowledgments We are greatly indebted to Axel Kleinschmidt and to Nassiba Tabti for fruitful discussions. One of us, François Englert, is grateful to Peter West for an interesting discussion on the possible extensions of E_8^{+++}. We would like to warmly thank Marc Henneaux, Vivian Scharager and Jorge Zanelli for having organized such friendly and magnificent ClaudioFest, which was made even more lively by the constant participation of Claudio himself.

Laurent Houart is a Research Associate of the Fonds de la Recherche Scientifique-FNRS, Belgium. This work was supported in part by IISN-Belgium (convention 4.4511.06 and convention 4.4505.86), and by the Belgian Federal Science Policy Office through the Interuniversity Attraction Pole P VI/11.

References

1. Angelantonj, C., Sagnotti, A.: Open strings. Phys. Rept. **371**, 1–150 (2002); Erratum-ibid. **376**, 339–405 (2003). `arXiv:hep-th/0204089`, and references therein.
2. Bergshoeff, E.A., De Baetselier, I., Nutma, T.A.: E_{11} and the embedding tensor. J. High Energy Phys. **0709**, 047 (2007). `arXiv:0705.1304 [hep-th]`.

3. Casher, A., Englert, F., Nicolai, H., Taormina, A.: Consistent superstrings as solutions of the D = 26 bosonic string theory. Phys. Lett. **B162**, 121 (1985).
4. Chattaraputi, A., Englert, F., Houart, L., Taormina, A.: The bosonic mother of fermionic D-branes. J. High Energy Phys. **0209**, 037 (2002). arXiv:hep-th/0207238.
5. Cremmer, E., Julia, B.: The $N = 8$ supergravity theory. 1. The lagrangian. Phys. Lett. **B80**, 48 (1978).
6. Cremmer, E., Julia, B.: The $SO(8)$ supergravity. Nucl. Phys. **B159**, 141 (1979).
7. Cremmer, E., Julia, B., Scherk, J.: Supergravity theory in 11 dimensions. Phys. Lett. **B76**, 409 (1978).
8. Cremmer, E., Scherk, J., Ferrara, S.: $SU(4)$ invariant supergravity theory. Phys. Lett. **B74**, 61 (1978).
9. Cremmer, E., Julia, B., Lu, H., Pope, C.N.: Higher-dimensional origin of D = 3 coset symmetries. (1999) (unpublished). arXiv:hep-th/9909099.
10. Damour, T., Henneaux, M.: E_{10}, BE_{10} and arithmetical chaos in superstring cosmology. Phys. Rev. Lett. **86**, 4749 (2001). arXiv:hep-th/0012172.
11. Damour, T., Henneaux, M., Julia, B., Nicolai, H.: Hyperbolic Kac–Moody algebras and chaos in Kaluza–Klein models. Phys. Lett. **B509**, 323 (2001). arXiv: hep-th/0103094.
12. Damour, T., de Buyl, S., Henneaux, M., Schomblond, C.: Einstein billiards and overextensions of finite-dimensional simple Lie algebras. J. High Energy Phys. **08**, 030 (2002). arXiv:hep-th/0206125.
13. Damour, T., Henneaux, M., Nicolai, H.: E_{10} and a 'small tension expansion' of M theory. Phys. Rev. Lett. **89**, 221601 (2002). arXiv:hep-th/0207267.
14. Damour, T., Henneaux, M., Nicolai, H.: Cosmological billiards. Class. Quant. Grav. **20**, R145 (2003). arXiv:hep-th/0212256.
15. de Wit, B., Nicolai, H., Samtleben, H.: Gauged supergravities, tensor hierarchies, and M-theory. J. High Energy Phys. **0802**, 044 (2008). arXiv:0801.1294 [hep-th].
16. Englert, F., Houart, L.: $\mathcal{G}^{+++}$ Invariant formulation of gravity and M-theories: exact BPS solutions. J. High Energy Phys. **0401**, 002 (2004). arXiv: hep-th/0311255.
17. Englert, F., Houart, L.: $\mathcal{G}^{+++}$ Invariant formulation of gravity and M-theories: Exact intersecting BPS solutions. J. High Energy Phys. **0405**, 059 (2004). arXiv: hep-th/0405082.
18. Englert, F., Nicolai, H., Schellekens, A.N.: Superstrings from 26 dimensions. Nucl. Phys. **274**, 315 (1986).
19. Englert, F., Houart, L., Taormina, A.: Brane fusion in the bosonic string and the emergence of fermionic strings. J. High Energy Phys. **0108**, 013 (2001). arXiv:hep-th/0106235.
20. Englert, F., Houart, L., Taormina, A., West, P.: The symmetry of M-theories. J. High Energy Phys. **0309**, 020 (2003). arXiv: hep-th/0304206.
21. Englert, F., Henneaux, M., Houart, L.: From very-extended to overextended gravity and M-theories. J. High Energy Phys. **0502**, 070 (2005). arXiv: hep-th/0412184.
22. Englert, F., Houart, L., Kleinschmidt, A., Nicolai, H., Tabti, N.: An E_9 multiplet of BPS states. J. High Energy Phys. **0705**, 065 (2007). arXiv: hep-th/0703285.
23. Ferrara, S., Scherk, J., Zumino, B.: Algebraic properties of extended supergravity theories. Nucl. Phys. **B121**, 393 (1977).
24. Gaberdiel, M.R., Olive, D.I., West, P: A class of Lorentzian Kac–Moody algebras. Nucl. Phys. **B645**, 403 (2002). arXiv:hep-th/0205068.
25. Goldhaber, A.S.: Spin and statistic connection for charge-monopole composites, Phys. Rev. Lett. **36**, 1122 (1976).
26. Hasenfratz, P., 't Hooft, G.: A Fermion-Boson puzzle in a Gauge theory. Phys. Rev. Lett. **36**, 1119 (1976).
27. Jackiw, R., Rebbi, C.: Spin from isospin in a gauge theory. Phys. Rev. Lett. **36**, 1116 (1976).
28. Julia, B.: Infinite Lie algebras in physics. Invited talk given at Johns Hopkins Workshop on Current Problems in Particle Theory, Baltimore, MD, May 25–27, (1981).
29. Julia, B.: Group disintegrations. Lecture in Applied Mathematics, Vol. 21 (1985) AMS-SIAM.
30. Klebanov, I.R., Tseytlin, A.A.: D-branes and dual gauge theories in type 0 strings. Nucl. Phys. **B546**, 155 (1999). arXiv:hep-th/9811035.

31. Kleinschmidt, A., Schnakenburg, I., West, P.: Very-extended Kac–Moody algebras and their interpretation at low levels. Class. Quant. Grav. **21**, 2493 (2004). arXiv:hep-th/0309198.
32. Lambert, N., West, P.: Coset symmetries in dimensionally reduced bosonic string theory. Nucl. Phys. **B615**, 117 (2001). arXiv:hep-th/0107209.
33. Narain, K.: New heteroric string theories in uncompactified dimensions <10. Phys. Lett. **B169**, 41 (1986).
34. Nicolai, H., Fischbacher, T.: Low level representations for E_{10} and E_{11}. Published in: Kac-Moody Lie algebra and related topics, eds. N. Sthanumoorthy and K.C. Misra, Contempary Mathematics 343, American Mathematical Society, 2004. arXiv: hep-th/0301017.
35. Nutma, T.: Software SimpLie. http://strings.fmns.rug.nl/SimpLie/.
36. Polchinski, J.: String Theory, Vol. 1. Cambridge University Press, Cambridge (1998).
37. Polchinski, J.: String Theory, Vol. 2. Cambridge University Press, Cambridge (1998).
38. Riccioni, F., West, P.: Dual fields and E_{11}. Phys. Lett. **B645**, 286 (2007). arXiv:hep-th/0612001.
39. Riccioni, F., West, P.: The E_{11} origin of all maximal supergravities. J. High Energy Phys. **0707**, 063 (2007). arXiv:0705.0752 [hep-th].
40. Schwarz, J.H., West, P.: Symmetries and transformations of chiral $N = 2$ $D = 10$ supergravity. Phys. Lett. **B126**, 301 (1983).
41. West, P.: E_{11} and M theory. Class. Quant. Grav. **18**, 4443 (2001). arXiv:hep-th/0104081.
42. West, P: Very extended E_8 and A_8 at low levels, gravity and supergravity. Class. Quant. Grav. **20**, 2393 (2003). arXiv: hep-th/0212291.

Why Does the Universe Inflate?

S.W. Hawking

Can you hear me?

It is a great pleasure for me to be back again in Chile, to celebrate the 60th birthday of an old friend, and esteemed colleague, Claudio Bunster, whom I have known for almost 40 years. Claudio has done so much for science in general, and for science in Chile in particular. Being in the city of Valdivia where CECS, the center he created, is located, is quite meaningful to me.

Twenty-five years ago, we held a Nuffield workshop on the Very Early Universe in Cambridge. The inflation scenario had just been proposed, by Guth and others, to account for many of the otherwise unexplained features of the Hot Big Bang model. The original Old Inflation proposal depicted in Fig. 1, of thin walled bubbles, forming in a meta-stable vacuum state, was shown not to work. If the bubble formation rate was high, the bubbles would be close together, and inflation would not last long enough. On the other hand if the bubble formation rate was low, the bubbles would be so far apart, that they never join up and thermalize.

Instead, a different scenario, shown in Fig. 2, called new inflation was proposed. I have got into trouble in the past, in assigning credit for new inflation, so I will just say I first encountered the idea when I visited Andrei Linde in Moscow in October 1981.

The essential ingredient of new inflation, and of nearly all later inflationary proposals, is that the scalar field that drives inflation, rolls slowly down an effective potential. This slow roll down mechanism, enabled me, and a number of other people, to calculate the perturbations that would be caused by quantum fluctuations in the very early universe. Again I'm not going to stir up a hornets nest by trying to assign credit for this. I leave that to the Nobel committee. At first, different people got different results, but I invited the major players to a workshop in Cambridge in the summer of 1982, and they nearly all came. After long and occasionally heated

S.W. Hawking
Department of Applied Mathematics and Theoretical Physics, Cambridge, UK

M. Henneaux, J. Zanelli (eds.), *Quantum Mechanics of Fundamental Systems: The Quest for Beauty and Simplicity*, DOI 10.1007/978-0-387-87499-9_10,

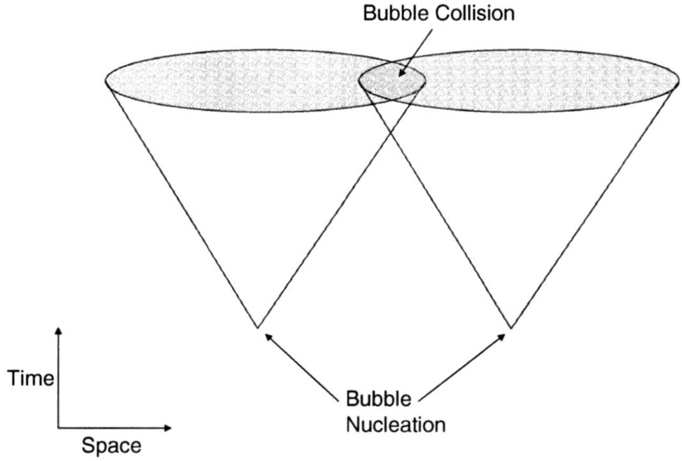

Fig. 1 Old inflationary scenario

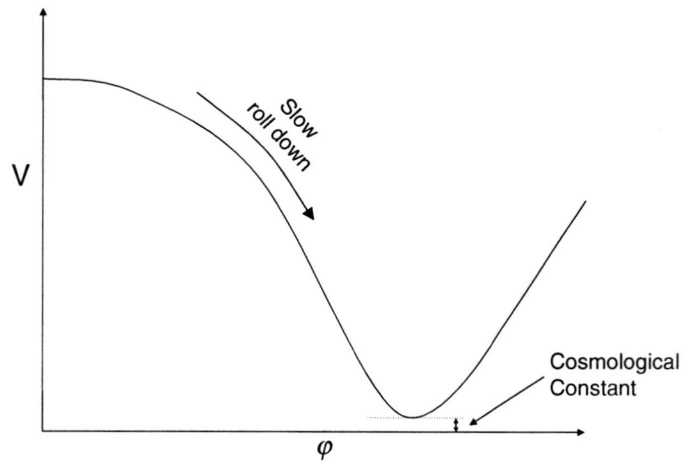

Fig. 2 New inflationary scenario

discussions, we all agreed on the amplitude and the spectrum of the perturbations that would be generated by slow roll inflation.

- Amplitude

$$\Delta^2(k) = \frac{1}{24\pi^2 M_P^4} \frac{V}{\varepsilon}$$

- Spectrum

$$\Delta^2(k) \alpha k^{n_s - 1},$$

where $n_s - 1 = -4\varepsilon + 2\eta$.

These predictions were spectacularly confirmed by Cobee 10 years later, and then in more detail by W-map. They rank in importance with the discovery of the expansion of the universe, and of the microwave background Radiation, because they would explain the origin of structure in the universe, and ultimately our own existence. It demonstrates that the early universe, is governed by quantum theory.

Slow roll inflation, is a mechanism for generating perturbations in a smooth background, that depends only on the shape of the effective potential, but is otherwise independent of what is giving rise to the potential. This is fortunate, because there is no agreement on the source of the potential. However whatever it is, one can ask,

Why did the universe inflate? Why did the scalar field start with a high value of the potential, and run down to a minimum. Why didn't it just start at the minimum?

In the original new inflation model, the universe was supposed to have started out as a hot big bang model, and to have cooled to a meta-stable state, in which the scalar field is delicately balanced at a maximum of the potential. This seemed implausible, and it was unsatisfactory to invoke one hot big bang stage, to explain another.

Instead, in this talk, I want to provide an answer to the question, that will apply to any model of the potential. For that, one needs a theory of initial conditions. The only well formulated theory I know, is the no boundary proposal. This says that in the Euclidean regime, space time is compact, and without boundary. It is the only natural boundary condition.

$$\Psi[h_{ij},\Sigma] = \int Dge^{-S[g]}$$

Sum over all metrics that have Σ as boundary
and where the induced metric on Σ is hij.

According to the no boundary proposal, the amplitude, Ψ, for a state with metric h, and matter field's ϕ, on a spacelike surface Σ, is given by the path integral over all no boundary metrics, with those values on the surface Σ. For simplicity, I shall consider only final states in which the metric is that of a three sphere of radius a, and the matter fields are a single homogeneous scalar field, ϕ. The amplitudes for slightly inhomogeneous final states, will be lower, and this, as Halliwell and I first showed, leads to the fluctuations in the microwave background, that we all discussed 25 years ago. My main concern in this talk however, will not be with these inhomogeneous fluctuations, but with the homogeneous and isotropic background, on which they occur. I will discuss large scale inhomogeneities at the end.

The amplitude, Ψ, is the wave function of the universe. It will obey the Wheeler DeWitt equation.

$$\left[-G_{ijkl}\frac{\delta^2}{\delta h_{ij}\delta h_{kl}} - {}^3R(h)h^{1/2} + 2\Lambda h^{1/2}\right]\Psi[h_{ij}] = 0$$

Where G_{ijkl} is the metric on superspace, $G_{ijkl} = \frac{1}{2}h^{-1/2}(h_{ik}h_{jl} + h_{il}h_{jk} - h_{ij}h_{kl})$ and ${}^3R$ is the scalar curvature of the intrinsic geometry of the three-surface.

In the case that the surfaces, Σ, are three spheres of radius a, and the matter is a single scalar field ϕ, this is a wave equation in the (a,ϕ) plane, with a playing the role of time.

$$\frac{1}{2}\left[\frac{\partial^2}{\partial a^2} - a^2 - \frac{1}{a^2}\frac{\partial^2}{\partial \phi^2} + a^4 V\right]\Psi(a,\phi 0 = 0$$

In the Euclidean region, $a^2V < 1$, there will be a real Euclidean solution of the field equations, and the wave function will be exponential. Outside this region, however, there will only be complex solutions, and the wave function will oscillate rapidly.

One can represent the wave function as the product of a rapidly varying phase, C, and slowly varying amplitude, B. Plugging this in the Wheeler DeWitt equation, one finds that C obeys the Hamilton Jacobi equation.

$$\Psi = Be^{iC}$$
$$\nabla C \cdot \nabla C + J = 0$$
$$\nabla B \cdot \nabla C = 0$$

One can therefore interpret the wave function by WKB, as corresponding to a family of Lorentzian solutions of the field equations. The trajectories of the solutions are given by the gradient of C, raised by the Wheeler DeWitt metric. The amplitude, B, obeys a conservation equation, which implies that the amplitudes of individual solutions are constant over the evolution of the solutions. This is shown in more detail in a paper by Jim Hartle, Thomas Hertog, and myself.

These Lorentzian solutions, start at different potentials, and roll down to the minimum. The amplitude for a solution that starts at a potential V_0, will be $e^{\frac{3}{4\pi V_0}}$. Clearly, this strongly favors starting at a low value of the potential, which would give little or no slow roll inflation (Fig. 3).

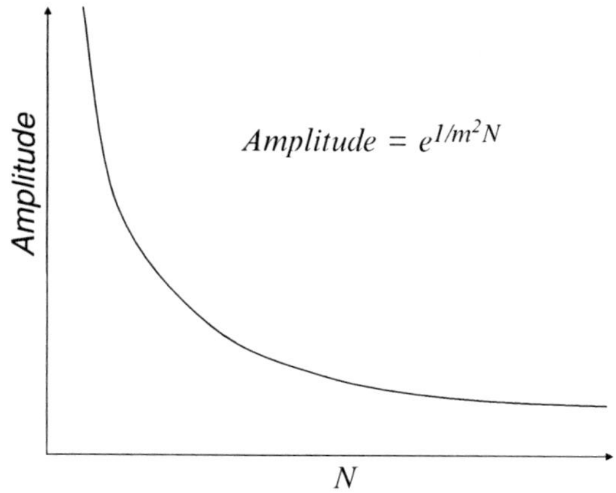

Fig. 3 No boundary amplitude

This has been recognized to be a problem with the no boundary proposal, for some time. I think the answer, is that there are two different probabilities involved. The amplitude gives the probability for the entire universe. However, one does not observe the entire universe, but only a Hubble volume around oneself. The number of such Hubble volumes at a given matter density, is proportional to the volume of the universe at that time, which in turn is proportional to e^{3N}, where N is the number of e-foldings of inflation. Thus on a frequency definition of probability, the probability of observing a Hubble volume of a given history, is proportional to the probability of that history, times e^{3N}.

The volume weighting transforms the probability distribution over N, the amount of inflation. It can more than compensate for the reduction in amplitude, due to a higher starting value of the potential, if the slow roll parameter, ε, at the start of inflation, is less than the potential, V_0, in Planck units. This is the same as the condition at which it is argued that quantum fluctuations will drive the volume averaged potential upwards, in the eternal inflation scenario. However, this derivation of eternal inflation, is not gauge invariant, and violates energy conservation and the constraint equations. The argument I have given for volume weighting, suffers from none of those difficulties (Fig. 4).

The condition, $\varepsilon < V$, was not satisfied at the time the microwave fluctuations we observe within our horizon were produced, because V/ε was about 10^{-5} at that time. This indicates that inflation must have started much earlier, and that N, the number of e-foldings of inflation, must have been much greater than 60.

It is reasonable to suppose that near the minimum, the potential is quadratic, plus a cosmological constant. If the potential, continued to be quadratic all the way up to the Planck density, the volume weighted probability will increase with increasing starting potential, only for starting potentials near the Planck density. Even at the Planck density, the volume weighted probability, will only be equal to that, for one e-folding of inflation.

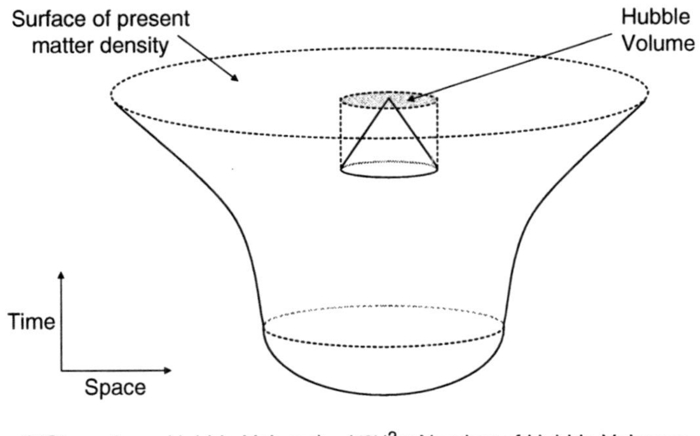

$$P(\text{Observing a Hubble Volume}) = |\Psi|^2 \times \text{Number of Hubble Volumes}$$
$$\propto |\Psi|^2 e^{3N}$$

Fig. 4 Volume weighting

This diagram, taken from the paper by Hartle, Hertog, and myself, shows the volume weighted probability as a function of the potential, for a quadratic potential with minimum at $\phi = 0$. On the other hand, for a solution that starts at a maximum or a saddle point, the probability distribution will favor very large N, if the curvature of the potential at the maximum is low. The dominant contribution is therefore likely to come from broad saddle points below the Planck density. The metrics will be well within the semi classical regime. They would start out with a Hawking–Moss instanton, a de Sitter like state which is unstable, and begins to run down the potential hill. The origin of the universe, is in the low energy regime of M theory, in which four dimensional general relativity, is a good approximation (Fig. 5).

The volume weighting implies that the universe starts at sufficiently high potential, that the eternal inflation condition, $V > \varepsilon$, is satisfied. But this is also the condition that the fluctuations in the scalar field are so large, that the surfaces of constant scalar field, don't foliate the spacetime. The usual calculation of inhomogeneous perturbations, uses the constant scalar field gauge. This will break down for modes that leave the horizon while the eternal inflation condition, is satisfied. For such modes, spatial gradients will not be important. Each mode will evolve independently, like a separate universe, and may fall into different vacua. The universe will be a mosaic of nearly homogeneous patches, like the structure predicted by eternal inflation. Unlike the eternal inflation scenario however, the no boundary amplitude with volume weighting, gives this result in a gauge invariant manner, without violation of the conservation or constraint equations. Any history is finite in duration, so inflation certainly is not eternal (Fig. 6).

I have shown that the no boundary proposal, with volume weighting, can explain why the universe inflated, why it started with a high potential, and ran down to a minimum. The volume weighting implies that the universe starts at sufficiently high potential, that different regions evolve independently. This gives the universe a mosaic structure.

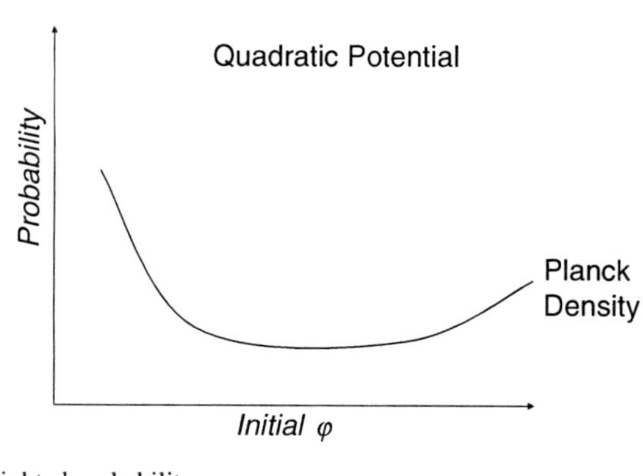

Fig. 5 Volume weighted probability

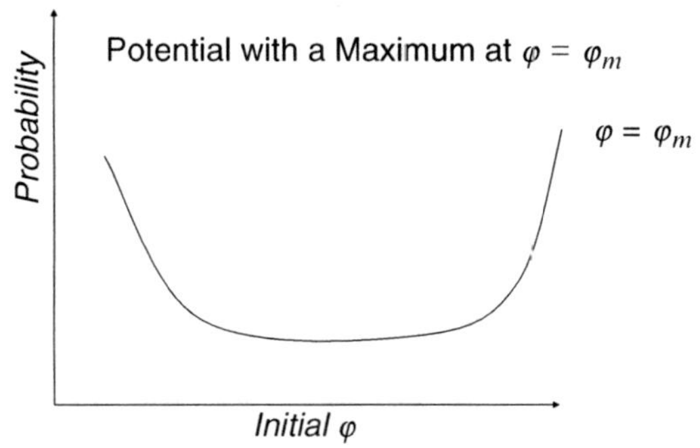

Fig. 6 Volume weighted probability

I would like to thank the organizers of this meeting again, for inviting me, to this beautiful country, which I discovered about 10 years ago. My stay in Chile is not over, and I look forward to the coming days.

Thank you for listening.

Kac–Moody Algebras and the Structure of Cosmological Singularities: A New Light on the Belinskii–Khalatnikov–Lifshitz Analysis

Marc Henneaux

Abstract The unexpected and fascinating emergence of hyperbolic Coxeter groups and Lorentzian Kac–Moody algebras in the investigation of gravitational theories in the vicinity of a cosmological singularity is briefly reviewed. Some open questions raised by this intriguing result, and some attempts to answer them, are outlined.

1 Introduction

As reported by Khalatnivov [46], the problem of cosmological singularities was considered by Lev Landau as one of the three most important problems of theoretical physics. While great breakthroughs occurred in the understanding of superconductivity and phase transitions – the other two important problems identified by Landau – taming singularities in gravity theory (understanding their structure and their possible resolution in an appropriate completion of gravity) remains to this day a challenge which raises a vast number of unanswered questions, in spite of the important advances achieved in the field in the last 50 years or so.

A major development in the area was the construction by Belinskii, Khalatnikov and Lifschitz ("BKL") [4,5] of a general solution of the gravitational field equations in the vicinity of a spacelike ("cosmological") singularity. Their work led recently to new investigations pointing to a fascinating and somewhat unexpected connection between gravity and Lorentzian Kac–Moody algebras [17,23]. The purpose of article is to briefly review these recent developments and to provide a guide to the literature on the subject.

It is a pleasure to dedicate this article to Claudio Bunster, long term collaborator and friend, on the occasion of his 60th birthday. The choice of cosmological

M. Henneaux
Physique théorique et mathématique, Université Libre de Bruxelles and International Solvay Institutes, ULB-Campus Plaine CP 231, B-1050 Bruxelles, Belgium
e-mail: henneaux@ulb.ac.be

M. Henneaux, J. Zanelli (eds.), *Quantum Mechanics of Fundamental Systems: The Quest for Beauty and Simplicity*, DOI 10.1007/978-0-387-87499-9_11,

singularities is a particularly appropriate subject since it is Claudio (then named "Claudio Teitelboim") who introduced some 30 years ago the author to the remarkable BKL analysis (interest in the BKL analysis was then motivated by the desire to understand the "zero signature" limit of gravity [35,37,56,57]).

2 Original BKL Analysis and Extension to Higher Dimensions

In their investigation of the generic dynamical behavior of the gravitational field in the vicinity of a cosmological (=spacelike) singularity, Belinskii, Khalatnikov and Lifschitz discovered the following remarkable features in four dimensions [4,5]:

- As one reaches the singularity, the spatial points decouple in the sense that the dynamical Einstein equations, which are partial differential equations, become ordinary differential equations with respect to time (one finite number of ODE's at each spatial point).
- In that limit, the off-diagonal components of the metric freeze (i.e. tend to definite limits) so that the non trivial dynamics is carried by the three independent scale factors that indicate how distances along three independent spatial directions change with time.[1]
- The dynamics of the scale factors exhibit a never-ending, oscillatory behavior of chaotic type with an infinite number of oscillations as one goes to the singularity (see also [50]).

This work was reformulated by Chitre and Misner in terms of a billiard motion in the 2-dimensional hyperbolic space of the dynamically independent scale factors (the three scale factors are related by the Hamiltonian constraint) [11,51]. Chaos is related in that picture to the finiteness of the volume of the billiard table. This reformulation turns out to be crucial for exhibiting the symmetries.

The extension to higher dimensions was started by Belinskii and Khalatnikov in five dimensions [3] and continued in [29,30,32] to all spacetime dimensions, with the surprising result that while the first two features (decoupling of spatial points and freezing of off-diagonal components) still hold, chaos disappear in spacetime dimensions ≥ 11. The infinite number of oscillations of the scale factors is replaced asymptotically by a monotonic Kasner regime. In particular, 11-dimensional pure gravity is non chaotic. However, if one includes the 3-form of 11-dimensional supergravity, chaos reappears [15].

The understanding of the higher dimensional dynamics in terms of a billiard motion was also achieved and led to the same picture of a cosmological billiard ball moving in an hyperbolic space of higher dimension [17,23,41,42]. This picture still holds if one includes dilatons and p-forms: the dilatons play the role of extra scale factors, while the p-form components play the role of extra off-diagonal components.

[1] See [4, 5, 23] for more information on the choice of slicing adapted to the singularity and the definition of scale factors.

It should be stressed that the emergence of chaos for those models that are chaotic is a statement valid for generic initial data. Chaos may be absent in models with particular spacetime symmetries, which form a set of measure zero. This corresponds to removing billiard walls and enlarging thereby the billiard table, making its originally finite volume infinite (see [16, 31]).

3 Emergence of Coxeter Groups and Kac–Moody Algebras

The billiard picture just described holds for any Lagrangian of the form

$$S[g_{\mu\nu},\phi,A^{(p)}] = \int d^D x \sqrt{-^{(D)}g}\left[R - \sum_i \partial_\mu \phi_i \partial^\mu \phi_i \right.$$
$$\left. -\frac{1}{2}\sum_p \frac{1}{(p+1)!} e^{l^{(p)}\phi} F^{(p)}_{\mu_1\cdots\mu_{p+1}} F^{(p)\,\mu_1\cdots\mu_{p+1}}\right] + \text{“more”} \quad (1)$$

where D is the spacetime dimension. The integer $p \geq 0$ labels the various p-forms $A^{(p)}$ present in the theory, with field strengths $F^{(p)}$ equal to $dA^{(p)}$,

$$F^{(p)}_{\mu_1\cdots\mu_{p+1}} = \partial_{\mu_1} A^{(p)}_{\mu_2\cdots\mu_{p+1}} \pm p \text{ permutations}\,. \quad (2)$$

In fact, the field strength could be modified by additional coupling terms of Yang–Mills or Chapline–Manton type [6, 10] (e.g. $F_C = dC^{(2)} - C^{(0)} dB^{(2)}$ for two 2-forms $C^{(2)}$ and $B^{(2)}$ and a 0-form $C^{(0)}$, as it occurs in ten-dimensional type IIB supergravity), but we include these additional contributions to the action in “more”. Similarly, “more” might contain Chern–Simons terms, as in the action for 11-dimensional supergravity [13]. The real parameters $l^{(p)}{}_i$ measure the strengths of the couplings to the dilatons.

However, a new feature emerges for theories which have the property that when reduced to three dimensions on a torus, their Lagrangian equals (after dualization to scalars of all the fields that can be dualized) the sum of the standard Einstein–Hilbert action plus the scalar Lagrangian of the non-linear sigma-model G/H, where G is some simple Lie group and H its maximal compact subgroup,

$$\mathscr{L} = \mathscr{L}_E + \mathscr{L}_{G/H}. \quad (3)$$

This class of theories include pure gravity in $D = d+1$ dimensions (for which the group G is $SL(d-1,R)$ and $H = SO(d-1)$ [9, 14]), or 11-dimensional supergravity, for which $G = E_{8,8}$ and $H = SO(16)$ [12, 49].

The crucial feature that emerges in that case is that the billiard table is then a Coxeter polyhedron and hence the billiard group (generated by the reflections in the billiard walls) is a Coxeter group [17]. This means that the angles of the billiard table are acute and equal to integer submultiples of π (see [38] for information on

Coxeter groups relevant to this context). Furthermore, the Coxeter polyhedron is a simplex and the matrix built out of the scalar products of the wall forms w_i defining the billiard

$$A_{ij} = 2\frac{(w_i|w_j)}{(w_i|w_i)} \tag{4}$$

turns out to be the Cartan matrix of the Lorentzian Kac–Moody algebra G^{++} [17, 21]. Here, G^{++} denotes the overextension [34,43,45] of the algebra G. Namely, it is obtained by adding a further simple root to the untwisted affine extension G^+ of G. That root, the "overextended root" is attached to the affine root by a single line. We give here a few examples.

- The algebra A_1^{++} relevant to pure, four-dimensional gravity.
- The algebra $E_8^{++} \equiv E_{10}$ relevant to 11-dimensional supergravity.
  Both A_1^{++} and E_{10} are hyperbolic, which means that if one removes any node from their Dynkin diagram, one obtains a Dynkin diagram which is either of finite or affine type. For instance, in the case of E_{10}, one gets successively (removing the node -1, 0, etc.): E_8^+ which is affine, and $A_1 \oplus E_8$, $A_2 \oplus E_7$, $A_3 \oplus E_6$, $A_4 \oplus D_5$, $A_5 \oplus A_4$, $A_6 \oplus A_1 \oplus A_2$, $A_8 \oplus A_1$, D_9, A_9, which are all of finite type (Figs. 1 and 2).
- The algebra $B_8^{++} \equiv BE_{10}$ relevant to $N = 1$ ten-dimensional supergravity with one vector multiplet.
  As shown in [36], "split symmetry controls chaos" and hence, it is the same billiard that controls the dynamics of ten-dimensional supergravity with k vector multiplets, in which case the symmetry algebra in three dimensions $so(8, 8+k)$ whose maximal split subalgebra $so(8, 9)$. The algebra BE_{10} is easily verified to be hyperbolic (Fig. 3).
- The algebra $D_8^{++} \equiv DE_{10}$ relevant to pure $N = 1$ ten-dimensional (Fig. 4).
  The algebra DE_{10} is also easily verified to be hyperbolic.

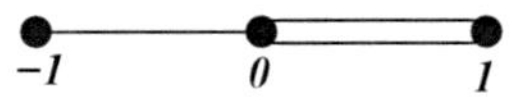

Fig. 1 The hyperbolic algebra A_1^{++}. The node 1 defines the Dynkin diagram of A_1, the nodes 1 and 0 form the Dynkin diagram of its affine extension, while the nodes 1, 0 and -1 define its overextension A_1^{++}

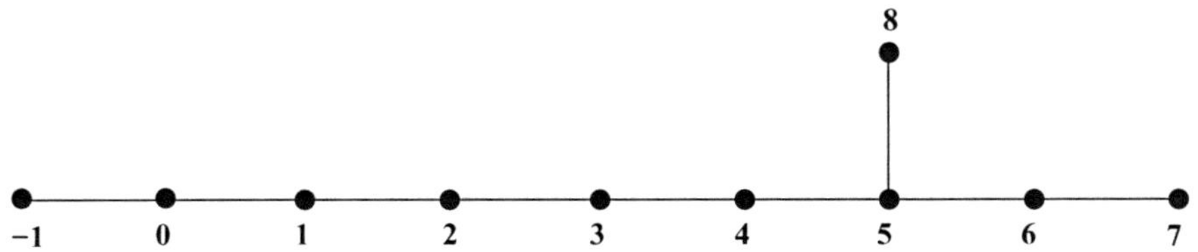

Fig. 2 The hyperbolic algebra E_{10}. The nodes labelled $1, \ldots, 8$ form the Dynkin diagram of E_8, the nodes $0, 1, \ldots, 8$ form the Dynkin diagram of its affine extension E_8^+, while the nodes $-1, 0, 1, \ldots, 8$ define its overextension $E_8^{++} \equiv E_{10}$

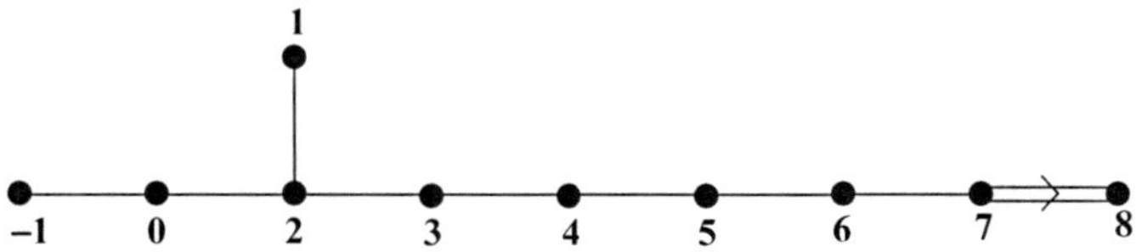

Fig. 3 The hyperbolic algebra BE_{10}. The nodes labelled $1,\dots,8$ form the Dynkin diagram of B_8, the nodes $0,1,\dots,8$ form the Dynkin diagram of its affine extension B_8^+, while the nodes $-1,0,1,\dots,8$ define its overextension $B_8^{++} \equiv BE_{10}$

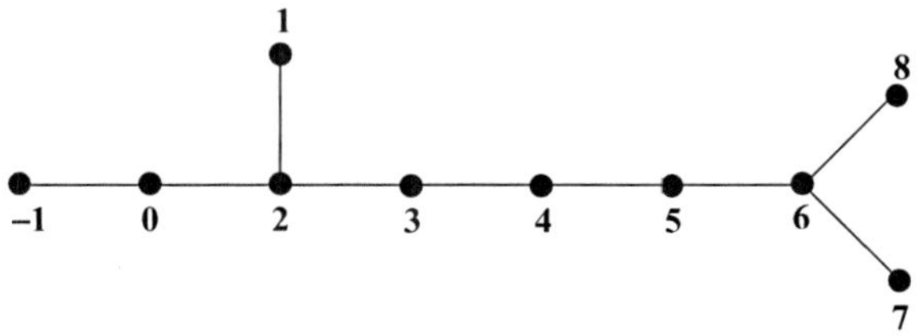

Fig. 4 The hyperbolic algebra DE_{10}. The nodes labelled $1,\dots,8$ form the Dynkin diagram of D_8, the nodes $0,1,\dots,8$ form the Dynkin diagram of its affine extension D_8^+, while the nodes $-1,0,1,\dots,8$ define its overextension $D_8^{++} \equiv DE_{10}$

We stress that the emergence of the Kac–Moody structure holds for all theories whose toroidal dimensional reduction to three dimensions has the properties indicated above, but it is by no means necessary to actually perform the reduction to three dimensions to get the billiard. This follows from the dynamics for *generic* initial conditions.

It also turns out that the billiard table is a fundamental domain for the action of the Weyl group on hyperbolic space (upper sheet of the unit hyperboloid). That it is acute-angled is crucial in this respect.

The fact that one gets the Weyl group of a Lorentzian Kac–Moody algebra is rather remarkable as it depends on the presence of all walls. As analysed in [39] removing some walls, as dictated for instance by spatial cohomology in a regime of intermediate asymptotics [58], might yield Coxeter groups that are not equal to Weyl groups. One might then get a billiard table that is not acute-angled, or which is not a simplex.[2] Getting the Weyl group of a Kac–Moody algebra is thus a quite non trivial phenomenon.

Finally, we note that hyperbolic Kac–Moody algebras exist only up to rank 10. In rank 10, there are four of them, namely, E_{10}, BE_{10}, DE_{10} already encountered, as well as CE_{10} which is the algebra dual to BE_{10} (its Dynkin diagram is obtained by reversing the arrow connecting the nodes 7 and 8). The Weyl group of rank 10 algebras act naturally on 9-dimensional hyperbolic space, and hyperbolicity translates itself in the property that the billiard table (fundamental domain) has finite volume. The disappearance of chaos for pure gravity as one increases the spacetime dimensions follows from the fact that the algebras A_k^{++} are hyperbolic up to A_7^{++}, while A_k^{++} is not hyperbolic for $k \geq 8$ [20].

[2] It should also be recalled that in hyperbolic space, Coxeter polyhedra need not be simplices.

For information we note that while simplex Coxeter groups exist in hyperbolic space up to dimension 9 as we have just recalled, non simplex Coxeter groups exist in hyperbolic space up to dimension 996 [59].

4 Motion in Cartan Subalgebra

One can precisely reformulate the billiard dynamics as a motion in the Cartan subalgebra of the associated Kac–Moody algebra. To that end, let us recall the basic features of a Kac–Moody algebra [45].

A Kac–Moody algebra is defined by a (generalized) Cartan matrix A_{ij} $(i,j = 1,\ldots,N)$, namely, a (square) matrix with the following properties:

- $A_{ii} = 2$
- $A_{ij} \in Z_{-}$ $(i \neq j)$
- $A_{ij} \neq 0 \Rightarrow A_{ji} \neq 0$.

The corresponding Kac–Moody algebra $\mathscr{A}$ is generated by $3N$ generators $\{h_i, e_i, f_i\}$ $(i = 1,\ldots,N = r+2)$ subject to the following relations

$$[h_i, h_j] = 0 \tag{5}$$

$$[h_i, e_j] = A_{ij} e_j, \;\; [h_i, f_j] = -A_{ij} f_j, \;\; [e_i, f_j] = \delta_{ij} h_i \tag{6}$$

$$\underbrace{[e_i,[e_i,[e_i,[\cdots,[e_i,e_j]]\cdots]}_{1-A_{ij}\text{ times}} = 0, \;\; \underbrace{[f_i,[f_i,[f_i,[\cdots,[f_i,f_j]]\cdots]}_{1-A_{ij}\text{ times}} = 0 \tag{7}$$

Relations (5) and (6) are the Chevalley relations, relations (7) are the Serre relations. The integer N is called the *rank* of the algebra.

A central feature of Kac–Moody algebra is the triangular decomposition,

$$\mathscr{A} = \mathscr{N}^- \oplus H \oplus \mathscr{N}^+$$

where (a) H is the Cartan subalgebra (linear combinations of h_i); (b) $\mathscr{N}^+$ contains the linear combinations of the "raising operators" e_i and their multiple commutators $[e_i, e_j]$, $[e_i, [e_j, e_k]]$, etc. not killed by the above relations; and (c) $\mathscr{N}^-$ contains the linear combinations of the "lowering operators" f_i and their multiple commutators $[f_i, f_j]$, $[f_i, [f_j, f_k]]$, etc. not killed by the above relations. This generalizes the well-known triangular decomposition of finite-dimensional simple Lie algebras, e.g.

$$\begin{pmatrix} a & b & c \\ d & e & f \\ g & h & -a-e \end{pmatrix} = \begin{pmatrix} 0 & 0 & 0 \\ d & 0 & 0 \\ g & h & 0 \end{pmatrix} + \begin{pmatrix} a & 0 & 0 \\ 0 & e & 0 \\ 0 & 0 & -a-e \end{pmatrix} + \begin{pmatrix} 0 & b & c \\ 0 & 0 & f \\ 0 & 0 & 0 \end{pmatrix}$$

for $sl(3)$.

One has

$$[h, e_i] = \alpha_i(h)\, e_i$$

where $\alpha_i \in H^\star$ are the *simple roots*. If

$$[e_{i_1}, [e_{i_2}, [\cdots, [e_{i_{m-1}}, e_{i_m}]] \cdots]] \neq 0,$$

then $\alpha_{i_1} + \alpha_{i_2} + \cdots \alpha_{i_m}$ is a (positive) root,

$$[h, [e_{i_1}, [e_{i_2}, [\cdots, [e_{i_{m-1}}, e_{i_m}]] \cdots]]] \\ = (\alpha_{i_1} + \alpha_{i_2} + \cdots \alpha_{i_m})(h)\, [e_{i_1}, [e_{i_2}, [\cdots, [e_{i_{m-1}}, e_{i_m}]] \cdots]]$$

(Jacobi identity). One has similar relations on the negative side. If the matrix A_{ij} is symmetrizable,

$$A_{ij} = 2d_i S_{ij}, \quad d_i > 0, \quad S_{ij} = S_{ji},$$

as we shall assume here, on can define a scalar product in the real linear span of the simple roots,

$$(\alpha_i|\alpha_j) = S_{ij}.$$

It is customary to normalize the scalar product such that the longest roots have length squared equal to 2.

One distinguishes three cases, according to which the scalar product is Euclidean (finite case), positive semi-definite (affine case) or of neither of these two types (indefinite case). It is only in the Euclidean case that the algebra is finite-dimensional. In the other two cases, it is infinite-dimensional.

In the affine and indefinite cases, a root can be real (=spacelike) or imaginary (=timelike or null). Simple roots are real; real roots are similar to roots of finite-dimensional, simple Lie algebras: they are non-degenerate and furthermore, if α is a real root, the only multiples of α that are roots are $\pm\alpha$. By contrast, the imaginary roots do not enjoy these properties and are poorly understood. The indefinite case with Lorentzian signature is called "Lorentzian".

The Weyl group of $\mathscr{A}$ is generated by *fundamental Weyl reflections*:

$$s_i : \lambda \to s_i(\lambda) = \lambda - 2\frac{(\lambda|\alpha_i)}{(\alpha_i|\alpha_i)}\alpha_i.$$

It is a discrete subgroup of $O(N-1,1)^+$ in the Lorentzian case (time orientation preserving elements of $O(N-1,1)$). It has a well defined action on the upper light cone and on the upper sheet of the unit hyperboloid ($(N-1)$-dimensional hyperbolic space). The fundamental Weyl chamber is defined in terms of the simple roots by $\alpha_i(h) \geq 0$. If it is completely contained within the light cone, the algebra is called hyperbolic. Its intersection with hyperbolic space has then finite volume and is a fundamental domain for the action of the Weyl group on hyperbolic space.

The dictionary between the billiard motion and the Kac–Moody algebra is given in the following table [17,20]

Gravity side		Kac–Moody side
Scale factors	$\leftrightarrow$	Cartan degrees of freedom
Billiard motion	$\leftrightarrow$	Lightlike motion in Cartan subalgebra
Walls	$\leftrightarrow$	Hyperplanes orthogonal to simple roots
Reflection against a wall	$\leftrightarrow$	Fundamental Weyl reflection
Finite volume of billiard table	$\leftrightarrow$	Hyperbolic algebra

5 Hidden Symmetries?

The intriguing emergence of the Weyl group of a Kac–Moody algebra in the BKL limit has prompted the conjecture that the Kac–Moody algebra itself might be a hidden symmetry of the corresponding gravitational theory (possibly augmented by new degrees of freedom) [22]. Part of the excitement regarding this conjecture is due to the fact that the same conjecture was made (earlier in the case of some of the approaches) following different lines in [33,40,43,44,52,60]. Attempts to substantiate the conjecture have been made based on the idea of non linear realizations, in which the conjectured symmetry is manifest [22, 60]. The problem is then to connect the dynamics of the non linear sigma model to the (super)gravity dynamics through an appropriate dictionary and to establish their equivalence (with the possible addition of new degrees of freedom on the gravity side). It is not the purpose here to review all the interesting work that has gone into studying various aspects of this conjecture. We shall only allude to the approach inspired by the cosmological billiards, which is the exclusive subject of this article.

The BKL analysis suggests to consider the $(1+0)$-non linear sigma model $G^{++}/K(G^{++})$ (geodesic motion on $G^{++}/K(G^{++})$) [22]. Namely, one goes beyond the dynamics in the Cartan subalgebra by including as dynamical variables the fields associated with the positive roots. This approach has met with spectacular successes at low "levels" [22] (see [18] for a systematic analysis) since its low level truncation reproduces the dynamics of supergravity consistently truncated to homogeneous modes (in a sense made precise in [22]). But no one has been able to push it systematically to higher levels so far to include the full supergravity theory. Further work (and new ideas) appears to be necessary. Part of the problem is that the dictionary between the sigma model variables and the supergravity fields is not understood. A better control of duality appears also to be necessary. The same success works, up to the same levels, if one includes the fermionic degrees of freedom, leading, on the sigma-model side, to a spinning particle action [24, 26–28]

Two other spectacular findings provide additional support to the conjecture:

- The quantum corrections to M-theory are compatible with the E_{10} algebraic structure, in the sense that they correspond to roots of E_{10} [19] (see also [1, 2, 25, 47, 48] for further discussions and developments).

- The massive deformation of type IIA supergravity corresponds to a level 4 root of E_{10} (or E_{11}) [21,53], a result that can be extended to other deformations of the theory in lower dimensions (see the original works [7,8,54,55] for entries into this fast growing literature).

6 Conclusions

We have reviewed the dynamics of the gravitational field in the vicinity of a cosmological singularity pioneered by the remarkable work of BKL and have shown that it can be described in terms of fascinating structures: hyperbolic Coxeter groups and Kac–Moody algebras. This seems to be the tip of an iceberg indicating an even more richer structure at a deeper level, yet to be discovered.

Acknowledgments It is a pleasure to thank the various people with whom I had the privilege to work on this problem over the years, and in particular the late Jacques Demaret with whom the analysis of BKL in higher dimensions was started, and Thibault Damour, with whom the BKL study was given a new impetus leading to results (Coxeter groups and Kac–Moody algebras), the significance of which are still being explored. Work supported in part by IISN-Belgium (conventions 4.4511.06 and 4.4514.08), by the Belgian National Lottery, by the European Commission FP6 RTN programme MRTN-CT-2004-005104, and by the Belgian Federal Science Policy Office through the Interuniversity Attraction Pole P6/11.

References

1. L. Bao, M. Cederwall and B. E. W. Nilsson, Class. Quant. Grav. **25**, 095001 (2008) [arXiv:0706.1183 [hep-th]].
2. L. Bao, J. Bielecki, M. Cederwall, B. E. W. Nilsson and D. Persson, JHEP **0807**, 048 (2008) [arXiv:0710.4907 [hep-th]].
3. V. A. Belinsky and I. M. Khalatnikov, Sov. Sci. Rev. A **3**, 555 (1981)
4. V. A. Belinsky, I. M. Khalatnikov and E. M. Lifshitz, Adv. Phys. **19**, 525 (1970).
5. V. A. Belinsky, I. M. Khalatnikov and E. M. Lifshitz, Adv. Phys. **31**, 639 (1982).
6. E. Bergshoeff, M. de Roo, B. de Wit and P. van Nieuwenhuizen, Nucl. Phys. B **195**, 97 (1982).
7. E. A. Bergshoeff, I. De Baetselier and T. A. Nutma, JHEP **0709**, 047 (2007) [arXiv:0705.1304 [hep-th]].
8. E. A. Bergshoeff, J. Gomis, T. Nutma and D. Roest, JHEP **02**, 069 (2008) [arXiv:0711.2035 [hep-th]].
9. P. Breitenlohner, D. Maison and G. W. Gibbons, Commun. Math. Phys. **120**, 295 (1988).
10. G. F. Chapline and N. S. Manton, Phys. Lett. B **120**, 105 (1983).
11. D. M. Chitre, "Investigations of Vanishing of a Horizon for Bianchi Type X (the Mixmaster) Universe", Ph.D. thesis, University of Maryland, College Park, MD.
12. E. Cremmer and B. Julia, Nucl. Phys. B **159**, 141 (1979).
13. E. Cremmer, B. Julia and J. Scherk, Phys. Lett. B **76**, 409 (1978).
14. E. Cremmer, B. Julia, H. Lu and C. N. Pope, (unpublished). arXiv:hep-th/9909099.
15. T. Damour and M. Henneaux, Phys. Rev. Lett. **85**, 920–923 (2000) [arXiv:hep-th/0003139].
16. T. Damour and M. Henneaux, Phys. Lett. B **488**, 108 (2000) [Erratum-ibid. B **491** (2000) 377] [arXiv:hep-th/0006171].

17. T. Damour and M. Henneaux, Phys. Rev. Lett. **86**, 4749 (2001) [arXiv:hep-th/0012172].
18. T. Damour and H. Nicolai, in 25th International Colloquium on Group Theoretical Methods in Physics (ICGTMP), Cocoyoc, Mexico, August 2004. arXiv:hep-th/0410245.
19. T. Damour and H. Nicolai, Class. Quant. Grav. **22**, 2849 (2005) [arXiv:hep-th/0504153].
20. T. Damour, M. Henneaux, B. Julia and H. Nicolai, Phys. Lett. B **509**, 323 (2001) [arXiv:hep-th/0103094].
21. T. Damour, S. de Buyl, M. Henneaux and C. Schomblond, JHEP **0208**, 030 (2002) [arXiv:hep-th/ 0206125].
22. T. Damour, M. Henneaux and H. Nicolai, Phys. Rev. Lett. **89**, 221601 (2002) [arXiv:hep-th/0207267].
23. T. Damour, M. Henneaux and H. Nicolai, Class. Quant. Grav. **20**, R145 (2003) [arXiv:hep-th/0212256].
24. T. Damour, A. Kleinschmidt and H. Nicolai, Phys. Lett. B **634**, 319 (2006) [arXiv:hep-th/0512163].
25. T. Damour, A. Hanany, M. Henneaux, A. Kleinschmidt and H. Nicolai, Gen. Rel. Grav. **38**, 1507 (2006) [arXiv:hep-th/0604143].
26. T. Damour, A. Kleinschmidt and H. Nicolai, JHEP **0608**, 046 (2006) [arXiv:hep-th/0606105].
27. S. de Buyl, M. Henneaux and L. Paulot, Class. Quant. Grav. **22**, 3595 (2005) [arXiv:hep-th/0506009].
28. S. de Buyl, M. Henneaux and L. Paulot, JHEP **0602**, 056 (2006) [arXiv:hep-th/0512292].
29. J. Demaret, M. Henneaux and P. Spindel, Phys. Lett. B **164**, 27 (1985).
30. J. Demaret, J. L. Hanquin, M. Henneaux, P. Spindel and A. Taormina, Phys. Lett. B **175**, 129 (1986).
31. J. Demaret, Y. De Rop and M. Henneaux, Phys. Lett. B **211**, 37 (1988).
32. J. Demaret, Y. De Rop and M. Henneaux, Int. J. Theor. Phys. **28**, 1067 (1989).
33. F. Englert, L. Houart, A. Taormina and P. West, JHEP **09**, 020 (2003) [arXiv:hep-th/0304206].
34. A. J. Feingold and I. B. Frenkel, Math. Ann. **263**, 87 (1983).
35. M. Henneaux, *Geometry of Zero Signature Space–Times*, Preprint PRINT-79-0606-Princeton (unpublished).
36. M. Henneaux and B. Julia, JHEP **0305**, 047 (2003) [arXiv:hep-th/0304233].
37. M. Henneaux, M. Pilati and C. Teitelboim, Phys. Lett. B **110**, 123 (1982).
38. M. Henneaux, D. Persson and P. Spindel, Living Rev. Rel. **11**, 1 (2008) [arXiv:0710.1818 [hep-th]].
39. M. Henneaux, D. Persson and D. H. Wesley, JHEP **0809**, 052 (2008) [arXiv:0805.3793 [hep-th]].
40. P. Henry-Labordere, B. Julia and L. Paulot, JHEP **0204**, 049 (2002) [arXiv:hep-th/0203070].
41. V. D. Ivashchuk and V. N. Melnikov, Class. Quant. Grav. **12**, 809 (1995).
42. V. D. Ivashchuk and V. N. Melnikov, J. Math. Phys. **41**, 6341 (2000) [arXiv:hep-th/9904077].
43. B. Julia, Proc. AMS-SIAM Chicago Meeting July 1982, *Lectures in Applied Mathematics*, AMS-SIAM, Vol. 21 (1985), p. 355.
44. B. Julia, "*Group Disintegrations*," Invited Paper Presented at Nuffield Gravity Workshop, Cambridge, England, Jun 22–Jul 12, 1980.
45. V. G. Kac, *Infinite Dimensional Lie Algebras*, third edition, Cambridge University Press, Cambridge (1990).
46. I. M. Khalatnikov and A. Y. Kamenshchik, arXiv:0803.2684 [gr-qc] (unpublished).
47. N. Lambert and P. West, Phys. Rev. D **74**, 065002 (2006) [arXiv:hep-th/0603255].
48. N. Lambert and P. West, Phys. Rev. D **75**, 066002 (2007) [arXiv:hep-th/0611318].
49. N. Marcus and J. H. Schwarz, Nucl. Phys. B **228**, 145 (1983).
50. C. W. Misner, Phys. Rev. Lett. **22**, 1071 (1969).
51. C. W. Misner, "The Mixmaster cosmological metrics" in Deterministic Chaos in General Relativity, D. Hobill, A. Burd and A. Coley, editors. Nato ASI Series **B** vol. 332, Plenum, 1994. [arXiv:gr-qc/9405068].
52. H. Nicolai, Phys. Lett. B **276**, 333 (1992).
53. I. Schnakenburg and P. C. West, Phys. Lett. B **540**, 137 (2002) [arXiv:hep-th/0204207].

54. F. Riccioni and P. West, JHEP **0707**, 063 (2007) [arXiv:0705.0752 [hep-th]].
55. F. Riccioni and P. West, JHEP **02**, 039 (2008) [arXiv:0712.1795 [hep-th]].
56. C. Teitelboim, *The Hamiltonian Structure of Space–Time*, Preprint PRINT-78-0682-Princeton (unpublished).
57. C. Teitelboim, In Held, A.(Ed.): *General Relativity and Gravitation*, Vol. 1, 195–225.
58. D. H. Wesley, P. J. Steinhardt and N. Turok, Phys. Rev. D **72**, 063513 (2005) [arXiv:hep-th/0502108].
59. Vinberg et al., eds., *Geometry II – Spaces of Constant Curvature*, Springer, Berlin (1993).
60. P. West, Class. Quant. Grav. **18**, 4443 (2001) [arXiv:hep-th/0104081].

Black Holes with a Conformally Coupled Scalar Field

Cristián Martínez

Abstract We consider gravity in presence of a cosmological constant in arbitrary spacetime dimensions with a conformally coupled scalar field and a self-interacting potential. The energy-momentum tensor is traceless when the constant appearing in front of the non-minimal coupling term and the power of the self-interacting potential are properly chosen. First, configurations with a constant scalar field are studied. In the general case, the spacetime is required to be an Einstein space. However, for a special value of the scalar field this condition can be relaxed and it is enough that the spacetime has a constant scalar curvature fixed by the cosmological constant. In this case the cosmological constant and the self-interacting coupling constant are related. The second part is devoted to searching black holes dressed with a conformally coupled scalar field in dimensions greater than four. Since the existence of a no-go theorem discarding static and asymptotically flat black holes in higher dimensions, we introduce a cosmological constant and a self-interacting potential in the action. Using a standard static ansatz for the metric, which includes both spherically symmetric as topological black holes, and a scalar field depending only on the radial coordinate, it is shown that there are no higher-dimensional counterparts of the known black holes in three and four dimensions.

1 Introduction

The organizers of Claudio's Fest have subtitled this volume as "The Quest for Beauty and Simplicity." It is a very good choice that describes the outstanding scientific contributions of Claudio Bunster. Trying to follow the spirit, the simplest form of matter interacting with a black hole that one could consider corresponds

C. Martínez
Centro de Estudios Científicos (CECS), Casilla 1469, and Centro de Ingeniería de la Innovación del CECS (CIN), Valdivia, Chile
e-mail: martinez@cecs.cl

M. Henneaux, J. Zanelli (eds.), *Quantum Mechanics of Fundamental Systems: The Quest for Beauty and Simplicity*, DOI 10.1007/978-0-387-87499-9_12,

to a single real scalar field. However, when this field is minimally coupled and the spacetime is asymptotically flat, the so called no-hair conjecture [1–3] indicates that this class of black hole would not exist. In the last 30 years, a large number of articles have studied this problem and recent works can be found in [4–14]. However, there are different ways to circumvent this conjecture as we show below.

In the 1970s a four-dimensional and spherically symmetric asymptotically flat black hole [15, 16] was reported. This black hole is dressed with a *conformally* coupled scalar field, i.e., when the corresponding stress-energy tensor is traceless. Conformally coupled scalar fields in General Relativity have been used to model quantum effects in semiclassical theories [17–23]. This model has a well posed initial value formulation [24], and it was shown to reproduce better – than the minimally coupled scalar field – the local propagation properties of Klein Gordon fields on Minkowski spacetime [25].

A black hole solution, where the scalar field is conformally coupled and *regular everywhere*[1] was found in [26]. In this three-dimensional black hole, the spacetime is asymptotically anti-de Sitter because a negative cosmological constant is included. This black hole solution was extended by considering a one parameter family of conformal self-interacting potential [27]. The presence of a cosmological constant allows to find exact four-dimensional black hole solutions, where the scalar field is regular on and outside the event horizon [28–31]. Numerical black hole solutions can also be found in four [32–36] and five dimensions [37]. Further exact solutions were found in [38].

Some interesting aspects of these black hole solutions are studied in [39–61]. Recently, using a combination of analytical and numerical methods, it was shown [62] that there no other four-dimensional static, spherically symmetric black hole solutions with a conformally coupled scalar field in presence of a positive cosmological constant, satisfying the dominant and strong energy condition between the event and cosmological horizon, besides the solution reported in [28].

In this contribution, we are interested in higher-dimensional solutions of the Einstein equations with a cosmological constant and a conformally coupled self-interacting scalar field. After introducing the model in Sect. 2, we analyze the constant scalar field configurations in Sect. 3. In the general case, the spacetime is required to be an Einstein space, however, for a special value of the scalar field this condition is relaxed and the spacetime must have constant scalar curvature, and the cosmological constant and the self-interacting coupling are related. Section 4 is devoted to search black hole solutions with non-constant conformally coupled scalar field in dimensions greater than four. It is known that in absence of a cosmological constant and a self-interacting potential a static, spherically symmetric black hole conformally dressed exist only in four dimensions. In this way, the quest of possible higher-dimensional black holes requires to include them in the action. Using a standard static ansatz for the metric and the scalar field, which include the possibility of topological event horizons, we show by direct integration of the field equations that there are no higher-dimensional black holes with a conformally coupled scalar field. Finally, concluding remarks are presented in Sect. 5.

[1] In the previous four-dimensional counterpart [15, 16] the scalar field diverges at the horizon.

2 The Conformal Scalar Coupling

In d spacetime dimensions the action for gravity, in presence of a cosmological constant Λ, with a non-minimally coupled scalar field ϕ is given by,

$$I[g_{\mu\nu},\phi] = \int d^d x\sqrt{-g}\left[\frac{R-2\Lambda}{2\kappa} - \frac{1}{2}g^{\mu\nu}\partial_\mu\phi\partial_\nu\phi - \frac{\xi}{2}R\phi^2 - \alpha\phi^p\right], \tag{1}$$

where κ and α are the gravitational and self-interaction coupling constants, respectively. The matter piece of the action is invariant under the conformal transformations $g_{\mu\nu} \to \Omega^2 g_{\mu\nu}$ and $\phi \to \Omega^{\frac{2-d}{2}}\phi$, provided that ξ and p take the following unique values:

$$\xi = \frac{1}{4}\frac{d-2}{d-1} \quad \text{and} \quad p = \frac{2d}{d-2}. \tag{2}$$

With these values in the action we say that the scalar field is *conformally coupled* to gravity. Hereafter, we will consider only this case.

The field equations derived from the action are

$$G_{\mu\nu} + \Lambda g_{\mu\nu} = \kappa T^{\phi}_{\mu\nu}, \tag{3a}$$

$$\Box\phi - \xi R\phi - p\alpha\phi^{p-1} = 0, \tag{3b}$$

where the energy-momentum tensor is

$$T^{\phi}_{\mu\nu} = \partial_\mu\phi\partial_\nu\phi - \frac{1}{2}g_{\mu\nu}g^{\alpha\beta}\partial_\alpha\phi\partial_\beta\phi - \alpha g_{\mu\nu}\phi^p + \xi\left[g_{\mu\nu}\Box - \nabla_\mu\nabla_\nu + G_{\mu\nu}\right]\phi^2. \tag{3c}$$

This energy-momentum tensor differs from the standard one by the last term, which is proportional to ξ, and that comes from the non-minimal coupling $R\phi^2$ of the action. Note that the trace of $T^{\phi}_{\mu\nu}$ vanishes on shell

$$T^{\phi\,\mu}_{\ \ \mu} = \left(\frac{d-2}{2}\right)\phi\left[\Box\phi - \xi R\phi - p\alpha\phi^{p-1}\right] = 0. \tag{4}$$

Thus, the trace of (3a) gives

$$R = \frac{2\Lambda d}{d-2}. \tag{5}$$

Then, all the possible solutions of this theory have a constant scalar curvature.

In general, (3b) can be derived from (3a). To see this, one can take the covariant derivative of Einstein equations (3a)

$$\nabla_\nu G^\nu_\mu + \Lambda\nabla_\nu\delta^\nu_\mu = \kappa\nabla_\nu T^{\phi\,\nu}_{\mu},$$

and since that $\nabla_\nu G^\nu_\mu$ and $\nabla_\nu\delta^\nu_\mu$ identically vanish,

$$\nabla_\nu T^{\phi\,\nu}_{\mu} = \partial_\mu\phi\left[\Box\phi - \xi R\phi - p\alpha\phi^{p-1}\right] = 0.$$

Thus, if ϕ is not a constant, the conservation of the energy-momentum tensor implies (3b). The case of a constant scalar field, where (3b) is independent, will be studied in the next section.

3 Solutions with a Constant Scalar Field

As it was noticed previously, it is interesting to comment on those solutions of the field equations (3) for which $\phi = \phi_o \neq 0$ is a constant. For this configuration, (3) reduce to:

$$\left(1 - \xi \kappa \phi_o{}^2\right) G_{\mu\nu} + (\Lambda + \kappa \alpha \phi_o{}^p) g_{\mu\nu} = 0, \tag{6a}$$

$$R + \alpha \frac{p}{\xi} \phi_o{}^{p-2} = 0. \tag{6b}$$

Compatibility between the trace of (6a) and (6b) implies

$$\phi_o{}^{p-2} = -\frac{\Lambda \xi}{\alpha} \quad (\alpha, \Lambda \neq 0). \tag{7}$$

Replacing this expression for ϕ_o in (6) we obtain

$$\left(1 - \kappa \xi \phi_o{}^2\right) \left[G_{\mu\nu} + \Lambda g_{\mu\nu}\right] = 0, \tag{8a}$$

$$R - \frac{2\Lambda d}{d-2} = 0. \tag{8b}$$

If $\kappa \xi \phi_o{}^2 \neq 1$, (8b) can be derived from (8a) and the general solution is given by any Einstein spacetime with a constant scalar field satisfying (7). In the case $\kappa \xi \phi_o{}^2 = 1$, the spacetime is not required to be an Einstein space, instead it must satisfy (8b), which is a weaker condition. However, since $\phi_o{}^{-2} = \kappa \xi$, the consistency with (7) implies that the ratio Λ / α is fixed in this case:

$$\frac{1}{\kappa} = \xi \left(-\frac{\Lambda}{\alpha} \xi\right)^{\frac{d-2}{2}}. \tag{9}$$

Note that for odd and four dimensions Λ and α should have opposite signs because the gravitational constant κ is a positive real number.

It is easy to check that in absence of a self-interacting potential, i.e., $\alpha = 0$, a constant scalar field is only possible for a vanishing cosmological constant, $\Lambda = 0$, and vice versa. Thus if $\kappa \xi \phi_o{}^2 \neq 1$, the line element verifies $G_{\mu\nu} = 0$ and the scalar field is an arbitrary constant. In the special case $\kappa \xi \phi_o{}^2 = 1$, the metric tensor must satisfy $R = 0$.

If one consider an additional *conformal* matter field, like the Maxwell field in four dimensions or the non-linear generalization proposed in [63], i.e., a matter field with a traceless energy-momentum tensor $T^c_{\mu\nu}$, (8a) now reads

$$\left(1-\kappa\xi\phi_o{}^2\right)\,\left[G_{\mu\nu}+\Lambda g_{\mu\nu}\right]=\kappa T^c_{\mu\nu}. \tag{10}$$

In the same way as in four dimensions [28, 62], (10) corresponds to the Einstein equation for the additional matter field with an effective gravitational constant $\kappa_{\text{eff}} = \left(1-\kappa\xi\phi_o{}^2\right)^{-1}\kappa$. Thus the case where $\left(1-\kappa\xi\phi_o{}^2\right)<0$, i.e., when κ_{eff} is negative, is weird because it is equivalent to having repulsive gravitational forces.

4 Black Hole Solutions

Now we are interested in searching static black hole spacetimes dressed with non-constant scalar fields. As it was mentioned before, in presence of a non-constant scalar field it is enough consider (3a), which can be written as

$$\begin{aligned}R^{\mu}{}_{\nu} = {}& \frac{\kappa}{1-\kappa\xi\phi^2}\Big[(1-2\xi)\partial^{\mu}\phi\partial_{\nu}\phi-2\xi\phi\nabla^{\mu}\nabla_{\nu}\phi \\ &+\frac{1}{d-1}\left(\alpha\phi^{p}-\frac{\xi}{2}R\phi^2-\frac{1}{2}\partial^{\rho}\phi\partial_{\rho}\phi\right)+\frac{2\Lambda}{\kappa(d-2)}\Big]\,\delta^{\mu}{}_{\nu}.\end{aligned} \tag{11}$$

In order to solve these equations we consider an extension of the general static and spherically symmetric line element in d spacetime dimensions

$$ds^2=-e^{h(r)}f^2(r)dt^2+f^{-2}(r)dr^2+r^2d\sigma^2_{d-2}\quad\text{and}\quad\phi=\phi(r), \tag{12}$$

where $d\sigma^2_{d-2}$ stands for the line element of the $d-2$-dimensional base manifold Σ, which is assumed to be an Einstein space with Ricci tensor $R^m{}_n(\Sigma)=(d-3)\gamma\delta^n_m$, where m and n denote the $d-2$ coordinates of the base manifold. For instance, in $d=4$, this means that Σ is a two-dimensional surface locally isometric to the sphere, flat space, or hyperbolic manifold for $\gamma=1,0,-1$, respectively. This ansatz allows accommodate possible topological black holes solutions. In principle, Σ is assumed to be compact for ensuring compact black hole horizons.

The presence of a Kronecker delta in (11) suggests to subtract the different components of the Ricci tensor. Moreover, the Ricci tensor is diagonal for the ansatz (12) and its base manifold components are all the same, which will be denote by $R^{\bar{m}}{}_{\bar{m}}$ (no summation over $\bar{m}$). Thus, the equations derived from $R^t{}_t-R^r{}_r$, $R^t{}_t-R^{\bar{m}}{}_{\bar{m}}$ and $R^{\bar{m}}{}_{\bar{m}}$ are respectively:

$$-\frac{d-2}{2r}h'=\frac{\kappa}{1-\kappa\xi\phi^2}\left(2\xi\phi\phi''+(2\xi-1)\phi'^2-\xi\phi\phi'h'\right) \tag{13a}$$

$$-\frac{1}{2}\left[f^{2\prime\prime}+\left(\frac{3}{2}h'+\frac{d-4}{r}\right)f^{2\prime}+\left(h''+\frac{h'^2}{2}+\frac{d-3}{r}h'\right)f^2\right]-\frac{d-3}{r^2}(\gamma-f^2)$$
$$=\frac{2\kappa\xi\phi\phi'}{1-\kappa\xi\phi^2}\left(\frac{f^2}{r}-\frac{f^{2\prime}}{2}-\frac{h'f^2}{2}\right) \tag{13b}$$

$$-\frac{f^{2\prime}}{r}-\frac{h'f^2}{2r}+\frac{d-3}{r^2}(\gamma-f^2)$$
$$=\frac{\kappa(d-1)^{-1}}{(1-\kappa\xi\phi^2)}\left(\alpha\phi^{\frac{2d}{d-2}}-\frac{d\Lambda\phi^2}{4(d-1)}+\frac{2(d-1)\Lambda}{\kappa(d-2)}-\frac{f^2}{2}\left(\phi'^2+\frac{(d-2)\phi\phi'}{r}\right)\right) \tag{13c}$$

It is convenient to divide the analysis of this equations in two cases: the first case considers $\Lambda=\alpha=0$, and in the second one Λ and α a priori have no restrictions.

4.1 Case $\Lambda=\alpha=0$

The first solution of these equations, in four dimensions and $\gamma=1$, was obtained by Bocharova, Bronnikov and Melnikov in 1970 and independently by Bekenstein in 1974 [15, 16]. The line element has the form

$$ds^2=-(1-M/r)^2dt^2+(1-M/r)^{-2}dr^2+r^2d\Omega^2,$$

and the scalar field is

$$\phi(r)=\sqrt{\frac{6}{\kappa}}\frac{M}{r-M}.$$

This solution (BBMB) is an extremal black hole, provided $M>0$, and the singularity of ϕ at the horizon $r=M$ should not be considered as a physical pathology since a particle does not feel infinite tidal forces when it is approaching the horizon [64].

One year before his tragic decease, Xanthopoulos in collaboration with Zannias found [70] the general static, spherically symmetric and asymptotically flat solution of (13). They shown that in this family of solutions there is only one with a smooth event horizon and it is the BBMB solution. Few months later Xanthopoulos, now in collaboration with Dialynas [71], using transformations that related the Einstein equations with a minimally coupled scalar field with the case of a conformally coupled scalar field, shown that the previous result also holds for *any spacetime dimension greater than three*. Thus the black hole solution only exists in four dimensions and it is the BBMB solution. Later Klimčík [72] confirmed this result by the direct integration of the field equations including the three-dimensional case. In what follow, we analyze how the presence of a cosmological constant and a self-interacting potential could change such a surprising result.

4.2 Case of Λ and α Not Simultaneously Null

As far as the author knows, the general solution of the involved equations (13) have not been discovered. In order to get a more manageable system, a simple choice is to set $h' = 0$, i.e., to choose the metric function $h(r)$ to be a constant. With this simplification the system (13) reads:

$$0 = 2\xi\phi\phi'' + (2\xi - 1)\phi'^2 \tag{14a}$$

$$-\frac{f^{2\prime\prime}}{2} - \frac{d-4}{2r}f^{2\prime} - \frac{d-3}{r^2}(\gamma - f^2) = \frac{2\kappa\xi\phi\phi'}{1-\kappa\xi\phi^2}\left(\frac{f^2}{r} - \frac{f^{2\prime}}{2}\right) \tag{14b}$$

$$-\frac{f^{2\prime}}{r} + \frac{d-3}{r^2}(\gamma - f^2) = \frac{\kappa(d-1)^{-1}}{(1-\kappa\xi\phi^2)}\left(\alpha\phi^{\frac{2d}{d-2}} - \frac{d\Lambda\phi^2}{4(d-1)} + \frac{2(d-1)\Lambda}{\kappa(d-2)} - \frac{f^2}{2}\left(\phi'^2 + \frac{(d-2)\phi\phi'}{r}\right)\right) \tag{14c}$$

The main simplification comes from (14a) since that it only involves the scalar field and it is integrable giving the simple expression

$$\phi(r) = \left(\frac{A}{r+B}\right)^{\frac{d-2}{2}}, \tag{15}$$

where A and B are integration constants.

In the case $h' = 0$, before dealing with (14b) and (14c), it is very useful to consider (5), which leads

$$R = -f^{2\prime\prime} - \frac{2(d-2)}{r}f^{2\prime} + \frac{(d-2)(d-3)}{r^2}(\gamma - f^2) = \frac{2\Lambda d}{d-2}. \tag{16}$$

This equation can be directly integrated and one obtains

$$f^2(r) = \gamma + cr^2 + \frac{a}{r^{d-3}} + \frac{b}{r^{d-2}}, \tag{17}$$

where a and b are integration constants, and $c = -2\Lambda[(d-1)(d-2)]^{-1}$. Now, replacing (15) and (17) in (14b) and (14c), we obtain

$$bd\left((r+B)^{d-1} - \kappa\xi A^{d-2}(r+B)\right) = (d-2)\kappa\xi A^{d-2}\left(2\gamma r^{d-1} + a(d-1)r^2 + bdr\right) \tag{18a}$$

$$2(2-d)(r+B)^2r^d\kappa\Lambda+8b(d-1)^2\left(A^{2-d}(r+B)^d-(r+B)^2\kappa\xi\right)$$
$$=\kappa\left((-2+d)(2B+r)\left((-2+d)(-1+d)\left(br+ar^2+r^{-1+d}\gamma\right)-2r^{1+d}\Lambda\right)\right)$$
$$+8\kappa A^2(d-1)\alpha r^d \tag{18b}$$

respectively. These equations can be written as vanishing polynomials of the variable r. This implies that each coefficient of these polynomials must be zero, producing relations between the integration constants and the coupling constants appearing in the action. In what follows an analysis of these equations, divided in the cases $d=3$, $d=4$ and $d>4$, will be presented.

4.2.1 Three-Dimensional Case

In the three dimensional case, $\gamma=0$ and (18a) and (18b) reduce to

$$\frac{3}{8}bB(8B-A\kappa)+b\left(6B-\frac{3A\kappa}{4}\right)r+\left(3b-\frac{aA\kappa}{4}\right)r^2=0, \tag{19}$$

$$\frac{1}{4}bB^2(8B-A\kappa)+\frac{3}{4}bB(8B-A\kappa)r+\left(6bB-\frac{3Ab\kappa}{8}-\frac{1}{4}aAB\kappa\right)r^2$$
$$+\left(2b-\frac{1}{8}A\kappa\left(a+B^2\Lambda+8A^2\alpha\right)\right)r^3=0, \tag{20}$$

respectively. These equations generate three different solutions:

1. *Stealth solution without self-interacting potential.* In this case $a=b=B=\alpha=0$, $\Lambda=-l^{-2}$ and A is an arbitrary positive constant. The metric and the scalar field take the form

$$ds^2=-\frac{r^2}{l^2}dt^2+\frac{l^2}{r^2}dr^2+r^2d\theta^2,\quad \phi=\sqrt{\frac{A}{r}}.$$

   This spacetime corresponds to a massless BTZ black hole endowed with a non-trivial scalar field whose energy-momentum tensor vanishes. This kind of matter field, named stealth field can be found also in flat space [65].
2. *Stealth solution with a self-interacting potential.* In this case $a=b=0$, $\Lambda=-l^{-2}$ and $A=\sqrt{\frac{B^2}{8\alpha l^2}}$. As the previous case, we have a massless BTZ black hole with a stealth field

$$\phi=\sqrt{\frac{|B|}{\sqrt{8\alpha l^2}(r+B)}}.$$

3. *Black hole solution dressed with a conformally coupled scalar field.* This class is characterized by $A=8B\kappa^{-1}, a=3B^2(\Lambda+512\alpha\kappa^{-2})$ and $b=2B^3(\Lambda+512\alpha\kappa^{-2})$. When the cosmological constant is negative $\Lambda=-l^{-2}$ and $B\geq 0$

we have the black hole solution defined with the metric function

$$f^2(r) = \frac{r^2}{l^2} - (1 - 512\alpha\kappa^{-2}l^2)\left(\frac{3B^2}{l^2} + \frac{2B^3}{l^2 r}\right),$$

and the scalar field

$$\phi = \sqrt{\frac{8B}{\kappa(r+B)}}.$$

This solution, without self-interacting potential, was found in [26], and it was generalized for the case $\alpha \neq 0$ in [27]. Remarkably the scalar field is regular everywhere. The entropy of this black hole is proportional to area of the event horizon, but the numerical factor is different to $1/4$. The case of $\alpha = 0$ is related with the back reaction of a conformal field on the BTZ black hole [66], and it was proved to be unstable against linear symmetric perturbations [67] like the BBMB solution [68].

4.2.2 Four-Dimensional Case

In four dimensions, (18a) and (18b) read

$$bB\left(4B^2 - \frac{2}{3}A^2\kappa\right) + 2b\left(6B^2 - A^2\kappa\right)r + \left(12bB - aA^2\kappa\right)r^2 + \left(4b - \frac{2}{3}A^2\gamma\kappa\right)r^3 = 0, \tag{21}$$

$$\begin{aligned} &3bB^2\left(6B^2 - A^2\kappa\right) + 12bBr\left(6B^2 - A^2\kappa\right) - 6r^2\left(-18bB^2 + A^2b\kappa + aA^2B\kappa\right) \\ &+ r^3\left(72bB - 3A^2(a + 2B\gamma)\kappa\right) + r^4\left(18b - A^2\kappa\left(6A^2\alpha + 3\gamma + B^2\Lambda\right)\right) = 0, \end{aligned} \tag{22}$$

respectively. The following solutions can be obtained:

1. *Stealth solutions*. This case occurs when $a = b = B = \gamma = 0$. The line element is

$$ds^2 = -\frac{r^2}{l^2}dt^2 + \frac{l^2}{r^2}dr^2 + r^2 d\sigma^2,$$

where $\Lambda = -3l^{-2}$ and $d\sigma^2$ is the line element of a surface locally isometric to the two-dimensional flat space. The scalar field is

$$\phi = \sqrt{\frac{1}{2l^2\alpha}}\frac{|B|}{r+B},$$

with $B \neq 0$ and $\alpha > 0$. In the case of a vanishing self-interacting coupling constant, $\alpha = 0$ the scalar field is $\phi = A/r$, where A is arbitrary. Note that these solutions are similar to those found in three dimensions with a vanishing energy-momentum tensor.

2. *Black hole solution dressed with a conformally coupled scalar field.* This class is defined by $A^2 = 6B^2\kappa^{-1}$, $a = 2\gamma B$, $b = \gamma B^2$, and also it requires a relation between α and Λ
$$\alpha = -\frac{\kappa}{36}\Lambda.$$
This relation is equivalent to the one described by (9) in the case of a constant scalar field.
The lapse function is
$$f^2(r) = -\frac{\Lambda}{3}r^2 + \gamma\left(1+\frac{B}{r}\right)^2,$$
and the scalar field
$$\phi = \sqrt{\frac{6}{\kappa}}\frac{|B|}{r+B}.$$
Then we have two different black holes. When the cosmological constant is positive, $\gamma = 1$ and $B < 0$ the solution corresponds to the de Sitter black hole reported in [28]. In the case of a negative cosmological constant, $\gamma = -1$ and $B > 0$ the solution is a topological black hole [30].

4.2.3 Higher-Dimensional Case

Using the binomial theorem in (18a)

$$\begin{aligned} bd&\left(r^{d-1} + (d-1)r^{d-2}B + \cdots + \frac{1}{2}(d-1)(d-2)B^{d-3}r^2\right. \\ &\left. + (d-1)B^{d-2}r + B^{d-1} - \kappa\xi A^{d-2}(r+B)\right) \\ &= (d-2)\kappa\xi A^{d-2}\left(2\gamma r^{d-1} + a(d-1)r^2 + bdr\right) \end{aligned} \tag{23}$$

For $d > 4$, the only possible solution of the previous equation is $a = b = \gamma = 0$. *Thus, there are no black holes solutions in higher dimensions*. This proves the conjecture proposed in [69]. Replacing these values in (18b) we obtain the single condition

$$A^2\alpha + B^2\xi\Lambda = 0. \tag{24}$$

This condition contains the horizonless cases found in three and four dimensions. Then it is possible to write a simple expression for any spacetime dimension: The metric is defined by
$$ds^2 = -\frac{r^2}{l^2}dt^2 + \frac{l^2}{r^2}dr^2 + r^2 d\sigma^2, \tag{25}$$

where $l^2 = -(d-1)(d-2)/(2\Lambda)$ and $d\sigma^2$ is the line element of $d-2$-dimensional Ricci flat manifold.[2] The stealth scalar field is given by

$$\phi(r) = \begin{cases} \left(\sqrt{\frac{(d-2)^2}{8l^2\alpha}}\frac{|B|}{r+B}\right)^{\frac{d-2}{2}} & \text{for } \alpha > 0, \\ \left(\frac{A}{r}\right)^{\frac{d-2}{2}} & \text{for } \alpha = 0. \end{cases} \tag{26}$$

5 Concluding Remarks

The main result presented here is somewhat negative. Choosing $h' = 0$ in the general static ansatz (12) it was possible to prove that there are no black holes with a conformally coupled scalar field in dimensions greater than four. However, the analysis of the field equations with a simplified version of the ansatz can be seen as a necessary step before dealing the involved general case, which will be our next target. Finally, I still remember the enormous emotion that I felt when the black hole solution [26] emerged from my notes. This happened at the very beautiful CECS' house in Santiago, where young students learnt about the *quest for beauty and simplicity*.

Acknowledgments As it can be seen along this contribution my work in this topic has largely benefited by the collaboration with my colleagues Eloy Ayón-Beato, Gustavo Dotti, Reinaldo Gleiser, Marc Henneaux, Ricardo Troncoso and Jorge Zanelli. This work was partially funded by FONDECYT (Chile) grants 1051056, 1061291, 1071125, 1085322. The Centro de Estudios Científicos (CECS) is funded by the Chilean Government through the Millennium Science Initiative and the Centers of Excellence Base Financing Program of Conicyt. CECS is also supported by a group of private companies which at present includes Antofagasta Minerals, Arauco, Empresas CMPC, Indura, Naviera Ultragas and Telefónica del Sur. CIN is funded by Conicyt and the Gobierno Regional de Los Ríos.

References

1. R. Ruffini and J. Wheeler, Phys. Today **24**(1), 30 (1971).
2. J. D. Bekenstein, Phys. Rev. D **5**, 1239 (1972).
3. C. Teitelboim, Phys. Rev. D **5**, 2941 (1972).
4. D. Sudarsky, Class. Quant. Grav. **12**, 579. (1995).
5. A. Saa, J. Math. Phys. **37**, 2346 (1996).
6. A. E. Mayo and J. D. Bekenstein, Phys. Rev. D **54**, 5059 (1996).
7. A. Saa, Phys. Rev. D **53**, 7377 (1996).

[2] In [69] $d\sigma^2$ is the metric of a $d-2$-dimensional maximally symmetric Einstein space, which is a more strong condition.

8. J. D. Bekenstein, "Black hole hair: Twenty-five years after". 2nd International Sakharov Conference on Physics, Moscow, Russia, 20–23 May, 1996, arXiv:gr-qc/9605059.
9. D. Sudarsky and T. Zannias, Phys. Rev. D **58**, 087502 (1998).
10. K. A. Bronnikov and G. N. Shikin, Grav. Cosmol. **8**, 107 (2002).
11. D. Sudarsky and J. A. González, Phys. Rev. D **67**, 024038 (2003).
12. E. Ayón-Beato, Class. Quant. Grav. **19**, 5465 (2002).
13. U. Nucamendi and M. Salgado, Phys. Rev. D **68**, 044026 (2003).
14. T. Hertog, Phys. Rev. D **74**, 084008 (2006).
15. N. Bocharova, K. Bronnikov and V. Melnikov, Vestn. Mosk. Univ. Fiz. Astron. **6**, 706 (1970).
16. J. D. Bekenstein, Ann. Phys. **82**, 535 (1974).
17. B. L. Hu and L. Parker, Phys. Rev. D **17**, 933 (1978) [Erratum-ibid. D **17**, 3292 (1978)].
18. P. R. Anderson, Phys. Rev. D **32**, 1302 (1985).
19. N. G. Phillips and B. L. Hu, Phys. Rev. D **55**, 6123 (1997) [arXiv:gr-qc/9611012].
20. B. L. Hu and Y. h. Zhang, Phys. Rev. D **37**, 2151 (1988).
21. M. V. Fischetti, J. B. Hartle and B. L. Hu, Phys. Rev. D **20**, 1757 (1979).
22. J. B. Hartle and B. L. Hu, Phys. Rev. D **20**, 1772 (1979).
23. J. B. Hartle and B. L. Hu, Phys. Rev. D **21**, 2756 (1980).
24. D. Noakes, J. Math. Phys. **24**, 1846 (1983).
25. S. Sonego and V. Faraoni, Class. Quant. Grav. **10**, 1185 (1993).
26. C. Martínez and J. Zanelli, Phys. Rev. D **54**, 3830 (1996).
27. M. Henneaux, C. Martínez, R. Troncoso and J. Zanelli, Phys. Rev. D **65**, 104007 (2002).
28. C. Martínez, R. Troncoso and J. Zanelli, Phys. Rev. D **67**, 024008 (2003).
29. C. Martínez, R. Troncoso and J. Zanelli, Phys. Rev. D **70**, 084035 (2004).
30. C. Martínez, J. P. Staforelli and R. Troncoso, Phys. Rev. D **74**, 044028 (2006).
31. C. Martínez and R. Troncoso, Phys. Rev. D **74**, 064007 (2006).
32. T. Torii, K. Maeda and M. Narita, Phys. Rev. **D64**, 044007 (2001).
33. E. Winstanley, Found. Phys. **33**, 111 (2003).
34. T. Hertog and K. Maeda, Phys. Rev. D **71**, 024001 (2005).
35. E. Winstanley, Class. Quant. Grav. **22**, 2233 (2005).
36. E. Radu and E. Winstanley, Phys. Rev. D **72**, 024017 (2005).
37. T. Hertog and K. Maeda, JHEP **0407**, 051 (2004).
38. K. G. Zloshchastiev, Phys. Rev. Lett. **94**, 121101 (2005).
39. M. Henneaux, C. Martínez, R. Troncoso and J. Zanelli, Phys. Rev. D **70**, 044034 (2004).
40. M. Henneaux, C. Martínez, R. Troncoso and J. Zanelli, Ann. Phys. **322**, 824 (2007).
41. A. J. Amsel and D. Marolf, Phys. Rev. D **74**, 064006 (2006) [Erratum-ibid. D **75**, 029901 (2007)].
42. M. Natsuume, T. Okamura and M. Sato, Phys. Rev. D **61**, 104005 (2000).
43. E. Ayón-Beato, A. García, A. Macías and J. M. Pérez-Sánchez, Phys. Lett. B **495**, 164 (2000).
44. G. Barnich, "Conserved charges in gravitational theories: contribution from scalar fields". 3rd International Sakharov Conference on Physics, Moscow, Russia, 24–29 June 2002. arXiv:gr-qc/0211031.
45. J. Gegenberg, C. Martínez and R. Troncoso, Phys. Rev. D **67**, 084007 (2003).
46. E. Ayón-Beato, C. Martínez and J. Zanelli, Gen. Rel. Grav. **38**, 145 (2006).
47. M. Hortacsu, H. T. Ozcelik and B. Yapiskan, Gen. Rel. Grav. **35**, 1209 (2003).
48. M. I. Park, Phys. Lett. B **597**, 237 (2004).
49. M. Bañados and S. Theisen, Phys. Rev. D **72**, 064019 (2005).
50. M. Salgado, Class. Quant. Grav. **20**, 4551 (2003).
51. A. Ashtekar, A. Corichi and D. Sudarsky, Class. Quant. Grav. **20**, 3413 (2003).
52. A. Ashtekar and A. Corichi, Class. Quant. Grav. **20**, 4473 (2003).
53. T. Hertog and S. Hollands, Class. Quant. Grav. **22**, 5323 (2005).
54. P. L. McFadden and N. G. Turok, Phys. Rev. D **71**, 086004 (2005).
55. A. Biswas and S. Mukherji, JCAP **0602**, 002 (2006).
56. I. Papadimitriou, JHEP **0702**, 008 (2007).
57. G. Koutsoumbas, S. Musiri, E. Papantonopoulos and G. Siopsis, JHEP **0610**, 006 (2006).
58. S. de Haro, I. Papadimitriou and A. C. Petkou, Phys. Rev. Lett. **98**, 231601 (2007).

59. A. M. Barlow, D. Doherty and E. Winstanley, Phys. Rev. D **72**, 024008 (2005).
60. T. J. T. Harper, P. A. Thomas, E. Winstanley and P. M. Young, Phys. Rev. D **70**, 064023 (2004).
61. G. Koutsoumbas, E. Papantonopoulos and G. Siopsis, JHEP **0805**, 107 (2008).
62. G. Dotti, R. J. Gleiser and C. Martínez, Phys. Rev. D **77**, 104035 (2008).
63. M. Hassaine and C. Martínez, Phys. Rev. D **75**, 027502 (2007).
64. J. D. Bekenstein, Ann. Phys. **91**, 75 (1975).
65. E. Ayón-Beato, C. Martínez, R. Troncoso and J. Zanelli, Phys. Rev. D **71**, 104037 (2005).
66. C. Martínez and J. Zanelli, Phys. Rev. D **55**, 3642 (1997).
67. C. Martínez, Phys. Rev. D **58**, 027501 (1998).
68. K. A. Bronnikov and Yu. N. Kireyev, Phys. Lett. A **67**, 95 (1978).
69. M. Nadalini, L. Vanzo and S. Zerbini, Phys. Rev. D **77**, 024047 (2008).
70. B. C. Xanthopoulos and T. Zannias, J. Math. Phys. **32**, 1875 (1991).
71. B. C. Xanthopoulos and T. E. Dialynas, J. Math. Phys. **33**, 1463 (1992).
72. C. Klimčík, J. Math. Phys. **34**, 1914 (1993).

Quantum Mechanics on Some Supermanifolds

Luca Mezincescu

Abstract Results of recent investigations on super Landau Models, are presented in a way streamlined as to outline the difficulties connected to the formulation of non relativistic motion on some supermanifolds and the subsequent solution of these difficulties.

1 Introduction

In what follows I will present some results of recent investigations [1–6] on super Landau Models, streamlined as to outline the difficulties connected to the formulation of non relativistic motion on some supermanifolds and the subsequent solution of these difficulties.

I will consider the quantum motion on some coset supermanifolds of the form $\mathcal{G}/\mathcal{H}$ where $\mathcal{G}$ is a supergroup while $\mathcal{H}$ can be just a group. More specifically I will consider 1-dimensional non relativistic sigma models, of Brink Schwarz type, where $\mathcal{G}$ acts in the target space. Such models have been considered only recently and I will explain the heuristic reason for this is. Given the superalgebra **g** corresponding to $\mathcal{G}$, we seek such **g**'s which have an involution which is also an automorphism of this algebra, and which upon quantization goes into the hermitian conjugation $\dagger$ with respect to the naturally defined invariant inner product. It may happen that some odd generators Q and $Q^{\dagger}$ have the following anticommutation relation:

$$\{Q, Q^{\dagger}\} = C. \tag{1}$$

If C is not a positive definite operator the corresponding relation implies that there must be negative norms in the corresponding quantum system, and this seems like the end of the story. It may happen however that the $\mathcal{G}$ invariant Hamiltonian H, has

L. Mezincescu
University of Miami, Coral Gables, FL 33124, USA
e-mail: mezincescu@miami.edu

M. Henneaux, J. Zanelli (eds.), *Quantum Mechanics of Fundamental Systems: The Quest for Beauty and Simplicity*, DOI 10.1007/978-0-387-87499-9_13,

nevertheless real eigenvalues and a complete system of eigenvectors whose norms as mentioned before are not positive definite:

$$H\,|\Phi_A\rangle = \lambda_A\,|\Phi_A\rangle\,, \tag{2}$$

$$\langle\Phi_A|\Phi_B\rangle = (-)^{g(A)}\,\delta_{AB}, \tag{3}$$

where $g(A) = \pm 1$ when the corresponding norm of the vector $|\Phi_A\rangle$ is positive respectively negative. We proceed in the following way: first we define the Hilbert space metric operator G by:

$$G\,|\Phi_A\rangle = (-)^{g(A)}\,|\Phi_A\rangle\,, \qquad G^2 = 1, \tag{4}$$

then we redefine the inner product[1] as:

$$\langle\langle\Phi_A|\Phi_B\rangle\rangle \equiv \langle G\Phi_A|\Phi_B\rangle = \delta_{AB}. \tag{5}$$

We can therefore introduce the $\ddagger$ Hermitian conjugation with respect to the new inner product

$$\langle\langle\Phi_A\,|\mathcal{O}|\,\Phi_B\rangle\rangle = \left\langle\left\langle\mathcal{O}^{\ddagger}\Phi_A|\Phi_B\right\rangle\right\rangle, \tag{6}$$

where

$$\mathcal{O}^{\ddagger} = G\mathcal{O}^{\dagger}G. \tag{7}$$

We end up with a space in which we have two Hermitian conjugations with respect to two inner products, the original inner product which was not positive definite but which was fixed by the requirement of its invariance under the transformations corresponding to the superalgebra g, and the new positive definite inner product just introduced. The fact that we have two operations of Hermitian conjugation, under which the Hamiltonian of the system is Hermitian, has the rather surprising consequence that the number of integrals of motion of this system is enhanced. Indeed consider an integral of motion $\mathcal{O}$, we have

$$[\mathcal{O},H] = 0, \tag{8}$$

taking now the $\dagger$ and the $\ddagger$ of this relation we obtain that

$$\mathcal{O}^{\ddagger} - \mathcal{O}^{\dagger} = G\left[\mathcal{O}^{\dagger},G\right], \tag{9}$$

is also an integral of motion, which is non vanishing if $[\mathcal{O},G] \neq 0$. We will organize these integrals of motion so as to have simple conjugation properties for each of them. We introduce the shift operator:

$$\mathcal{O}_G = G\left[\mathcal{O}^{\dagger},G\right], \qquad \mathcal{O}_G^{\ddagger} = -\mathcal{O}_G^{\dagger}, \tag{10}$$

[1] This step is similar to [7], see also [8].

and the tilde operator:

$$\widetilde{\mathcal{O}} = \mathcal{O} + \frac{1}{2}\mathcal{O}_G^{\dagger}, \qquad \widetilde{\mathcal{O}}^{\ddagger} = \widetilde{\mathcal{O}}^{\dagger}. \tag{11}$$

Therefore the restoration of the positive definite inner product, exhibits additional integrals of motion of the corresponding model. The standard lore is that if you have a quantum model with a certain unitary symmetry, by changing the inner product you are bound to loose this unitary symmetry. We are going to analyze in the next sections, how in specific examples, the additional integrals of motion combine to re-obtain a unitary symmetry of starting model.

2 Fermionic Landau Model

The fermionic Landau model [4, 9] is obtained from the ordinary Landau model by replacing the corresponding complex commuting coordinate with a complex anti-commuting one. We first establish the notation for the bosonic Landau model.

Consider a motion in the complex plane **C** with the Lagrangian,

$$L_b = |\dot{z}|^2 - i\kappa\,(\dot{z}\bar{z} - \dot{\bar{z}}z)\,. \tag{12}$$

This model can be viewed as a non linear realization of the magnetic translations group

$$[P, P^{\dagger}] = 2\kappa, \tag{13}$$

over the corresponding central charge. The kinetic term and the connection term in (12), are obtained from the Lie algebra element:

$$g^{-1}dg = i\omega_P P + i\bar{\omega}_P P^{\dagger} + A_{2\kappa}\kappa, \tag{14}$$

where g belongs to the corresponding coset.

In the quantum version of this model we have:

$$H_b = a^{\dagger}a + \kappa, \quad [a, a^{\dagger}] = 2\kappa, \quad E = 2\kappa\left(n + \frac{1}{2}\right), \tag{15}$$

where

$$a = i\,(\partial_{\bar{z}} + \kappa z)\,, \qquad a^{\dagger} = i\,(\partial_z - \kappa\bar{z})\,. \tag{16}$$

The generators of the magnetic translations are:

$$P = -i\,(\partial_z + \kappa\bar{z})\,, \qquad P^{\dagger} = -i\,(\partial_{\bar{z}} - \kappa z)\,. \tag{17}$$

Replacing now the commuting complex coordinate z, with the anticommuting complex coordinate ζ:

$$L_f = \dot{\zeta}\dot{\bar{\zeta}} - i\kappa\left(\dot{\zeta}\bar{\zeta} + \dot{\bar{\zeta}}\zeta\right), \tag{18}$$

note the change of sign in the terms linear in the time derivatives, which follows from reality condition. This Lagrangian also follows from a coset space construction of the supermagnetic translation group

$$\{\Pi^{\dagger}, \Pi\} = 2\kappa, \tag{19}$$

and because of the term quadratic in time derivatives one expects to have ghosts upon quantization [10]. Indeed this is what happens, a canonical quantization of this model leads to the following result:

$$\hat{H}_f = -\alpha^{\dagger}\alpha - \kappa, \quad \{\alpha, \alpha^{\dagger}\} = -2\kappa, \quad E = \pm\kappa, \tag{20}$$

where

$$\alpha = \left(\partial_{\bar{\zeta}} - \kappa\zeta\right), \qquad \alpha^{\dagger} = \left(\partial_{\zeta} - \kappa\bar{\zeta}\right). \tag{21}$$

It is clear that one will have problems with the positivity of the norm because of the wrong sign in the anticommutator of the odd creation and annihilation operators $\alpha^{\dagger}$ and α, ($\kappa > 0$). Indeed the wave vector for this system is

$$\psi\left(\zeta, \bar{\zeta}\right) = \mathcal{A} + \zeta\mathcal{B} + \bar{\zeta}\mathcal{C} + \bar{\zeta}\zeta\mathcal{D}, \tag{22}$$

and independently of the Grassmann parities of the coefficients in the wave vector, the creation and annihilation operators of this system will be hermitian conjugates to each other under the inner product:

$$\langle \Psi_1, \Psi_2 \rangle = \partial_{\zeta}\partial_{\bar{\zeta}}\left(\Psi_1^{*}\Psi_2\right), \tag{23}$$

under which the Hamiltonian is also hermitian. Under this inner product also the conserved magnetic translations

$$\Pi = \partial_{\zeta} + \kappa\bar{\zeta}, \qquad \Pi^{\dagger} = \partial_{\bar{\zeta}} + \kappa\zeta. \tag{24}$$

are hermitian conjugate to each other. We are therefore in the situation that the Hamiltonian admits a complete system of eigenvectors which even if orthogonal among themselves do not have a positive definite norm. The Hilbert space metric operator is extremely easy to find, as it is just proportional to the Hamiltonian

$$G = -\kappa^{-1}H_f. \tag{25}$$

Because of this the integrals of motion will not change their hermitian conjugates when one converts to the new inner product. The negative norms result from the wrong sign of the anticommutator of the odd creation and annihilation operators, and the change of the norm, results in a change of the hermitian conjugates of these operators. This renders the system into a usual 4-state system:

$$\alpha^{\ddagger} = \alpha^{\dagger}, \quad \{\alpha, \alpha^{\ddagger}\} = 2\kappa, \quad \hat{H}_f = \alpha^{\ddagger}\alpha - \kappa. \tag{26}$$

This is therefore our simplest example where upon quantization of a system, the naive norm has to be modified in order to obtain a sound quantum mechanical system. In the following two sections we will present further examples, in increasing order of difficulty, where such situations arise.

3 Superplane Landau Model

This model [4, 9] is obtained by taking

$$L = L_b + L_f, \tag{27}$$

where L_b is given by (12), and L_f is given by (18). It represents a motion in $\mathbf{C}^{(1|1)}$ with the coordinates (z, ζ). This model can be viewed as the coset space construction of $ISU(1|1)/[SU(1|1) \times Z]$. $SU(1|1)$ has the odd generators

$$Q = z\partial_\zeta - \bar{\zeta}\partial_{\bar{z}}, \qquad Q^{\dagger} = \bar{z}\partial_{\bar{\zeta}} + \zeta\partial_z, \tag{28}$$

whose anticommutators gives the central charge

$$\{Q, Q^{\dagger}\} = C = z\partial_z + \zeta\partial_\zeta - \bar{z}\partial_{\bar{z}} - \bar{\zeta}\partial_{\bar{\zeta}}. \tag{29}$$

The supergroup $ISU(1|1)$ is obtained by adjoining the magnetic translations P and $P^{\dagger}$ given by (17), and supermagnetic translations Π and $\Pi^{\dagger}$ given by (24). The nonvanishing brackets are

$$[P, P^{\dagger}] = 2\kappa, \quad \{\Pi^{\dagger}, \Pi\} = 2\kappa, \quad [Q, P] = i\Pi, \tag{30}$$

$$\{Q^{\dagger}, \Pi\} = iP, \quad [C, P] = -P, \quad [C, \Pi] = -\Pi, \tag{31}$$

together with the corresponding hermitian conjugate relations.

The quantization of this model leads to the quantum Hamiltonian

$$H = a^{\dagger}a - \alpha^{\dagger}\alpha, \tag{32}$$

acting on the space spanned by the wave functions $\Psi(z, \bar{z}, \zeta, \bar{\zeta})$. The involutions in the algebra are realized as Hermitian conjugates within the inner product

$$\langle \Psi_1, \Psi_2 \rangle = \int dz d\bar{z} \partial_\zeta \partial_{\bar{\zeta}} (\Psi_1^* \Psi_2), \tag{33}$$

and all the eigenstates are constructed in the usual way with the help of a, $a^\dagger$, α, $\alpha^\dagger$. As in the Fermionic Landau model case we will have negative norms and because of the direct product structure of this model the metric operator will be the same $G = -\kappa^{-1}H_f$. This operator commutes with all the generators of $ISU(1|1)$, less the odd generators Q and $Q^\dagger$. Therefore according to our established strategy we will compute the resulting integrals of motion according to (10, 11). We obtain

$$Q_G = -\frac{\iota}{\kappa}S, \quad S = a^\dagger\alpha, \quad \widetilde{Q} = -\frac{\iota}{2\kappa}P^\dagger\Pi, \tag{34}$$

together with the corresponding Hermitian conjugate operators. The upshot is that the unitary symmetry of the model is the following: The charges

$$P, \Pi, \widetilde{Q}; P^\dagger, \Pi^\dagger, \widetilde{Q}^\dagger; \widetilde{C} = C + \frac{1}{2\kappa}H, \tag{35}$$

form a $ISU(1|1)$ with commutation relations like (29–31), with the obvious correspondence of the notation. The surprising feature is that the remaining operators, whose graded brackets with the novel $ISU(1|1)$ vanish, form a supersymmetry algebra

$$\left\{S, S^\ddagger\right\} = H, \tag{36}$$

which is quite puzzling as the Hamiltonian is related to time translations. Now it is well known that the Landau electron is related to the planar limit of the particle on the sphere in the field of a monopole at the origin. For the particle on the sphere in the field of the monopole, Dirac's quantization condition applies so that the electric charge is an integer multiple $2N$ of a minimal allowed charge. The planar Landau model is then found by taking the limit in which the radius of the sphere $R \to \infty$, $N \to \infty$ with $\frac{N}{R^2}$ fixed. In the next section we are going present an example of a motion on a supermanifold which has the sphere as its body, and the subsequent redefinition of the norm and rearrangement of its symmetry.

4 Super-Flag

We will be seeking to find a minimal superalgebra which contains the $SU(2)$ algebra, and which has an involution which upon quantization goes into the Hermitian conjugate operation. The first candidate will be $Osp(1|2)$, but this algebra does not have the involution we seek. It has an involution which is a pseudoreality condition for the spinor generators, but this involution is not compatible with a standard Berezin like involutions, which is what we seek. This algebra has been considered in [11], and these authors do not mention any clash between the positive definiteness and invariance of the norm. The next algebra will be that of $SU(2|1)$ for which it is possible to find a standard involution. The Lie superalgebra $su(2|1)$ is spanned by even charges $(F, J_3, J_\pm)$, satisfying the commutation relations of $U(2)$, and a $U(2)$

doublet of odd charges (Π, Q); we write the complex conjugate charges as $(\Pi^{\dagger}, Q^{\dagger})$ since we want to exhibit the mentioned involution. The non-zero commutators of the even charges are

$$[J_+, J_-] = 2J_3, \qquad [J_3, J_\pm] = \pm J_\pm. \tag{37}$$

The non-zero commutators of the odd generators with the even generators are

$$\begin{aligned} [J_+, \Pi] &= iQ, \qquad [J_-, Q] = -i\Pi, \\ [J_3, \Pi] &= -\frac{1}{2}\Pi, \qquad [J_3, Q] = \frac{1}{2}Q, \\ [F, \Pi] &= -\frac{1}{2}\Pi, \qquad [F, Q] = -\frac{1}{2}Q, \end{aligned} \tag{38}$$

and their Hermitian conjugates. Finally, the non-zero anti-commutators of the odd charges are

$$\begin{aligned} \{\Pi, \Pi^{\dagger}\} &= -J_3 + F, \qquad \{Q, Q^{\dagger}\} = J_3 + F, \\ \{\Pi, Q^{\dagger}\} &= iJ_-, \qquad \{\Pi^{\dagger}, Q\} = -iJ_+. \end{aligned} \tag{39}$$

The relevant cosets are $SU(2|1)/U(1|1) \equiv CP^{(1|1)}$ (the bosonic generator in $U(1|1)$ is $J_3 + F$), this is what we call the SuperSphere [2,6], and $SU(2|1)/[U(1) \times U(1)]$, which we call the Super-Flag [3,6]. We have considered both these cases but I will concentrate on the Super-Flag in what follows, because here the results are sharper. For Super-Flag the formula analogous to (14), tells us that we have three Cartan forms and two $U(1)$ connection terms. We will use only one Cartan form, and in an appropriate parametrization of the coset space through coordinates $Z^M = z, \xi^i$, $\bar{Z}_M = \bar{z}\ \bar{\xi}_i, i = 1,2$ we will have

$$E^+ = K_1^{-\frac{1}{2}} K_2^{-1} \left[dz - K_1^{-1} \left(d\xi^1 - z d\xi^2\right)\left(\bar{\xi}_2 + z\bar{\xi}_1\right)\right], \tag{40}$$

where:

$$K_1 = 1 + \bar{\xi}_1\xi^1 + \bar{\xi}_2\xi^2, \quad K_2 = 1 + \bar{z}z + \left(\xi^1 - z\xi^2\right)\left(\bar{\xi}_1 - \bar{z}\bar{\xi}_2\right). \tag{41}$$

Our charge will move in the *magnetic* fields corresponding to the two $U(1)$, Kahler connections:

$$\mathcal{A} = -idZ^M \partial_M \ln K_2 + c.c., \qquad \mathcal{B} = idZ^M \partial_M \ln K_1 + c.c.. \tag{42}$$

By the standard rules of non linear realizations the Lagrangian:

$$L = |\omega^-|^2 + NA + MB, \tag{43}$$

where ω^+, A, B, are the pullbacks of the corresponding forms, will be invariant under the isometry transformations

$$\begin{aligned}\delta z &= a+\bar{a}z^2-(\bar{\varepsilon}_2+z\bar{\varepsilon}_1)\left(\xi^1-z\xi^2\right),\\ \delta\xi^1 &= a\xi^2+\varepsilon^1+(\bar{\varepsilon}\cdot\xi)\xi^1,\\ \delta\xi^2 &= -\bar{a}\xi^1+\varepsilon^2+(\bar{\varepsilon}\cdot\xi)\xi^2.\end{aligned} \tag{44}$$

As it can be easily seen the kinetic terms for the odd variables will come from the connection terms and will be first order in time. This model has been quantized using the so called Gupta–Bleuer method, and exhibits the interesting feature that the constraints become first class for a given energy [12], however we can stay away from this, by restricting the parameters N, M so that $N>0, -N-\frac{1}{2}<M<0$. The quantum Hamiltonian for this system is:

$$\hat{H}=H_N=-K_2^2K_1\nabla_z^{(N)}\nabla_{\bar{z}}^{(N)}+N, \tag{45}$$

where N is the strength of the charge monopole interaction, and it is quantized, that is $2N$ is a positive integer, and

$$\nabla_z^{(N)}=\partial_z-N\partial_z\ln K_2, \qquad \nabla_{\bar{z}}^{(N)}=\partial_{\bar{z}}+N\partial_{\bar{z}}\ln K_2. \tag{46}$$

The physical wave superfields are defined as solutions to the equations:

$$\hat{\varphi}^i|\Psi\rangle=0 \qquad (i=1,2), \tag{47}$$

where $\hat{\varphi}^i$ are some odd differential operators, the solutions of which are

$$\Psi_{(N,M)}=K_1^MK_2^{-N}\Phi\left(z,\bar{z}_{sh},\xi^1,\xi^2\right) \tag{48}$$

where $\bar{z}_{sh}$ is the "shifted" coordinate defined by

$$\bar{z}_{sh}=\bar{z}-(\xi^2+\bar{z}\xi^1)(\bar{\xi}_1-\bar{z}\bar{\xi}_2). \tag{49}$$

The Hamiltonian (45), is diagonalized by the set of physical eigenvectors:

$$\Psi_{(N,M)}{}^{(\ell)}=K_2{}^{-N}K_1{}^{M}\nabla_z{}^{2(N+1)}\ldots.\nabla_z{}^{2(N+\ell)}\Phi^{\left(N+\ell,M-\frac{\ell}{2}\right)}\left(z,\xi^i\right), \tag{50}$$

with

$$\nabla_z^{2N}=\partial_z-\frac{2\,(N)\,\bar{z}_{sh}}{1+z\bar{z}_{sh}}, \tag{51}$$

$$H\Psi_{(N,M)}{}^{(\ell)}=[(2\ell+1)N+\ell(\ell+1)]\,\Psi_{(N,M)}{}^{(\ell)}, \qquad \ell=0,1,2,\ldots \tag{52}$$

One can see that the higher level eigenfunctions are determined in terms of analytic superfields,

$$\Phi^{\left(N+\ell,\, M-\frac{\ell}{2}\right)}\left(z,\xi^i\right) = A^{\left(N+\ell,\, M-\frac{\ell}{2}\right)} + \xi^i \psi_i^{\left(N+\ell,\, M-\frac{\ell}{2}\right)} + \xi^1\xi^2 F^{\left(N+\ell,\, M-\frac{\ell}{2}\right)}, \quad (53)$$

like the ones which determine the ground state but with different external charges, with the help of a corresponding sequence of covariant derivatives. It is convenient to express the invariant norm in terms of these functions, and it is also convenient to find the images of physical operators of interest on these analytic functions. The upshot is very simple: to get the level ℓ norm, operator image (as we are interested in the action of conserved charges their action does not change the level), replace in the ground state norm, or operator image, the numbers N, M by $N+\ell, M-\frac{\ell}{2}$. This is a remarkable simplification as it reduces the study of the higher levels norms, or operators to those of the ground state with corresponding external charges. The $SU(2|1)$ invariant norm of the ground state of external charges N, M is

$$\begin{aligned} ||\Psi_{(N,\, M)}{}^{(0)}||^2 = 2\int \frac{dz d\bar{z}}{(1+z\bar{z})^{2(N++1)}} \Big\{ & M\,(2M+2N+1)\bar{A}A + \frac{1}{2}\bar{F}F \\ & + M\left(\bar{\psi}^1\psi_1 + \bar{\psi}^2\psi_2\right) + \frac{N+1}{1+z\bar{z}}\left(\bar{\psi}^2 + \bar{z}\bar{\psi}^1\right)\left(\psi_2 + z\psi_1\right) \Big\}. \end{aligned} \quad (54)$$

All the functions appearing in the norm are polynomials in z whose maximum degree is determined from the convergence of the corresponding integral. This norm can be diagonalized, however important clues about its nature come from considering the images of the $SU(2)$ generators

$$iJ_+ = J_{\bar{z}} = -2Nz + z^2\partial_z - \xi^1 \frac{\partial}{\partial \xi^2}, \quad (55)$$

$$iJ_- = J_z = \partial_z + \xi^2 \frac{\partial}{\partial \xi^1}, \quad (56)$$

$$J_3 = z\partial_z - N + \frac{1}{2}\left(\xi^1 \frac{\partial}{\partial \xi^1} - \xi^2 \frac{\partial}{\partial \xi^2}\right), \quad (57)$$

acting on the analytic superfields. Indeed solving the equation:

$$J_- \Phi^{(N,M)}\left(z,\xi^i\right) = 0, \quad (58)$$

one obtains that the $SU(2)$ representation content in the analytic superfield $\Phi^{(N,\, M)}$, is four representations with angular momentum, N, $N \pm \frac{1}{2}, N$. The corresponding norms, (for the range mentioned before: $N > 0$, $-N - \frac{1}{2} < M < 0$), are negative for the states with angular momentum N, $N - \frac{1}{2}$, while the states with angular momentum $N + \frac{1}{2}$, N, have positive norm. The standard angular momentum procedure of acting with rising operator on the lowest weight states establishes that all the states in a given representation of $SU(2)$ contribute the same sign to the norm. The Hilbert space metric operator is

$$G = \frac{1}{N+\frac{1}{2}}\left[J^2 + (B-2M)^2 - 2\left(N+\frac{1}{2}\right)^2\right], \tag{59}$$

and it commutes with the even charges of $SU(2|1)$ $\boldsymbol{J}$, and B. It does not commute with the odd charges $\Pi, Q, \Pi^\dagger, Q^\dagger$. Therefore according to our procedure outlined before we expect a doubling of the odd generators of the original $SU(2|1)$ algebra that is we expect the new generators $\Pi_G, \widetilde{\Pi}, Q_G, \widetilde{Q}$, together with the corresponding conjugates. We should stress that all these new generators are constructed on the enveloping algebra of the original algebra, therefore they are automatically conserved quantities. Next we will outline the closure of these new generators for this particular model. As a sample relation we take the anticommutators of Π_G with $\Pi_G^\ddagger$

$$\{\Pi_G, \Pi_G^\ddagger\} = \frac{-8M}{2N+1}\left(J_3 + F_G\right), \quad F_G = 2M + 2N + 1 - F, \tag{60}$$

this formula is to be compared with the original commutation relations of the odd generators

$$\left\{\Pi, \Pi^\dagger\right\} = -J_3 + F, \qquad \left\{Q, Q^\dagger\right\} = J_3 + F, \tag{61}$$

which suggests to rescale the generators like

$$Q_G' = \sqrt{\frac{2N+1}{-8M}}\,\Pi_G^\ddagger. \tag{62}$$

The numerical factors appearing in these rescalings can be expressed in terms of the Casimir operators of the $SU(2|1)$ algebra. For a given irreducible representation of $SU(2|1)$acting on $\Phi^{(N,\,M)}\left(z, \xi^i\right)$ the action of the Casimir operators is

$$C_2 = -2M\left(2M + 2N + 1\right), \quad C_3 = \left(4M + 2N + 1\right)C_2. \tag{63}$$

and then from these two relations one can express the numbers N, M in terms of the corresponding Casimirs. Now when one considers higher levels, the rule is to replace the charges N, M by $N + \ell, M - \frac{\ell}{2}$ and ℓ gets reabsorbed into the corresponding redefinition of the image of the Casimir for that level, rendering a level free definition of the generators, albeit a complicated one. The upshot is that the Π_G', Q_G' and their conjugates generate a new $SU(2|1)$ algebra, which is the the analogue of the supersymmetry algebra obtained in the Planar Landau model considered in a previous section. There remain the other operators $\widetilde{\Pi}\cdots$ a sample of which is

$$\widetilde{\Pi} = \Pi + \frac{1}{2}\Pi_G = \Pi + \sqrt{\frac{-2M}{2N+1}}Q_G'^{\ddagger}, \tag{64}$$

whose anticommutator $\left\{\widetilde{\Pi}, \Pi_G'^{\ddagger}\right\}$, leads to a new even operator. With additional rescalings this operator together with its Hermitian conjugate generate a new $SU(2)$ algebra. The final result is that the closure of the algebra is an $SU(2|2)$ algebra with a central charge $N + \ell + \frac{1}{2}$, which can also be expressed in terms of the $SU(2|1)$ Casimirs.

5 Conclusions

We have therefore shown that the change of the norm resurrects some Quantum Mechanical models on some supermanifolds. Contrary to the usual belief that the change of the norm leads to the loss of the unitary symmetries, in the examples considered, this change of norm lead to a rearrangement of the symmetries to which the enveloping algebra of the original algebra played a fundamental role. This may be a purely superalgebra phenomena as it appears that when one ends up with an enhanced symmetry, this is due to the fact that, that the $SU(2|1)$ representations appearing in the spectra of the Super-Flag can be reinterpreted as short representations of $SU(2|2)$, which can be unitarily represented. The importance of the short representations in supersymmetric models with extended supersymmetry suggests that such constructions may be relevant. It is a challenging goal to try to give a strictly geometrical construction of these type of models bypassing the dynamical aspect of the redefinition of the norm.

Acknowledgments I wish to thank Andrey Beylin, Thomas L. Curtright, Evgeny Ivanov, the late Anatoly Pasnev, and Paul K. Townsend, my coauthors in the results outlined here. I wish to thank Marc Henneaux and Jorge Zanelli for organizing this exciting event. I acknowledge partial support from the National Science Foundation under grant No. 555603.

References

1. E. Ivanov, L. Mezincescu, A. Pashnev and P. K. Townsend, Odd coset quantum mechanics, Phys. Lett. B **566** (2003) 175 [arXiv:hep-th/0301241].
2. E. Ivanov, L. Mezincescu and P. K. Townsend, Fuzzy $CP(n|m)$ as a quantum superspace, in *Symmetries in Gravity and Field Theory*, eds. V. Aldaya, J.M. Cerveró and P. García, Ediciones Universidad de Salamanca, Salamanca, Spain, 2004, pp. 385–408; arXiv:hep-th/0311159.
3. E. Ivanov, L. Mezincescu and P. K. Townsend, A super-flag Landau model, in *From Fields to Strings: Circumnavigating Theoretical Physics*, eds. M. Shifman, A. Vainshtein and J. Wheater, World Scientific, Singapore, 2004; arXiv:hep-th/0404108.
4. E. Ivanov, L. Mezincescu and P. K. Townsend, Planar super-Landau models, JHEP **0601** (2006) 143 [arXiv:hep-th/0510019].

5. T. Curtright, E. Ivanov, L. Mezincescu and P. K. Townsend, Planar super-Landau models revisited, JHEP **0704** (2007) 020 [arXiv:hep-th/0612300].
6. A. Beylin, T. L. Curtright, E. Ivanov, L. Mezincescu and P. K. Townsend, Unitary spherical super-Landau models, arXiv:0806.4716 [hep-th].
7. C. M. Bender, Introduction to PT-symmetric quantum theory, Contemp. Phys. **46** (2005) 277 [arXiv:quant-ph/0501052].
8. T. Curtright and L. Mezincescu, Biorthogonal quantum systems, J. Math. Phys. **48** (2007) 092106 [arXiv:quant-ph/0507015].
9. K. Hasebe, Quantum Hall liquid on a noncommutative superplane, Phys. Rev. D **72** (2005) 105017 [arXiv:hep-th/0503162].
10. D. V. Volkov and A. I. Pashnev, Supersymmetric Lagrangian for particles in proper time, Theor. Math. Phys. **44** (1980) 770 [Teor. Mat. Fiz. **44** (1980) 321].
11. K. Hasebe and Y. Kimura, Fuzzy supersphere and supermonopole, Nucl. Phys. B **709** (2005) 94 [arXiv:hep-th/0409230].
12. M. Banados, L. J. Garay and M. Henneaux, The dynamical structure of higher dimensional Chern-Simons theory, Nucl. Phys. B **476** (1996) 611 [arXiv:hep-th/9605159].

John Wheeler's Quest for Beauty and Simplicity

Charles W. Misner

Abstract Some recollections and comments are given about John A. Wheeler as a mentor, as a teaching researcher, and as a driving force in the revitalization of general relativity in the 1950s and 1960s. His relationship to Hugh Everett and the "Many worlds" interpretation of quantum mechanics is also discussed.

It's a great pleasure to be here on this occasion, Claudio's 60th birthday Fest. I'm going to talk about John Wheeler. Both Claudio and I were Wheeler's Ph.D. students. That means we could call Wheeler our doktorfather which makes us brothers in science.

This talk is essentially historical. It describes the atmosphere in which Claudio and I were trying to advance the state of gravitational physics a long time ago when it was not at the forefront of physics the way it has been in recent times. At the times I started with Wheeler in the 1950s and even into the later 1960s when Claudio came there, gravitation physics was considered a very suspect and distant piece of physics that was not relevant to anything that was going on in the real physicists' world. John Wheeler began the movement of gravitation physics into the forefront of physics in the United States, and contributed strongly to this transition in the rest of the world. My talk here today is mainly concerned with this era in Wheeler's career.

But this is just one piece of John Wheeler. He actually talks about going through three different periods of his life spending about 20 years on each of them. I won't concentrate on his earlier work where he took the viewpoint that particles are fundamental and would try to explain everything in terms of particles. This lasted through about 1955 and includes all his work on nuclear physics and some field theory. During that period he also did important work on nuclear fission and contributed to the development of atomic bombs during World War II and to H-bombs at the beginning of the Cold War. He was involved in that for a couple years just before I arrived at Princeton in 1953. I will talk mostly about his years at Princeton before he went

C.W. Misner
University of Maryland, College Park, MD, USA
e-mail: misner@umd.edu

M. Henneaux, J. Zanelli (eds.), *Quantum Mechanics of Fundamental Systems: The Quest for Beauty and Simplicity*, DOI 10.1007/978-0-387-87499-9_14,

to Texas. This period he describes as one where he was trying to explain particles (and everything) in terms of fields. Although his approach was initially out of the mainstream, it led to great developments in general relativity and its astrophysical applications. When he moved to Texas after retiring from Princeton in 1976 his interests went more to the fundamentals of quantum mechanics and he had many important students there. His motto for that period was "It from Bit" and he has referred to that period as "everything is information." He retired again in 1986 and returned to a quieter life in the neighborhood of Princeton where he's been living since. [A few months after this Claudio Fest meeting, John Archibald Wheeler died on April 13, 2008 at age 96.]

1 Wheeler and Teaching

John Wheeler was very interested in teaching and devoted himself to teaching all his life. As one aspect important for physics, he thought that the best way to learn anything is to teach it, and that's in the way he got into the general relativity. When he decided that it was safe to let Ph.D. students work on relativity, he asked to teach it as a graduate course and taught it the first time a year before I arrived in Princeton. When teaching, John focuses on inspiration before he focuses on content. Every so often he would teach elementary courses (perhaps the honors course for freshman physics majors or perhaps a more general course). In any course his first lecture would talk about something he was very enthusiastic about. This was usually a research project he or his students were working on currently. He would give an impression of the questions that were in the forefront and then explain how they were attacking the problem. Then he would then slowly morph that into whatever he needed to teach after the first lecture and begin to get down to brass tacks. In an elementary course this might be velocity, Newton's laws, vectors or whatever. This way he would show the enthusiasm he had from current research. Another favorite theme he pressed on students was that, more important than solving a problem, was getting the right question. He also felt students were essential to his research, as their questions would suggest new ways of approaching a problem.

Many insights into John's interactions with students can be found in a privately distributed volume of letters and career comments prepared at the instigation of Peter Putnam (one of John's lesser known students) on the occasion of John's 65th birthday. At a talk similar to this given at the opening of the "John Archibald Wheeler International School of Relativistic Astrophysics" in Erice, Sicily in June 2006 I presented many selections from that book, so these are not repeated in this report of my talk at the Claudio Fest. They can be found in the Proceeding of that School which are edited by Matzner and Ciufolini and published by World Scientific in 2008. But I am happy to repeat a quotation from Claudio's contribution to that book. Claudio found that John worked tremendously in trying to help him improve his presentations of all the work he was doing but always with a sense of humor. One advice to Claudio was that "the central idea should always stand out very clearly, sharply, just as in 'Cuba si, Yankee no'."

2 John A Wheeler and the Recertification of GR as True Physics

This section of my ClaudioFest talk can also be found in the 2006 Erice report (Matzner and Ciufolini, World Scientific 2008), where it is explained that one of John's themes was daring conservativism. That meant exploring the equations of well established theories while looking for predictions for possibilities others dared not consider. In this line Wheeler treated wormholes in spacetime, bundles of radiation held together by gravitation to form configuration with mass but containing no massive particles, etc. These explorations and other developments led Wheeler with many students, including Claudio, to reinvigorating general relativity in the United States and preparing for its still continuing resurgence under the impact of new astrophysical data.

3 John A Wheeler and Hugh Everett III

In view of the emphasis of the ClaudioFest as distinct from the Erice 2006 talk, I here present another example of daring conservativism which was recognized as important by Wheeler, but which in this case left him conflicted as he hoped it need not be so daring. This was the dissertation of my roommate Hugh Everett III which Wheeler edited and supported for Hugh's Ph.D., but from which he eventually withdrew full support.

Although Hugh was making great progress in game theory which he had already studied as an undergraduate, he was provoked by informal discussions in Princeton with Niels Bohr's assistant Aage Petersen to think about the interpretation of quantum mechanics. This he approached in a combination of Wheelerish daring conservativism and his own inclination to enjoy arguments where he thought he could be more logical than his counterparty in the discussion. Knowing of my decision to work with Wheeler who was optimistic that a thesis could be short and quick, Hugh also decided to get a Ph.D. with Wheeler. But anyway, he liked to argue and when he heard Bohr's assistant Aage Petersen describing the Bohr interpretation he just thought that sounded like nonsense. So, wanting to prove his point, he began working up his ideas and, as Wheeler would say, just following the equations. Let them tell us something instead of us assuming that they will verify what we know, or give us a numerical correction. And so Hugh worked out this interpretation of quantum mechanics which says that the Schrödinger equation continues to work all the time. It never stops. There is no collapse of the wave function. The wave function that includes both the observer and the equipment just goes its own way according to the Schrödinger equation. This was attractive for Wheeler, because it was daring conservativism that says "push ahead with the equations." And Everett could be a very very logical. But Bohr was Wheeler's principal mentor and a man he highly respected and emulated in many ways. John got Everett his Ph.D. – I think Wheeler forced him to tone it down a bit in the writing before he presented to the committee – but he got him a degree. But then, a year or two later, he sent Hugh off to try to argue it all out with Bohr so the questions would all get straightened out. Well that was unlikely to happen and didn't. Everett went to Copenhagen and spent several weeks there. Even before Hugh finished his Ph.D. he had gone to the Pentagon and got a good job doing game theory kind of things. Hugh was also very clever and accomplished there, and so by the time he came to Copenhagen he was making such a good salary that he could stay at the Hotel d'Angleterre, which is the most expensive and fancy hotel in Copenhagen. I think the Bohr Institute had never had a visitor stay there before. But anyway he had two or three afternoons trying to talk to Bohr with the expected result that they couldn't hear each other.

There is more to learn about Hugh Everett this year, as there is activity spurred on by the fact that 2007 was 50 years since Hugh's dissertation was published. In addition to conferences on that occasion there is a BBC television hour. That program is about his son trying to discover his father and what his father had done. It was, from the son's point of view, such a dysfunctional family that he never knew his father was a scientist. The son is a rock star and I think that's the reason that the BBC could do the program. Having a rock star to follow around as he researches his father makes the show much more attractive to general audiences.

Bibliography

- J. A. Wheeler with K. Ford, "Geons, Black Holes and Quantum Foam" (Norton, New York, 1998)
- K. S. Thorne, "Black Holes and Time Warps", (Norton, New York, 1994)
- Anon., ed., "Family Gathering", 2 vols. (privately distributed 1977: AIP, Princeton, TX)
- Wheeler obituary:
  http://www.nytimes.com/2008/04/14/science/14wheeler.html
- R. Matzner and I. Ciufolini, editors, "The John Archibald Wheeler International School of Relativistic Astrophysics: Erice 2006" (World Scientific, Singapore, 2008)
- M. Bartusiak, "Einstein's Unfinished Symphony" (J Henry Press, Washington, DC, 2000)
- D. Kennefick, "Quest for Gravitational Waves" (Princeton University Press, Princeton, NJ, 2007)
- Misner – Beckedorff on black holes:
  http://hdl.handle.net/1903/4280
- Everett biography:
  http://space.mit.edu/home/tegmark/everett/
  http://www.broadcastnow.co.uk/printPage.html?pageid=217829
- Everett conferences:
  http://users.ox.ac.uk/ everett
  http://www.perimeterinstitute.ca/manyworlds/

Magnetic Monopoles in Electromagnetism and Gravity

Rubén Portugueś

Abstract Magnetic monopoles have rivaled black holes as the most beautiful and elusive entities in the world of theoretical physics since they were first considered in the early twentieth century. They naturally arise in electromagnetism and have immediate implications on the underlying symmetries, both dynamical and internal, at both the classical and quantum level. In this article we will recap some facts about magnetic monopoles in electromagnetism, in particular notions regarding their contribution to the angular momentum of a system and how these concepts are related to the quantization condition for the product of the charges of a fundamental electric pole and magnetic pole. We then proceed to consider magnetic poles in general relativity and briefly address similar considerations in this theory.

1 Introduction

In many ways Dirac can be considered the father, or at least a very influential parent, of modern theoretical physics: the importance of his contributions is unquestionable. He sought to bestow upon theoretical physics a sense of beauty and simplicity which he believed to be realized through the power of mathematics and its unequivocal elegance and unity. In his paper of 1931 [1] he writes:

> "*The most powerful method of advance that can be suggested at present is to employ all the resources of pure mathematics in attempts to perfect and generalize the mathematical formalism that forms the existing basis of theoretical physics, and* after *each success in this direction, to try to interpret the new mathematical features in terms of physical entities.*"

He was extremely successful himself following this philosophy with the prediction in 1930 of the existence of positrons which earned him the Nobel Prize in

R. Portugueś
Department of Molecular and Cellular Biology, Harvard University, BioLabs 2073, 16 Divinity Avenue, Cambridge, MA 02138, USA
e-mail: rportugues@mcb.harvard.edu

M. Henneaux, J. Zanelli (eds.), *Quantum Mechanics of Fundamental Systems: The Quest for Beauty and Simplicity*, DOI 10.1007/978-0-387-87499-9_15,

1933. In the 1931 paper Dirac predicted yet another physical entity: the magnetic pole. Magnetic poles have not yet been observed in nature, yet their beauty and importance in modern theoretical physics are only rivaled by those of another elusive entity, the black hole. This may not be a coincidence. Consider the following wild argument. The quantization condition for the product of a fundamental electric charge q and a magnetic charge g, which will be the main subject of this article,

$$\frac{qg}{2\pi\hbar} \in \mathbb{Z}, \tag{1}$$

can be used to infer that the fundamental magnetic charge may actually be rather large. If we further assume some notion of supersymmetry and the fact that a magnetic pole may satisfy a BPS-like bound, it is not completely out of the question that objects carrying magnetic charge may be very massive and undergo gravitational collapse. Hopefully the arguments outlined in this brief paper will add weight to the belief that the physics of magnetic monopoles is beautiful and important and may help shed light upon our understanding of quantum effects in gravity theories.

2 Magnetic Poles in Classical Electromagnetism

With the conventions we will use throughout, Maxwell's equations are

$$\begin{aligned} \nabla \cdot \mathbf{E} &= \rho, \\ \nabla \cdot \mathbf{B} &= 0, \\ \nabla \times \mathbf{E} &= -\frac{\partial \mathbf{B}}{\partial t}, \\ \nabla \times \mathbf{B} &= \mathbf{J} + \frac{\partial \mathbf{E}}{\partial t}. \end{aligned} \tag{2}$$

These give the electric field $\mathbf{E}$ and magnetic field $\mathbf{B}$ in terms of the sources, namely the charge density ρ and the current density $\mathbf{J}$. The second equation allows us to write the magnetic field in terms of a vector potential $\mathbf{A}$

$$\mathbf{B} = \nabla \times \mathbf{A}, \tag{3}$$

and then the third equation allow us to express the electric field in terms of $\mathbf{A}$ and a scalar potential Φ:

$$\mathbf{E} = -\nabla\Phi - \dot{\mathbf{A}}. \tag{4}$$

This potential $\mathbf{A}$ is often said to be the electromagnetic potential "in the electric picture" meaning that its definition allows only for the existence of electric sources.

The motion for a particle of charge q in the background of these fields is described by the Lorentz force law:

$$\mathbf{F} = q\left(\mathbf{E} + \dot{\mathbf{r}} \times \mathbf{B}\right). \tag{5}$$

In what follows we will often use the solutions to Maxwell's equations corresponding to a point electric particle with charge q placed at a point with position vector $\mathbf{a}$

$$\mathbf{E}(\mathbf{r}) = \frac{q\,(\mathbf{r}-\mathbf{a})}{4\pi|r-a|^3}, \tag{6}$$

and the corresponding scalar potential

$$\Phi(\mathbf{r}) = \frac{q}{4\pi|r-a|}. \tag{7}$$

By analogy with (6), if the second of Maxwell's equations in (2) was modified to allow for magnetic sources, we would expect the magnetic field around a point particle with magnetic charge g placed at $\mathbf{a}$ to be given by

$$\mathbf{B}(\mathbf{r}) = \frac{g\,(\mathbf{r}-\mathbf{a})}{4\pi|r-a|^3}. \tag{8}$$

In this section we will start by considering simple implications of the existence of a monopole magnetic field, especially on the angular momentum of a particle moving in this field. Then, in preparation for the quantum theory, we will look at electromagnetic potentials and see Dirac strings emerge already in the classical theory.

2.1 The Field Angular Momentum

Let us now consider an electric charge q of mass m placed at a point $\mathbf{a}$ in the background field of a magnetic pole g placed at the origin. The background magnetic and dynamical electric fields are given by (8) with $\mathbf{a} = 0$ and (6) respectively. As is well known, the angular momentum stored in the electromagnetic field does not vanish. It can be calculated from the standard symmetric energy-momentum tensor of the electromagnetic field and is given by

$$\mathbf{L}_{field} = \int_V \mathbf{r}\times(\mathbf{E}\times\mathbf{B}) = \frac{qg}{4\pi}\frac{\mathbf{a}}{|a|}. \tag{9}$$

This expression is independent of the magnitude of $\mathbf{a}$ even though it depends on its direction, and therefore the limit $\mathbf{a}\to 0$ cannot be taken continuously. For example, if we place the charge above the pole the angular momentum will be negative, whereas it will be positive if placed below. Whatever the sign, if we assume the angular momentum to be quantized in integer multiples of $\hbar/2$, we obtain the quantization condition

$$\frac{qg}{2\pi\hbar} \in \mathbb{Z}. \tag{10}$$

This was first noted by Fierz [2] and is the statement that quantization of the field angular momentum results in the Dirac quantization condition. The fact that there exists a shift in the angular momentum by $\hbar/2$ when considering charge monopole composites without fermions in sight has been lucidly analyzed in [3]. We will return to this in Sect. 3.3.

2.2 A Test Electric Charge in the Background of a Magnetic Pole (I)

Let us consider a charged particle moving in the magnetic field of a pole placed at the origin. In particular, following [4,5], let us examine its angular momentum. The torque τ experienced by the particle is given by

$$\tau = \mathbf{r} \times \mathbf{F}, \tag{11}$$

where the force is given by the Lorentz force law (5). We find that the change in angular momentum is

$$\frac{d\mathbf{L}}{dt} = \tau = q\left(\mathbf{r} \times \dot{\mathbf{r}} \times \frac{g\,\mathbf{r}}{4\pi r^3}\right) = \frac{d}{dt}\left(\frac{qg}{4\pi}\frac{\mathbf{r}}{|r|}\right). \tag{12}$$

We can therefore deduce that the conserved angular momentum is not the usual $\mathbf{L} = m\mathbf{r} \times \dot{\mathbf{r}}$, but that there is an extra term:

$$\mathbf{L} = m\mathbf{r} \times \dot{\mathbf{r}} - \frac{qg}{4\pi}\frac{\mathbf{r}}{|r|}. \tag{13}$$

The same result was derived in [6] by demanding that the velocity should transform as a vector under rotations. We will see below that there are at least two more ways of deriving this result.

2.3 The Electromagnetic Potential and the Appearance of the Dirac String

We know that the magnetic field due to a point source at the origin will be given by (8) with $\mathbf{a} = 0$:

$$\mathbf{B}(\mathbf{r}) = \frac{g\,\mathbf{r}}{4\pi r^3}. \tag{14}$$

What is the electromagnetic potential that gives rise to this field? We find that, mathematically, the potential giving rise to this field in the sense of (3) can be written as the 1-form

$$\mathbf{A} = \frac{g}{4\pi}\,(k - \cos\theta)\,\mathbf{d}\phi, \tag{15}$$

where k is an arbitrary constant. We will now see that this (alone) is not the potential for a magnetic pole. The monopole field (14) is manifestly spherically symmetric. We can calculate the magnetic flux through a sphere surrounding the pole and we obtain

$$\int_{S^2} \mathbf{B}\cdot\mathbf{dS} = g\left(=\int_V \nabla\cdot\mathbf{B}\right). \tag{16}$$

We can in fact remove an infinitesimally small cap on the sphere (take the cap $\mathscr{S}$ to be on the North pole surrounded by the curve $\mathscr{C}$) and the value of this surface integral will be unchanged. This configuration is shown on the left hand side of Fig. 1.

According to the argument just given, the outward magnetic flux through this cap should be zero. We may try to check this explicitly:

$$\begin{aligned}\int_{\mathscr{S}} \mathbf{B}\cdot\mathbf{dS} &= \int_{\mathscr{S}} (\nabla\times\mathbf{A})\cdot\mathbf{dS}\\ &= \oint_{\mathscr{C}} \mathbf{A}\cdot\mathbf{dl}\\ &= \frac{g}{2}(k-1),\end{aligned} \tag{17}$$

where we have used Stokes' theorem. We find in fact that when $k=-1$ there is actually an inward flux equal to the pole strength g across the cap at the North pole.[1] We arrive at the conclusion that the potential (15) does not fully correspond to the magnetic field (14), but that it describes a magnetic pole everywhere, except on the positive z-axis (when $k=-1$), on the negative z-axis (when $k=1$) or on the whole of the z-axis (for all other values of k). This ought to be expected because a potential in the electric picture, as mentioned in the introduction to Sect. 2, should not allow

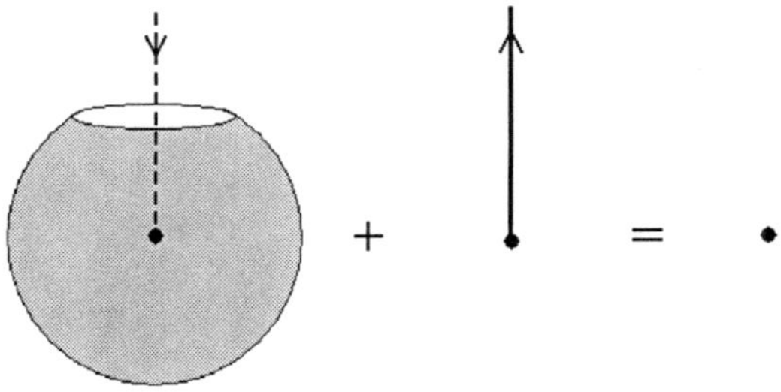

Fig. 1 A magnetic pole, shown on the *right*, is described by the potential (15), shown here for $k=-1$ on the *left hand side* plus a Dirac string shown in the *middle* which must be added to cancel the inward flux that the potential has along the positive z-axis

[1] Similarly, it can be checked that when $k=1$ there will exist an inward magnetic flux across the South pole cap. For all other values of k there is magnetic flux across both caps.

for magnetic sources. We can however truly describe a magnetic monopole in the electric picture: we must add a string carrying total positive magnetic flux $+g$ and superimpose it on the configuration described by (15) in order to cancel these north and/or south pole contributions. This is the Dirac string, and as shown schematically in Fig. 1, by introducing it we are able to describe magnetic monopoles when considering electromagnetic potentials. In a very similar way, electric charges can be described with potentials in the magnetic picture by introducing dual Dirac strings.

2.4 The Dirac String and Rotations

The electric field due to a point charge is spherically symmetric and so is the magnetic field due to a magnetic pole (14). The vector potential (15) is manifestly not so. From the discussion in the last section it is clear that rotating the vector potential will rotate the location of the Dirac string. This rotation can nevertheless be gauged away. Consider the vector potential with $k = 1$ and the string along the negative z-axis:

$$\mathbf{A}_{k=1} = \frac{g}{4\pi}(1-\cos\theta)\,\mathbf{d}\phi. \tag{18}$$

After a rotation by π radians along the x-axis the angular coordinates change as $\phi \to \phi' = 2\pi - \phi$ and $\theta \to \theta' = \pi - \theta$ such that

$$\mathbf{A}_{(k=1)} \to \mathbf{A}' = \frac{g}{4\pi}(-1-\cos\theta)\,\mathbf{d}\phi = \mathbf{A}_{(k=-1)}, \tag{19}$$

as expected. We can accompany this with the gauge transformation $\mathbf{A} \to \mathbf{A}' = \mathbf{A} + \mathbf{d}\lambda$ with $\lambda = \frac{g\phi}{2\pi}$ which transforms $\mathbf{A}_{(k=-1)}$ back to $\mathbf{A}_{(k=1)}$, so the rotation can be undone by a gauge transformation. Let us see how this works in general.

The usual Killing vectors which generate SO(3) rotations on the two-sphere are given by

$$\begin{aligned}
\xi_X &= -\sin\phi\,\frac{\partial}{\partial\theta} - \cos\phi\cot\theta\,\frac{\partial}{\partial\phi},\\
\xi_Y &= \cos\phi\,\frac{\partial}{\partial\theta} - \sin\phi\cot\theta\,\frac{\partial}{\partial\phi},\\
\xi_Z &= \frac{\partial}{\partial\phi}.
\end{aligned} \tag{20}$$

They satisfy the algebra

$$[\xi_A, \xi_B] = -\varepsilon_{ABC}\xi_C. \tag{21}$$

Under an infinitesimal rotation generated by one of these vectors, the vector potential changes by the Lie derivative along the vector field:

$$\mathscr{L}_{\xi_B}A_i = \xi_B^j\partial_j A_i + A_j\partial_i\xi_B^j = \xi_B^j F_{ji} + \partial_i(\xi_B^j A_j). \tag{22}$$

For the monopole vector potential (15) this Lie derivative is non-zero and hence usual rotational symmetry would appear to be broken. As illustrated above however, we can undo its effect if we accompany the Lie derivative by a gauge transformation $A_i \to A_i + \partial_i \Lambda$ such that the total change is

$$\delta_{\xi_B} A_i = \mathscr{L}_{\xi_B} A_i + \partial_i \Lambda_B = 0. \tag{23}$$

The gauge transformations that we find associated with the three Killing vector fields are

$$\begin{aligned}
\Lambda_X &= -\frac{g}{4\pi}\cos\phi\sin\theta - \xi_X^j A_j = \frac{gk}{4\pi}\cot\theta\cos\phi - \frac{g}{4\pi}\frac{\cos\phi}{\sin\theta} \\
\Lambda_Y &= -\frac{g}{4\pi}\sin\phi\sin\theta - \xi_Y^j A_j = \frac{gk}{4\pi}\cot\theta\sin\phi - \frac{g}{4\pi}\frac{\sin\phi}{\sin\theta} \\
\Lambda_Z &= -\frac{g}{4\pi}\cos\theta - \xi_Z^j A_j = -\frac{gk}{4\pi}.
\end{aligned} \tag{24}$$

We note that Λ_Z is a constant gauge parameter, yet it needs to be considered in order that the new rotation generators defined below satisfy the algebra.

We can combine the gauge transformations Λ_B and the naive rotation operators ξ_B to create new rotation operators $\hat{\xi}_B = \xi_B + \Lambda_B \frac{\partial}{\partial\lambda}$ by using the fact that the generator of gauge transformations is the electric charge q, which is in turn conjugate to the coordinate on the U(1) fiber which we call λ, so we can write:

$$\begin{aligned}
\hat{\xi}_X &= -\sin\phi\frac{\partial}{\partial\theta} - \cos\phi\cot\theta\frac{\partial}{\partial\phi} + \left(\frac{gk}{4\pi}\cos\phi\cot\theta - \frac{g}{4\pi}\frac{\cos\phi}{\sin\theta}\right)\frac{\partial}{\partial\lambda}, \\
\hat{\xi}_Y &= \cos\phi\frac{\partial}{\partial\theta} - \sin\phi\cot\theta\frac{\partial}{\partial\phi} + \left(\frac{gk}{4\pi}\sin\phi\cot\theta - \frac{g}{4\pi}\frac{\sin\phi}{\sin\theta}\right)\frac{\partial}{\partial\lambda}, \\
\hat{\xi}_Z &= \frac{\partial}{\partial\phi} - \frac{gk}{4\pi}\frac{\partial}{\partial\lambda}.
\end{aligned} \tag{25}$$

These operators satisfy the SO(3) algebra just as (20). We have therefore seen that when considering rigid rotations in the presence of magnetic poles we are automatically led to consider gauge transformations. This, which is a priori unexpected, points to an intricate relationship between spacetime and internal symmetries.

2.5 *A Test Electric Charge in the Background of a Magnetic Pole (II)*

Let us go back to considering a test electric charge in the background of a magnetic pole. We will now see that it is possible to get direct access to the angular momentum in the field by analyzing the angular momentum of this charge.

There is an alternative way to arrive at this result which we believe is worth recalling. The motion of the charge is described by the action

$$S = \int \left(\frac{1}{2} m\dot{x}_i \dot{x}^i - qA_i \dot{x}^i \right) dt. \tag{26}$$

We may apply the standard Noether procedure to obtain that for a variation δx the action changes by

$$\delta S = \int \frac{d}{dt} \left[(m\dot{x}_i - qA_i) \delta x^i \right] \tag{27}$$

on-shell, i.e., when the equations of motion hold. The transformation will generate a conserved charge if the action is invariant off-shell up to the integral of a total derivative. Let us consider what the case is for a rotation generated by $\delta x^i = \xi_B^i$. The first term in (26) is the norm of a vector and therefore manifestly invariant under rotations. The variation of the second term becomes

$$\begin{aligned} \delta S &= -q \int \left(\dot{x}^i \delta_{\xi_B} A_i + A_i \delta_{\xi_B} \dot{x}^i \right) \\ &= -q \int \left(\mathcal{L}_{\xi_B} A_i \right) \dot{x}^i, \end{aligned} \tag{28}$$

where we have used $\delta_{\xi_B} A_i = (\partial_j A_i)\, \delta_{\xi_B} x^j$. If, as is the case here (see (23)), the Lie derivative of A_i is given by a total derivative $\mathcal{L}_{\xi_B} A_i = -\partial_i \Lambda_B$ this becomes

$$\delta S = q \int \frac{d}{dt} \left[\Lambda_B \right]. \tag{29}$$

We can now equate (27) and (29) to obtain the charge J_B conserved under rotations

$$J_B = (m\dot{x}_i - qA_i)\, \xi_B^i - q\Lambda_B. \tag{30}$$

In fact, the Λ_B are the gauge transformations we computed in (24). For example, for rotations about the z-axis, we obtain that

$$J_Z = m\dot{\phi} + \frac{qg}{4\pi} \cos\theta, \tag{31}$$

and in general the result (13).

From the expression (13) it is clear that the angular momentum does not vanish when the particle is at rest. This apparent paradox is explained by comparing (13) with (9). The extra piece is the angular momentum of the electromagnetic field, which must be included because the electric field of the test particle is dynamical and hence the angular momentum of the electromagnetic field changes with time. Only the sum of the standard orbital angular momentum and the field angular momentum is conserved.

Expressing the angular momentum in terms of the (non gauge invariant) conjugate momenta $p_\phi = m\dot{\phi} - qA_\phi$ one obtains, for the angular momentum about the z-axis,

$$J_Z = p_\phi + k\frac{qg}{4\pi}. \tag{32}$$

In the gauge $k = 1$ and when the particle is on the positive z-axis, the field angular momentum is just equal to the extra term appearing in J_Z, besides the usual p_ϕ. This is a way to identify the field angular momentum if one knows the total angular momentum, as we'll see below is the case for gravity, and as we saw in Sect. 2.1, this may then be quantized to obtain the Dirac quantization condition. Note that in the gauge $k = 0$, p_ϕ is equal to the total angular momentum.

3 Onwards: Magnetic Monopoles and Quantum Theory

We have seen that the consideration of magnetic monopoles in electromagnetism requires the introduction of Dirac strings for a consistent mathematical description. We have also seen that usual notions of angular momentum need to be modified in their presence. The most surprising implication however is yet to come and that is the quantization condition written above in (1). This obviously requires that we should consider the quantum theory, and we start by looking at the argument as first put forward by Dirac.

3.1 Dirac's Original Argument

Dirac originally considered a particle whose motion is represented by a complex wave function Ψ. It is straightforward to deduce that the value of the phase of the wave function is unobservable, but that changes in the phase along curves in space must be the same for wave functions of the same representation if the theory is to have an unambiguous physical interpretation. It is possible to write Ψ as

$$\Psi = \Psi_1 e^{i\beta}, \tag{33}$$

where Ψ_1 has a definite phase at each point in space and any indeterminacy in the phase is absorbed into β. This means that β is not necessarily a well defined function of x, y, z, t but that its derivatives are well defined at each point (and obviously need not satisfy the integrability conditions). We therefore have that

$$-i\hbar\nabla\Psi = e^{i\beta}\left(-i\hbar\nabla\Psi_1 + \hbar\nabla\beta\right)\Psi_1. \tag{34}$$

We recognize this as a minimal coupling term: the wave function Ψ_1 now describes a particle of arbitrary charge q moving in an electromagnetic field with potentials

$$\mathbf{A} = \frac{\hbar}{q}\nabla\beta, \qquad A_0 = -\frac{\hbar}{q}\frac{\partial\beta}{\partial t}. \tag{35}$$

Therefore, non-integrable derivatives of the wave function can be naturally interpreted in terms of potentials of the electromagnetic field. The change in phase along a closed curve $\mathscr{C}$ which bounds a surface $\mathscr{S}$ will be given by

$$\begin{aligned}\oint_{\mathscr{C}=\delta\mathscr{S}}(\nabla\beta)\cdot dl &= \int_{\mathscr{S}}(\nabla\times\nabla\beta)\cdot\mathbf{dS} \\ &= \frac{q}{\hbar}\int_{\mathscr{S}}(\nabla\times\mathbf{A})\cdot\mathbf{dS} \\ &= \frac{q}{\hbar}\int_{\mathscr{S}}\mathbf{B}\cdot\mathbf{dS}.\end{aligned} \tag{36}$$

Note that from the point of view of the mathematics of the theory, this phase is defined up to integral multiples of 2π. We will now try to interpret this mathematical leniency physically. First consider a very small closed curve $\mathscr{C}$. Schrödinger's equation implies that the wave function will be continuous and hence the change in phase around the curve will be small and cannot differ by an integral multiple of 2π for different wave functions. An exception to this argument occurs when the wave function vanishes as then the phase is not defined and continuity considerations will no longer apply. The vanishing of a complex quantity like the wave function in three dimensions will generally occur along a line, which Dirac called a "nodal line" and we now recognize as a Dirac string. If the curve $\mathscr{C}$ considered above encircles a nodal line, the change in the phase of the wave function could indeed possibly be close to an integral multiple of 2π; this multiple n will be a characteristic of the nodal line under consideration and its sign will confer upon the nodal line a direction. The total change in the phase when encircling $\mathscr{C}$ will be

$$2\pi\sum_i n_i + \frac{q}{\hbar}\int_{\mathscr{S}}\mathbf{B}\cdot\mathbf{dS}, \tag{37}$$

where the sum is over nodal lines which are indexed by the integer i. If the surface $\mathscr{S}$ is closed then this change must be zero. Only nodal lines that originate within the surface will contribute to the sum as every other line will puncture the surface an even number of times and its contributions will cancel. We therefore obtain the result that the end points of the nodal lines are points of singularity in the electromagnetic field. The total flux that crosses the closed surface $\mathscr{S}$ in this case is $g = \int\mathbf{B}\cdot\mathbf{dS} = \frac{2\pi n\hbar}{q}$, which leads to the well known Dirac quantization condition

$$\frac{qg}{2\pi\hbar} \in \mathbb{Z}. \tag{38}$$

By relentlessly pursuing the mathematical implications of quantum mechanics we have been led to the deduction that magnetic poles can exist and that the product of the magnetic and electric charges must obey the quantization condition (38). Along the way we have also learnt that the wave function must vanish along a Dirac string. In fact, in terms of the physical fields it is as if the Dirac string had been removed entirely and the physics was defined on $\mathbb{R}^3$ – string(s).

3.2 An Argument by Wilzcek

Another derivation of the quantization condition was presented by Wilzcek in [7]. Consider a solenoid extending along the z axis and a particle of charge q orbiting around it. When no current is flowing through the solenoid the orbital angular momentum will be quantized in units $L_z \in \hbar\mathbb{Z}$. When there is a current flowing through the solenoid the electric field the particle experiences is given by Faraday's law (the third of Maxwell's equations in (2)):

$$\mathbf{E}(\mathbf{r}) = -\frac{\hat{z} \times \mathbf{r}\dot{\Phi}}{2\pi(x^2+y^2)}, \tag{39}$$

where Φ is the magnetic flux through the solenoid, dot denotes time derivative and hat denotes a unit vector. The change in angular momentum is given by the torque

$$\dot{L}_z = [\mathbf{r} \times \mathbf{F}]_z = -\frac{q}{2\pi}\dot{\Phi}, \tag{40}$$

where we have used the Lorentz force law $\mathbf{F} = q(\mathbf{E} + \mathbf{v} \times \mathbf{B})$, so that the angular momentum is now quantized in units

$$L_z \in \hbar\mathbb{Z} - \frac{q\Phi}{2\pi}. \tag{41}$$

If you consider the charge to be orbiting infinitesimally above a magnetic pole of strength g, the flux threaded through the orbit will be $g/2$. The angular momentum will be quantized in units

$$L_z \in \hbar\mathbb{Z} - \frac{qg}{4\pi}. \tag{42}$$

If the particle was orbiting just below the magnetic pole the same argument would lead to a quantization of the form

$$L_z \in \hbar\mathbb{Z} + \frac{qg}{4\pi}. \tag{43}$$

If these two spectra are to be equivalent, the difference between them must be an integer times $\hbar$, which leads yet again to the quantization condition

$$\frac{gq}{2\pi\hbar} \in \mathbb{Z}. \tag{44}$$

Notice that for an odd integer, this changes the spectrum of L_z from integer values to half integer values as was mentioned at the end of Sect. 2.1. The question that now arises is whether the wave function will be periodic or antiperiodic, an issue which we will now investigate.

3.3 Considerations on the (Anti-)periodicity of the Wave Function

We saw in Sect. 2.5 that by applying Noether's procedure to the action describing an electric charge moving in the field of a magnetic monopole we could obtain the conserved angular momentum (13) and in particular the expression (32) for the angular momentum about the z-axis in terms of the conjugate momentum:

$$J_Z = p_\phi + k\frac{qg}{4\pi}. \tag{45}$$

It follows from our discussion in the preceding section, and has been shown in [3,6], that the difference $J_Z - qg/4\pi$ has integer eigenvalues. We may rewrite this assertion in the notation used here as the statement that in the gauge $k = 1$ the operator $p_\phi = \frac{\hbar}{i}\frac{\partial}{\partial\phi}$ has integer eigenvalues and therefore the wave function is periodic in ϕ.

However the periodicity of the wave function depends on k. This can be traced to the fact that differentiating with respect to ϕ

$$p_\phi = \frac{\hbar}{i}\frac{\partial}{\partial\phi}|_\lambda \tag{46}$$

at constant λ (where λ is the coordinate along the U(1) fiber) is not invariant under the gauge transformation

$$\lambda \to \lambda - \frac{g\phi}{4\pi}, \tag{47}$$

which brings k from 1 to 0. As (47) itself shows, the wave function picks up a phase factor $e^{-i\frac{qg}{4\pi}\phi}$. This in turn implies that in the gauge $k = 0$ the wave function is antiperiodic in ϕ when $qg/4\pi$ is half integer. This anti-periodicity is permissible because as discussed at the end of Sect. 3.1 in the gauge $k = 0$ an infinite line is removed from $\mathbb{R}^3$ and the resulting configuration space for the electron is therefore not simply connected.

Another interesting manifestation of the lack of simple connectedness due to the infinite line formed by the two strings when $k = 0$ is the following: the closed path traveled in configuration space when the string sweeps out a closed surface around the electric charge is not contractible to the identity, whereas it is so when the string is half infinite (see [8,9]). For this reason the string wave function changes sign after the turn and one obtains the same quantization condition as when $k = 1$, in which case the path is contractible and the wave function returns to its original value.

4 Beyond the Linear Theory for Spin Two

We have seen that magnetic monopoles in electromagnetism give rise to interesting physics and bring together ideas involving spatial rotations and gauge transformations and that, when considered in the quantum theory, they lead to the aforementioned quantization condition. It is natural to ask whether magnetic sources can be introduced in gravity and if so whether any of the arguments presented in Sects. 2 and 3 carry through to the gravity case.[2]

The gravitational analog of the point electric charge is the Schwarzschild solution, in the sense that it is a solution of the vacuum field equations regular everywhere except for at the origin. It is also spherically symmetric and the Killing vectors (20) act on sections of constant (t,r) and generate two-spheres. It is in fact the unique solution with all these properties. In what follows we will consider another spacetime: the Taub-NUT solution [11]. This is an exact solution of the vacuum Einstein theory describing a gravitational dyon. The quantization condition on the energy of a particle moving in the Taub-NUT geometry is a well known result which has been discussed by many authors [12–16] and which can be viewed as a consequence of the existence of closed timelike curves [17].

In what follows we shall rederive the quantization condition along new lines, from the quantization of the angular momentum stored in the gravitational field as was done for electromagnetism above.

4.1 The Gravitating Magnetic Pole

The Taub-NUT metric is given by

$$ds^2 = -V(r)[dt + 2N(k - \cos\theta)\,d\phi]^2 + V(r)^{-1}dr^2 + (r^2 + N^2)(d\theta^2 + \sin^2\theta\, d\phi^2), \tag{48}$$

with

$$V(r) = 1 - \frac{2(N^2 + Mr)}{(r^2 + N^2)} = \frac{r^2 - 2Mr - N^2}{r^2 + N^2}, \tag{49}$$

where N and M are referred to as the magnetic and electric masses. A pure magnetic mass has $M = 0$. The number k can be changed according to

$$k \to k' = k - \alpha \tag{50}$$

by performing a t coordinate transformation

$$t \to t' = t + 2N\alpha\phi. \tag{51}$$

---

[2] This question was studied in [10], where the linearized case was considered in detail for spins 2 and higher and then the spin 2 case was considered along the lines presented here. We refer the reader to that paper for a full list of references.

The metric (48) is singular on the z-axis. This singularity is most easily seen by calculating $|\nabla t|^2$ and can be interpreted as a singularity of the metric or of the t coordinate. It is known in the literature as the Dirac–Misner string singularity, due to its analogy with the Maxwell case. Its location depends on the value of k: for $k = 1$ the singularity is at $\theta = \pi$ and for $k = -1$ it is at $\theta = 0$. For all other values of k both string singularities exist. The choice $k = 0$ makes the North and South poles play a symmetrical role. In his paper [17], Misner showed that the singularity is a coordinate singularity and that the metric describes a non-singular manifold provided that the time coordinate t is taken to be periodic with period $8\pi N$.

4.2 Spatial Rotations and Quantization Condition

Even though the metric contains a $dtd\phi$ term, it is spherically symmetric. However, the rotation group acts on spacetime in an unconventional way [17] and the rotation Killing vectors differ from those of flat space by extra terms. One can understand the origin of these additional terms by comparing the Taub-NUT solution with the standard electromagnetic magnetic monopole solution as presented in Sect. 2.4 and recalling that for stationary metrics, the mixed time-space metric components are naturally interpreted as the component of an electromagnetic vector potential (gravitomagnetism): $g_{0i} \sim A_i$. In this spirit, the above metric component g_{0i} would correspond to a monopole potential

$$A_\phi = -2N(k - \cos\theta) \tag{52}$$

with magnetic charge N, as has been observed by many authors.

To understand the form of the Killing vectors, we start with the generator of rotations around the z-axis. It is recalled in Sect. 2.4 that in the electromagnetic case, rotations of the electromagnetic potential of a magnetic pole must be accompanied by gauge transformations. When the g_{0i} metric components are interpreted as the components of a vector potential, the electromagnetic gauge transformations lift to diffeomorphisms along the time direction. We therefore expect the gauge parameter which accompanies a rotation to lift to a component along $\partial/\partial t$.

This expectation turns out to be correct. The metric (48) has four Killing vectors, given in [17] for $k = 1$, which we display as

$$\begin{aligned}
\xi_t &= \frac{\partial}{\partial t}, \\
\xi_x &= -\sin\phi\,\frac{\partial}{\partial\theta} - \cos\phi\cot\theta\,\frac{\partial}{\partial\phi} + \left(2Nk\cos\phi\cot\theta - 2N\frac{\cos\phi}{\sin\theta}\right)\frac{\partial}{\partial t}, \\
\xi_y &= \cos\phi\,\frac{\partial}{\partial\theta} - \sin\phi\cot\theta\,\frac{\partial}{\partial\phi} + \left(2Nk\sin\phi\cot\theta - 2N\frac{\sin\phi}{\sin\theta}\right)\frac{\partial}{\partial t}, \\
\xi_z &= \frac{\partial}{\partial\phi} - 2Nk\frac{\partial}{\partial t}.
\end{aligned} \tag{53}$$

These satisfy the commutation relations

$$[\xi_a, \xi_b] = -\varepsilon_{abc}\xi_c, \qquad [\xi_a, \xi_t] = 0, \tag{54}$$

where $a,b,c = x,y,z$ and $\varepsilon_{xyz} = 1$, which constitute the standard su(2) $\times$ u(1) Lie algebra. Notice the similarities between (53) and the vectors displayed in (25). For $N = 0$, which is the Schwarzschild case, the three Killing vectors ξ_a generate space-like two-spheres. In the case of Taub-NUT, although the algebra is the same, the action of the group on the manifold is, as pointed out by Misner, different. In fact, the ξ_a which satisfy the su(2) algebra now generate the r =constant three-spheres. These three-spheres have a Lorentzian metric.

The three rotation Killing vectors ξ_a are invariant under reflections with respect to the origin ($t \to -t$, $\theta \to \pi - \theta$, $\phi \to \phi + \pi$, $N \to -N$ and $k \to -k$). This invariance is expected because the Killing vectors, being the generators of rotations, are pseudo-vectors.

One also observes that ξ_z has a component along $\partial/\partial t$ proportional to the magnetic mass a fact which we will now use to show a different derivation of the quantization condition.

Consider an electric mass following a geodesic in the Taub-NUT spacetime, with four-momentum $p_\mu = mu_\mu$. There exists a conserved charge associated with every Killing vector field, and in particular a charge J_Z associated with ξ_z:

$$J_Z = m\xi_z^\mu u_\mu = p_\phi - 2Nmu_0 k. \tag{55}$$

We see that just as in the electromagnetic case, the angular momentum about the z-axis has an additional term besides the standard p_ϕ. This extra piece comes from the angular momentum in the field, which varies as the particle moves and which must be taken into account in the conservation law. In fact, the extra term coincides with the angular momentum in the field when $k = 1$ and the particle is on the positive z-axis. Thus the angular momentum in the field is equal to $2Nmu_0$. Requiring this angular momentum to be quantized in multiples of $\hbar/2$ yields the quantization condition

$$4Nmu_0 \in \hbar\mathbb{Z}. \tag{56}$$

Note that when the particle is on the negative z-axis, the angular momentum in the field coincides with the extra term in J_Z when $k = -1$ and therefore changes sign.

Formula (56) agrees with the condition that comes from periodicity in time of the wave function. Suppose indeed that the electric mass is described by a wave function ψ. The time dependence of ψ is given by

$$\psi \propto e^{-\frac{iEt}{\hbar}} \tag{57}$$

where $E = mu_0$ is the energy of the particle. Recalling that time is periodic with period $\Delta t = 8\pi N$, and requiring single valuedness of the wave function, we obtain $E\Delta t = 2n\pi\hbar$ which implies, yet again, that

$$4Nmu_0 \in \hbar\mathbb{Z}. \tag{58}$$

Single-valuedness is required because the closed timelike curves are contractible.

Equations (56) and (58) derived above are particular cases of the more general covariant result derived for the linearized theory of spin ≥ 2 in [10], which followed an adaptation of the arguments presented by Dirac in another great paper [18].

Acknowledgments I am extremely grateful to Claudio Bunster and Marc Henneaux who have taught me all I know about magnetic monopoles and to everyone at CECS for the wonderful times I enjoyed in Valdivia.

References

1. P. A. M. Dirac, "Quantised singularities in the electromagnetic field," Proc. R. Soc. Lond. A **133** (1931) 60.
2. M. Fierz, Helv. Phys. Acta. **17** (1944) 27.
3. A. S. Goldhaber, "Spin and statistics connection for charge – monopole composites," Phys. Rev. Lett. **36** (1976) 1122.
4. J. S. Schwinger, "Sources and magnetic charge," Phys. Rev. **173** (1968) 1536.
5. J. S. Schwinger, "A magnetic model of matter," Science **165** (1969) 757.
6. S. R. Coleman, "The magnetic monopole fifty years later," HUTP-82/A032 *Lectures given at International School of Subnuclear Physics, Erice, Italy, Jul 31–Aug 11, 1981, at 6th Brazilian Symposium on Theoretical Physics, Jan 7–18, 1980, at Summer School in Theoretical Physics, Les Houches, France, and at Banff Summer Institute on Particle and Fields, Aug 16–28, 1981.*
7. F. Wilczek, Phys. Rev. Lett. **48** (1982) 1144.
8. S. Deser, A. Gomberoff, M. Henneaux and C. Teitelboim, "Duality, self-duality, sources and charge quantization in abelian N-form theories," Phys. Lett. B **400** (1997) 80 [arXiv:hep-th/9702184].
9. S. Deser, A. Gomberoff, M. Henneaux and C. Teitelboim, "p-Brane dyons and electric-magnetic duality," Nucl. Phys. B **520** (1998) 179 [arXiv:hep-th/9712189].
10. C. W. Bunster, S. Cnockaert, M. Henneaux and R. Portugues, Phys. Rev. D **73** (2006) 105014 [arXiv:hep-th/0601222].
11. E. Newman, L. Tamburino and T. Unti, "Empty space generalization of the Schwarzschild metric," J. Math. Phys. **4** (1963) 915.
12. A. Zee, "Gravitomagnetic pole and mass quantization," Phys. Rev. Lett. **55** (1985) 2379 [Erratum-ibid. **56** (1986) 1101].
13. S. Ramaswamy and A. Sen, "Comment on 'Gravitomagnetic pole and mass quantization.'," Phys. Rev. Lett. **57** (1986) 1088.
14. J. Samuel and B. R. Iyer, "Comment on 'Gravitomagnetic pole and mass quantization.'," Phys. Rev. Lett. **57** (1986) 1089.
15. M. T. Mueller and M. J. Perry, "Constraints on magnetic mass," Class. Quant. Grav. **3** (1986) 65.
16. A. Magnon, "Global techniques, dual mass, and causality violation," J. Math. Phys. **27** (1986) 1066.
17. C. W. Misner, "The flatter regions of Newman, Unti, and Tamburino's generalized Schwarzschild space," J. Math. Phys. **4** (7) (1963) 924.
18. P. A. M. Dirac, "The theory of magnetic poles," Phys. Rev. **74** (1948) 817.

The Census Taker's Hat

Leonard Susskind

Abstract If the observable universe really is a hologram, then of what sort? Is it rich enough to keep track of an eternally inflating multiverse? What physical and mathematical principles underlie it? Is the hologram a lower dimensional quantum field theory, and if so, how many dimensions are explicit, and how many "emerge?" Does the Holographic description provide clues for defining a probability measure on the Landscape?

The purpose of this lecture is first, to briefly review a proposal for a holographic cosmology by Freivogel, Sekino, Susskind, and Yeh (FSSY), and then to develop a physical interpretation in terms of a "Cosmic Census Taker:" an idea introduced in [1]. The mathematical structure – a hybrid of the Wheeler-DeWitt formalism and holography – is a boundary "Liouville" field theory, whose UV/IR duality is closely related to the time evolution of the Census Taker's observations. That time evolution is represented by the renormalization-group flow of the Liouville theory.

Although quite general, the Census Taker idea was originally introduced in [1], for the purpose of counting bubbles that collide with the Census Taker's bubble. The "Persistence of Memory" phenomenon discovered by Garriga, Guth, and Vilenkin, has a natural RG interpretation, as does slow roll inflation. The RG flow and the related C-theorem are closely connected with generalized entropy bounds.

1 Introduction

Of all the "String Inspired" cosmological scenarios, only one seems to me to have an element of inevitability to it. The facts and principles that drive it are as follows:

- Observational evidence supports the existence of a period of slow-roll inflation during which the universe exponentially expanded by a factor no less than e^{50}.

L. Susskind
Department of Physics, Stanford University, Stanford, CA 94305-4060, USA
e-mail: susskind@stanford.edu

M. Henneaux, J. Zanelli (eds.), *Quantum Mechanics of Fundamental Systems: The Quest for Beauty and Simplicity*, DOI 10.1007/978-0-387-87499-9_16,

The universe grew to a size which is at least 1,000 times larger (in volume) than the portion which is observable.

- A small residual vacuum energy of order $10^{-123}M_p^4$ remained at the end of inflation and now dominates the energy density of the universe. If this situation persists, then not only is the universe at least 1,000 larger than what can be seen; it is 1,000 larger than what can *ever* be seen [2, 3].
- String Theory apparently gives rise to an immense Landscape of de Sitter vacua [4–7] with a very dense "discretuum" of vacuum energies. None of these vacua are absolutely stable: each can decay to vacua with smaller cosmological constant.
- Black Hole (or Observer) Complementarity [8–10], and the Holographic Principle [11, 12], have been confirmed by string theory, at least in a certain wide class of backgrounds [13–15]. The implication is twofold. On the one hand, observer complementarity requires the identification of a causal patch; conventional quantum mechanics only makes sense within such a patch. The Holographic Principle requires that a region of space be described by boundary degrees of freedom whose number does not exceed the area, measured in Planck Units.
- Inflation, if it lasts long enough, has a tendency [16–21] to populate the Landscape with a great diversity of nucleated "pocket universes."

The first two items imply that all of observable cosmology consisted of a roll from one value of the vacuum energy (probably no bigger than $10^{-14}M_p^4$), to its final current value. How and why the universe began with such an unnatural energy density is not explained by any standard theory, but the Landscape suggests the following guess: At some point in the remote past the universe occupied a point on the Landscape with a much higher vacuum energy, perhaps of order one in Planck units. Rolling, unimpeded, to a vacuum energy of 10^{-14} without getting stuck in a local minimum is unlikely. (Think of rolling a bowling ball from the top of Mount Everest to sea level.) It is far more likely that the universe would get stuck in many minima, and have to tunnel [22] multiple times, before arriving at the very small vacuum energy required by conventional slow-roll inflation. We will not dwell on Anthropic issues in this paper, but I would point out that a long period of conventional inflation appears to be required for structure formation [23]. The argument is similar to the well-known Weinberg argument concerning the cosmological constant.

These considerations strongly suggest that the period of conventional slow-roll inflation was preceded by a tunneling event from a previous neighboring vacuum. In other words, the observed universe evolved by a sudden bubble nucleation from an *"Ancestor"* vacuum, once removed on the Landscape. It seems obvious that one of the next big questions for cosmology will be to find the theoretical and observational tools to confirm or refute the past existence of an Ancestor, and to find out as much as we can about it. If we are lucky and the amount of slow-roll inflation that followed bubble nucleation is as small as observational evidence allows, then we have a chance of seeing features of the Ancestor imprinted on the sky [23]. The two smoking guns would be:

- Negative spatial curvature: bubble nucleation leads to a negatively curved, infinite, FRW universe.
- Tensor modes in the CMB, but only in the lowest harmonics. Although the vacuum energy subsequent to tunneling (during conventional slow-roll inflation) was almost certainly too small to create observable tensor modes, the cosmological constant in the Ancestor was probably much larger. During the Ancestor epoch, large tensor fluctuations would be created by rapid inflation. A tail (diminishing rapidly with l) of those fluctuations could be visible if the number of slow-roll e-foldings is minimal.

If the observational evidence for an Ancestor is weak, so is the current theoretical framework. To many of us, eternal inflation, bubble nucleation, and a multiverse, seem all but inevitable, but it is also true that they have inspired what Bjorken[1] has called "the most extravagant extrapolation in the history of physics." Eternal inflation leads to an uncontrolled infinity of "pocket universes" which we have no good idea how to regulate – the inevitable has led to the preposterous. In my opinion, this situation reflects serious confusion, and perhaps even a crisis.

Eternal inflation is not the only extravagance that we have had to tame in recent decades. I have in mind the fact that a naive but very compelling interpretation of black holes seemed, at one time, to imply that a black hole can absorb an infinite amount of information behind its horizon [24, 25]. By feeding a black hole with coherent energy at the same rate that it evaporates, it would seem that an infinity of bits could be lost to the observable world.

I believe these two crises may be related. In both cases the infinities result from "cutting across horizons" and attempting to describe global space-like surfaces with *independent* degrees of freedom at each location. The cure is to focus attention on a single causal region, and to describe it by a Holographic set of degrees of freedom [8, 26].

In FSSY [27] the authors described one such holographic framework – call it holography in a hat – based on mathematical ideas that have become familiar from String Theory. At the same time, Shenker and collaborators [1] have developed an intuitive "gedanken observational" approach based on a fictitious observer called the "Census Taker." My purpose in this lecture is to explain the close connection between these ideas.

2 The Census Bureau

Let us begin with a precise definition of a causal patch. Start with a cosmological space–time and assume that a future causal boundary exists. For example, in flat Minkowski space the future causal boundary consists of $\mathscr{I}^+$ (future light like

[1] James Bjorken, private communication.

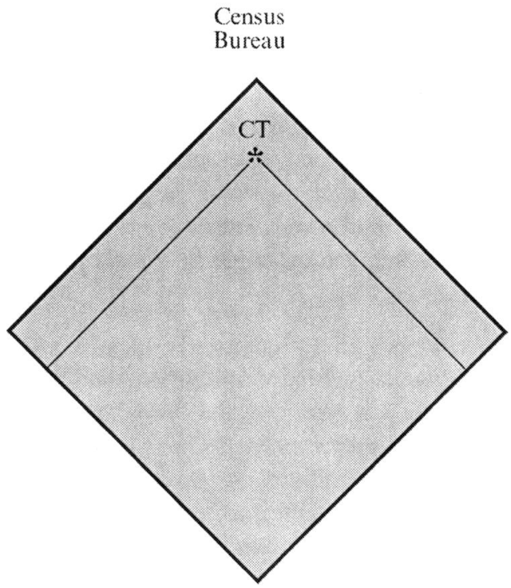

Fig. 1 Conformal diagram for ordinary flat Minkowski space. The causal patch associated with the "Census Bureau" is the entire space–time. A Census Taker and his past light-cone are also shown

infinity) and a single point, time-like-infinity. For a (non-eternal) Schwarzschild black hole, the future causal boundary has an additional component: the singularity.

A causal patch is defined in terms of a point **a** on the future causal boundary. I'll call that point the "Census Bureau.[2]" The causal patch is, by definition, the causal past of the Census Bureau, bounded by its past light cone. For Minkowski space, one usually picks the Census Bureau to be time-like-infinity. In that case the causal patch is all of Minkowski space as seen in Fig. 1.

In the case of the Schwarzschild geometry, **a** can again be chosen to be time-like-infinity, in which case the causal patch is everything outside the horizon of the black hole. There is no clear reason why one can't choose **a** to be on the singularity, but it would lead to obvious difficulties.

The term "Census Taker" was introduced [1] to denote an observer, at a point inside a causal patch, who looks back into the past and collects data. He can count galaxies, other observers, hydrogen atoms, colliding bubble-universes, civilizations, or anything else within his own causal past. As time elapses the Census Taker sees more and more of the causal patch. Eventually all Census takers within the causal patch arrive at the Census Bureau where they can compare data.

De Sitter space has the well known causal structure as shown in Fig. 2. In this case all points at future infinity are equivalent: the Census Bureau can be located

[2] This term originated during a discussion between myself and Steve Shenker in a Palo Alto Cafe. Neither of us will admit to having coined it first, but it wasn't me.

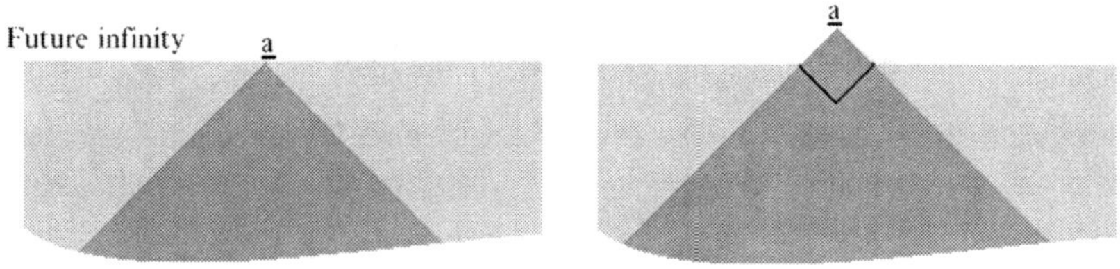

Fig. 2 Conformal diagrams for eternal and metastable de Sitter space. The grey areas are causal patches associated with the points **a**. In the metastable case the causal patch is associated with the tip of a hat

at any of them. However, String Theory and other considerations suggest that de Sitter minima are never stable. After a series of tunneling events they eventually end in terminal vacua with exactly zero or negative cosmological constant. The entire distant future of de Sitter space is replaced by a fractal of terminal bubbles.

Decay to negative cosmological constant always leads to a singular crunch. Barring governmental stupidity, this seems an unlikely place for a Census Bureau. The disadvantages (or advantages) of locating a government agency at a crunch are the same as at a black hole singularity.

Terminal vacua with zero cosmological constant seem more promising; the bubble then evolves to an open, negatively curved, FRW geometry, bounded by a "hat" [27]. The Census Bureau is at the tip of the hat.

In the case of the black hole, the degrees of freedom beyond the horizon, i.e., outside the causal patch, are redundant descriptions of degrees of freedom within the patch: they should not be double-counted. We assume that the same is true of the causal patch of a hat. In both cases the conventional rules of quantum mechanics are expected to apply *only* within the causal patch. Furthermore the rules should respect the Holographic Principle.

The reader may wonder about the relationship between hatted terminal geometries, and observational cosmology with a non-zero cosmological constant. There are two answers: the first is that for many purposes, the current cosmological constant is so small that it can be set to zero. Later we will argue that the conformal field theory description of the approximate hat which results from non-zero cosmological constant is an ultraviolet cut-off version of the type of field theory that describes a hat.

The second answer was emphasized by Shenker et al. [1] who argued that because our present de Sitter vacuum will eventually decay, a Census Taker can look back into our current vacuum from a point at or near the tip of a hat, and gather information. In principle the Census Taker can peek back, not only into the Ancestor vacuum (our vacuum in this case), but also into bubble collisions with other vacua of the Landscape. Much of this paper is about the gathering of information as the Census Taker's time progresses, and how it is encoded in the renormalization-group (RG) flow of a holographic field theory.

3 Open FRW and Euclidean ADS

The classical space–time in the interior of a Coleman De Luccia bubble, has the form of an open infinite FRW universe, Let $\mathscr{H}_3$ represent a hyperbolic geometry with constant negative curvature.

$$d\mathscr{H}_3^2 = dR^2 + \sinh^2 R \;\; d\Omega_2^2. \tag{1}$$

The metric of open FRW is

$$ds^2 = -dt^2 + a(t)^2 d\mathscr{H}_3^2, \tag{2}$$

or in terms of conformal time T (defined by $dT = dt/a(t)$)

$$ds^2 = a(T)^2(-dT^2 + d\mathscr{H}_3^2) \tag{3}$$

Note that in (1) the radial coordinate R is a hyperbolic angle and that the symmetry of the spatial sections is the non-compact group $O(3,1)$. This $O(3,1)$ symmetry plays a central role in what follows.

If the vacuum energy in the bubble is zero, i.e., no cosmological constant, then the future boundary of the FRW region is a hat. The scale factor $a(t)$ then has the early and late-time behaviors

$$a(T) \sim t \sim De^T. \tag{4}$$

For early time when $T \to -\infty$ the constant D is conveniently chosen to be the Hubble scale of the Ancestor, H^{-1}.

$$a(T) = H^{-1}e^T \quad (\mathrm{T} \to -\infty) \tag{5}$$

At late time it is always larger. In the simplest thin-wall case D is given by the Ancestor Hubble-length at all times.

In Fig. 3, a conformal diagram of FRW is illustrated, with surfaces of constant T and R shown in red and blue. The green region represents the de Sitter Ancestor vacuum. Figure 4 shows the Census Taker, as he approaches the tip of the hat, looking back along his past light cone.

Part of the inspiration for FSSY was the geometry of the spatial slices of constant T. Each slice, taken by itself, is a three dimensional, negatively curved, hyperbolic plane. It is very familiar to relativists and string theorists, being identical to 3-D Euclidean *anti de Sitter space*. The best way that I know of for becoming familiar with the hyperbolic plane is to study Escher's drawing "Limit Circle IV." It is both a drawing of Euclidean ADS and also a fixed-time slice of open FRW. In Fig. 5, the green circle is the intersection of Census Takers past light cone with the time-slice. As the Census Taker advances in time, the green circle moves out, ever closer to the boundary.

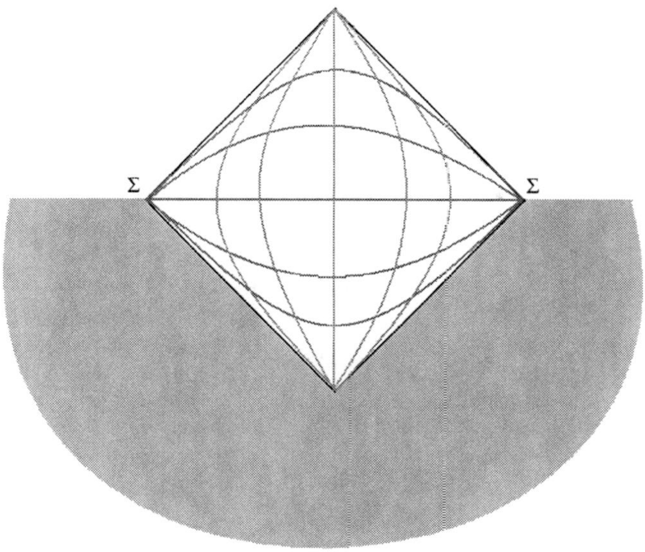

Fig. 3 A Conformal diagram for the FRW universe created by bubble nucleation from an "Ancestor" metastable vacuum. The Ancestor vacuum is shown in grey. The time-like and space-like curves are surfaces of constant T and R. The two-sphere at spatial infinity is indicated by Σ

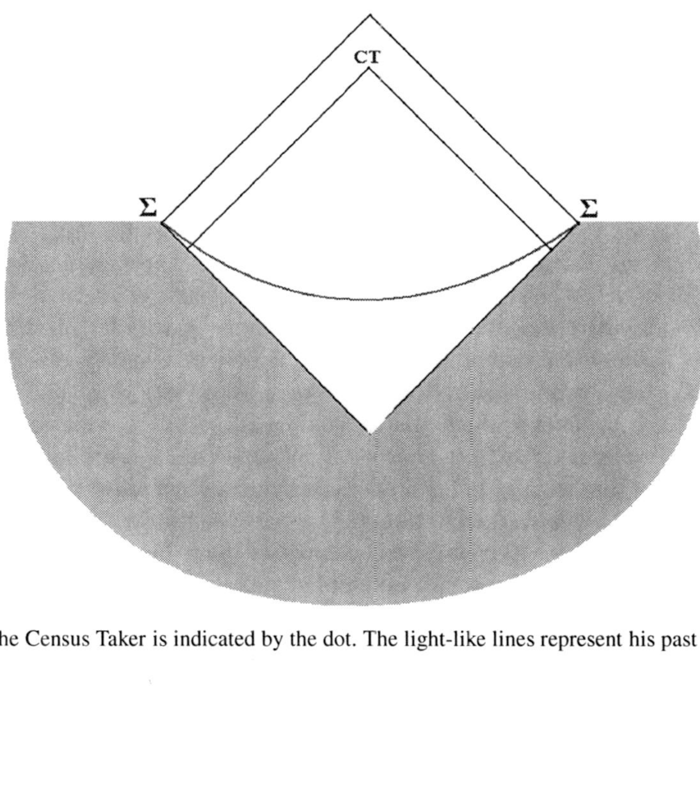

Fig. 4 The Census Taker is indicated by the dot. The light-like lines represent his past light-cone

Fig. 5 Escher's drawing of the Hyperbolic Plane, which represents Euclidean anti de Sitter space or a spatial slice of open FRW. The circle near the boundary shows the intersection of the Census Taker's past light-cone, which moves toward the boundary with Census-Taker-time

A fact (to be explained later) which will play a leading role in what follows, concerns the Census Taker's angular resolution, i.e., his ability to discern small angular variation. If the time at which the CT looks back is called T_{CT}, then the smallest angle he can resolve is of order $\exp(-T_{CT})$. It is as if the CT were looking deeper and deeper into the ultraviolet structure of a quantum field theory on Σ.

The boundary of anti de Sitter space plays a key role in the ADS/CFT correspondence, where it represents the extreme ultraviolet degrees of freedom of the boundary theory. The corresponding boundary in the FRW geometry is labeled Σ and consists of the intersection of the hat $\mathscr{I}^+$, with the space-like future boundary of de Sitter space. From within the interior of the bubble, Σ represents *space-like infinity*. It is the obvious surface for a holographic description. As one might expect, the $O(3,1)$ symmetry which acts on the time-slices, also has the action of two dimensional conformal transformations on Σ. Whatever the Census Taker sees, it is very natural for him to classify his observations under the conformal group. Thus, the apparatus of (Euclidean) conformal field theory, such as operator dimensions, and correlation functions, should play a leading role in organizing his data.

In complicated situations, such as multiple bubble collisions, Σ requires a precise definition. The asymptotic light-cone $\mathscr{I}^+$ (which is, of course, the limit of the Census Takers past light cone), can be thought of as being formed from a collection of light-like generators. Each generator, at one end, runs into the tip of the hat, while the other end eventually enters the bulk space–time. The set of points where the generators enter the bulk define Σ.

4 The Holographic Wheeler-DeWitt Equation

Supposedly, String Theory is a quantum theory of gravity, and indeed it has proved to be a remarkably powerful one, but only in certain special backgrounds. As effective as it is in describing scattering amplitudes in flat (supersymmetric) space–time, and black holes in anti de Sitter space, it is an inflexible tool which at present is close to useless for formulating a mathematical framework for cosmology. What is it that is so special about flat and ADS space that allows a rigorous formulation of quantum gravity, and why are cosmological backgrounds so difficult?

The problem is frequently blamed on *time-dependence*. But time-dependent deformations of anti de Sitter space or Matrix Theory are easy to describe. Something else is the culprit. There is one important difference between the usual String Theory backgrounds and more interesting cosmological backgrounds. Asymptotically-flat and anti de Sitter backgrounds have a property that I will call *asymptotic coldness*. Asymptotic coldness means that the boundary conditions require the energy density to go to zero at the asymptotic boundary of space.[3] Similarly, the fluctuations in geometry tend to zero. This condition is embodied in the statement that all physical disturbances are composed of normalizable modes. Asymptotic coldness is obviously important to defining an S-matrix in flat space–time, and plays an equally important role in defining the observables of anti de Sitter space.

But in cosmology, asymptotic coldness is never the case. Closed universes have no asymptotic boundary, and homogeneous infinite universes have matter, energy, and geometric variation out to spatial infinity; under the circumstances an S-matrix cannot be formulated. String theory at present is ill equipped to deal with *asymptotically warm* geometries. To put it another way, there is a conflict between a homogeneous cosmology, and the Holographic Principle which requires an isolated, cold, boundary.

The traditional approach to quantum cosmology – the Wheeler-DeWitt equation – is the opposite of string theory; it is very flexible from the point of view of background dependence – it doesn't require any definite boundary condition, it can be formulated for a closed universe, a flat or open FRW universe, de Sitter space, or for that matter, flat and anti de Sitter space space–time – but it is not a consistent quantum theory of gravity. It is based on an obsolete approach – local quantum field theory – that fails to address the problems that String Theory and the Holographic Principle were designed to solve: the huge over-counting of degrees of freedom implicit in a local field theory.

FSSY suggested a way out of dilemma: synthesize the Wheeler-DeWitt philosophy with the Holographic Principle to construct a Holographic Wheeler-DeWitt theory. We will begin with a review of the basics of conventional WDW; For a more complete treatment, especially of infinite cosmologies, see [28].

---

[3] Note that asymptotic coldness refers only to conditions at spatial infinity. A violation of asymptotic coldness does not imply that the temperature remains finite as the time goes to infinity, although even this is a problem in geometries that contain de Sitter boundary conditions. Hats are somewhat better in that they become cold at late time.

The ten equations of General Relativity take the form

$$\frac{\delta}{\delta g_{\mu\nu}} I = 0 \tag{6}$$

where I is the Einstein action for gravity coupled to matter. The canonical formulation of General Relativity makes use of a time–space split [29]. The six space–space components are more or less conventional equations of motion, but the four equations involving the time index have the form of constraints. These four equations are written,

$$H^{\mu}(x) = 0. \tag{7}$$

They involve the space–space components of the metric g_{nm}, the matter fields Φ, and their conjugate momenta. The time component $H^0(x)$, is a local Hamiltonian which "pushes time forward" at the spatial point x. More generally, if integrated with a test function,

$$\int d^3x\, f(x)\, H^0(x) \tag{8}$$

it generates infinitesimal transformations of the form

$$t \to t + f(x). \tag{9}$$

Under certain conditions H^0 can be integrated over space in order to give a global Hamiltonian description. Since H^0 involves second space derivatives of g_{nm}, it is necessary to integrate by parts in order to bring the Hamiltonian to the conventional form containing only first derivatives. In that case the ADM equations can be written as

$$\int d^3x\, H = E. \tag{10}$$

The Hamiltonian density H has a conventional structure, quadratic in canonical momenta, and the energy E is given by a Gaussian surface integral over spatial infinity. The conditions which allow us to go from (7) to (10) are satisfied in asymptotically cold flat-space–time, as well as in anti de Sitter space; in both cases global Hamiltonian formulations exist. Indeed, in anti de Sitter space the Hamiltonian of the Holographic boundary description is identified with the ADM Energy, but, as we noted, cosmology, at least in its usual forms, is never asymptotically cold. The only recourse for a canonical description, is the local form of equations (7).

When we pass from classical gravity to its quantum counterpart, the usual generalization of the canonical equations (7) become the Wheeler-DeWitt equations,

$$H^{\mu}|\Psi\rangle = 0 \tag{11}$$

where the state vector $|\Psi\rangle$ is represented by a wave functional that depends only on the space components of the metric g_{mn}, and the matter fields Φ.

The first three equations

$$H^m|\Psi\rangle = 0 \quad (m = 1,\ 2,\ 3.) \tag{12}$$

have the interpretation that the wave function is invariant under spatial diffeomorphisms,

$$x^n \to x^n + f^n(x^m) \tag{13}$$

In other words $\Psi(g_{mn},\ \Phi)$ is a function of spatial invariants. These equations are usually deemed to be the easy Wheeler-DeWitt equations.

The difficult equation is the time component

$$H^0|\Psi\rangle = 0. \tag{14}$$

It represents invariance under local, spatially varying, time translations. Not only is (14) difficult to solve; it is difficult to even formulate: the expression for H^0 is riddled with factor ordering ambiguities. Nevertheless, as long as the equations are not pushed into extreme quantum environments, they can be useful.

4.1 Wheeler-DeWitt and the Emergence of Time

Asymptotically cold backgrounds come equipped with a global concept of time. But in the more interesting asymptotically warm case, time is an approximate, derived, concept [28, 30], which emerges from the solutions to the Wheeler-DeWitt equation. The perturbative method for solving (14) that was outlined in [28], can be adapted to the case of negative spatial curvature. We begin by decomposing the spatial metric into a constant curvature background, and fluctuations. Since we will focus on open FRW cosmology, the spatial curvature is negative, the space metric having the form,

$$ds^2 = a^2\left(dR^2 + \sinh^2 R\ \ (d\theta^2 + \sin^2\theta d\phi^2)\right) + a^2 h_{mn} dx^m dx^n \tag{15}$$

In (15) a is the usual FRW scale factor and the x's are $(R,\ \ \theta,\ \phi)$.

The first approximation, in which all fluctuations are ignored, is usually called the mini-superspace approximation, but it really should be seen as a first step in a semiclassical expansion. In lowest order, the Wheeler-DeWitt wave function depends only on the scale factor a. To carry out the leading approximation in open FRW it is necessary to introduce an infrared regulator which can be done by bounding the value of R,

$$R < R_0 \ \ (R_0 \gg 1). \tag{16}$$

Lets also define the total dimensionless coordinate-volume within the cutoff region, to be V_0.

$$V_0 = 4\pi \int dR \sinh^2 R \approx \frac{1}{2}\pi e^{2R_0}. \tag{17}$$

The first (mini-superspace) approximation is described by the action,

$$L = \frac{-aV_0\dot{a}^2 - V_0 a}{2} \tag{18}$$

Defining P to be the momentum conjugate to the scale factor a,

$$P = -aV_0\dot{a} \tag{19}$$

the Hamiltonian H^0 is given by,[4]

$$H^0 = \frac{1}{2V_0} P \frac{1}{a} P + \frac{1}{2} V_0 a \tag{20}$$

Finally, using $P = -i\partial_a$, the first approximation to the Wheeler-DeWitt equation becomes,

$$-\partial_a \frac{1}{a} \partial_a \Psi - V_0^2 a \Psi = 0. \tag{21}$$

The equation has the two solutions,

$$\Psi = \exp(\pm i V_0 a^2), \tag{22}$$

corresponding to expanding and contraction universes; to see which is which we use (19). The expanding solution, labeled Ψ_0 is

$$\Psi_0 = \exp(-iV_0 a^2). \tag{23}$$

From now on we will only consider this branch.

There is something funny about (23). Multiplying V_0 by a^2 seems like an odd operation. $V_0 a^3$ is the proper volume, but what is $V_0 a^2$? The answer in flat space is that it is junk, but in hyperbolic space its just the proper area of the boundary at R_0. One sees from the metric (3) that the coordinate volume V_0, and the coordinate area A_0, of the boundary at R_0, are (asymptotically) equal to one another, to within a factor of 2.

$$A_0 = 2V_0. \tag{24}$$

Thus the expression in the exponent in (23) is $-\frac{1}{2}iA$, where A is the proper area of the boundary at R_0.

$$\Psi_0 = \exp(-2iA). \tag{25}$$

This is a suggestive indication of a Wheeler-DeWitt boundary-holography of open FRW.

---

[4] The factor ordering in the first term is ambiguous. I have chosen the simplest Hermitian factor ordering.

To go beyond the mini-superspace approximation one writes the wave function as a product of Ψ_0, and a second factor $\psi(a,h,\Phi)$ that depends on the fluctuations.

$$\Psi(a,h,\Phi) = \Psi_0 \;\; \psi(a,h,\Phi) = \exp(-iV_0 a^2) \;\; \psi(a,h,\Phi). \tag{26}$$

By integrating the Wheeler-DeWitt equation over space, and substituting (26), an equation for ψ can be obtained.

$$i\partial_a \psi + \frac{1}{aV_0}\partial_a \frac{1}{a}\partial_a \psi = H_m \psi \tag{27}$$

In this equation H_m has the form of a conventional Hamiltonian (quadratic in the momenta) for both matter and metric fluctuations.

In the limit of large scale factor the term $\frac{1}{aV_0}\partial_a \frac{1}{a}\partial_a \psi$ becomes negligible and (27) takes the form of a Schrodinger equation.

$$i\partial_a \psi = H_m \psi \tag{28}$$

Evidently the role of a is not as a conventional observable, but a parameter representing the unfolding of cosmic time. One does not calculate its probability, but instead constrains it – perhaps with a delta function or a Lagrange multiplier. As Banks has emphasized [30], in this limit, and maybe *only* in this limit, the wave function ψ has a conventional interpretation as a probability amplitude.

4.2 Holographic WDW

All of this brings us to the central question of this lecture: what form does the correct holographic theory take in asymptotically warm cosmological backgrounds? The answer suggested in FSSY was a holographic version of the Wheeler-DeWitt theory, living on the space-like boundary Σ.

As we have described it, the Wheeler-DeWitt theory is a throwback to an older view of quantum gravity based on the existence of bulk, space-filling degrees of freedom. It has become clear that this is a tremendous overestimate of the capacity of space to contain quantum information. The correct (holographic) counting of degrees of freedom is in terms of the area of the boundary of space [11, 12]. In the present case of open FRW, the special role of the boundary is played by the surface Σ at $R = \infty$.

Just as in the ADS/CFT correspondence [15], it is useful to define a regulated boundary, Σ_0, at $R = R_0$. In principle R_0 can depend on angular location on Ω_2. In fact later we will discuss invariance under gauge transformations of the form

$$R \to R + f(\Omega_2). \tag{29}$$

(The notation $f(\Omega_2)$. indicating that f is a function of location on Σ_0.)

The conjecture of FSSY is that the correct Holographic description of open FRW is a Wheeler-DeWitt equation, but one in which the degrees of freedom are at the boundary of space, i.e., on Σ, instead of being distributed throughout the bulk.

Thus we assume the existence of a set of boundary fields, that include a two dimensional spatial metric on Σ_0. The induced spatial geometry of the boundary can always be described in the conformal gauge in terms of a Liouville field $U(\Omega_2)$.

$$ds^2 = e^{2U(\Omega_2)} e^{2R_0(\Omega_2)} d\Omega_2^2 \tag{30}$$

U may be decomposed into a homogeneous term U_0, and a fluctuation; obviously the homogeneous term can be identified with the FRW scale factor by

$$e^{U_0} = a. \tag{31}$$

In Sect. 8 we will give a more detailed definition of the Liouville degree of freedom.

In addition we postulate a collection of boundary "matter" fields. The boundary matter fields, y, are not the limits of the usual bulk fields Φ, but are analogous to the boundary gauge fields in the ADS/CFT correspondence. In this paper we will not speculate on the detailed form of these boundary matter fields.

4.3 The Wave Function

In addition to U and y, we assume a local Hamiltonian $H(x^i)$ that depends only on the boundary degrees of freedom (the notation x^i refers to coordinates of the boundary Σ), and a wave function $\Psi(U,y)$,

$$\Psi(U,y) = e^{-\frac{1}{2}S + iW}. \tag{32}$$

At every point of Σ, Ψ satisfies

$$H(x^i)\ \Psi(U,y). = 0 \tag{33}$$

In (32), $S(U,y)$ and $W(U,y)$ are real functionals of the boundary fields. For reasons that will become clear, we will call S the action. However, S should not in any way be confused with the four-dimensional Einstein action.

The local Hamiltonian $H(x^i)$, and the imaginary term W in the exponent, play important roles in determining the expectation values of canonical momenta, as well as the relation between scale factor and ordinary time. In this paper H and W will play secondary roles.

We make the following three assumptions about S and W:

- Both S and W are invariant under conformal transformations of Σ. This follows from the symmetry of the background geometry: open FRW.

- The leading (non-derivative) term in the regulated form of W is $-\frac{1}{2}A$ where A is the proper area of Σ_0,

$$W = -\frac{1}{2}\int_{\Sigma} e^{2R_0} e^{2U} + \ldots \tag{34}$$

This follows from (25).
- S and W have the form of *local* two dimensional Euclidean actions on Σ. In other words they are integrals, over Σ, of densities that involve U, y, and their derivatives with respect to x^i.

The first of these conditions is just a restatement of the symmetry of the Coleman De Luccia instanton. Later we will see that this symmetry is spontaneously broken by a number of effects, including the extremely interesting "Persistence of Memory" discovered in by Garriga et al. [31].

The second condition follows from the bulk analysis described earlier in (25). It allows us to make an educated guess about the dependence of the local Hamiltonian $H(x^i)$ on U. A simple form that reproduces (25) is

$$H(x) = \frac{1}{2}e^{-2U}\pi_U^2 - 2e^{2U} + \ldots \tag{35}$$

where π_U is the momentum conjugate to U. It is easily seen that the solution to the equation $H\Psi = 0$ has the form (25).

The highly nontrivial assumption is the third item – the locality of the action. As a rule quantum field theory wave functions are not local in this sense. That the action S is local is far from obvious. In our opinion it is the strongest (meaning the weakest) of our assumptions and the one most in need of confirmation. At present our best evidence for the locality is the discrete tower of correlators, including a transverse, traceless, dimension-two correlation function, described in the next section. In principle, much more information can be obtained from bulk multi-point functions, continued to Σ. For example, correlation functions of h_{ij} would allow us to study the operator product expansion of the energy-momentum tensor.

As I said, the assumption that S is local is a very strong one, but I mean it in a rather weak sense. One of the main points of this lecture is that there is a natural RG flow in cosmology (see Sect. 6). By locality I mean only that S is in the basin of attraction of a local field theory. If it is true, locality would imply that the measure

$$\Psi^*\Psi = e^{-S} \tag{36}$$

has the form of a local two dimensional Euclidean field theory with action S, and that the Census Taker's observations could be organized not only by conformal invariance but by conformal field theory.

5 Data

The conjectured locality of the action S is based on data calculated by FSSY. The background geometry studied in [27] was the Minkowski continuation of a thin-wall Coleman De Luccia instanton, describing transitions from the Ancestor vacuum to a hatted vacuum. For a number of reasons such a background cannot be a realistic description of cosmology. First of all, there is a form of spontaneous breaking of the $O(3,1)$ symmetry that Garriga, Guth, and Vilenkin call "The Persistence of Memory." In Sect. 8 we will see that these type of effects are "dual" to effects expected in the theory of RG-flows.

More importantly, we do not live in a universe with zero cosmological constant. Observational cosmology has come close to ruling out vanishing cosmological constant, but also theoretical considerations rule it out; in the Landscape of String Theory the only vacua with exactly vanishing cosmological constant are supersymmetric. Nevertheless, hatted geometries are interesting in that they are the simplest versions of asymptotically warm geometries.

In FSSY, correlation functions were computed in the thin-wall, Euclidean, Coleman De Luccia instanton, and then continued to Minkowski signature. The more general situation, including the possibility of slow-roll inflation after tunneling, is presently under investigation with Ben Freivogel, Yasuhiro Sekino, and Chen Pin Yeh. Here we will mostly confine ourselves to the thin-wall case.

We begin by reviewing some facts about three-dimensional hyperbolic space and the solutions of its massless Laplace equation. An important distinction is between normalizable modes (NM) and non-normalizable modes (NNM); a scalar minimally coupled field χ is sufficient to illustrate the important points.

The norm in hyperbolic space is defined in the obvious way:

$$\langle \chi | \chi \rangle = \int dR d\Omega_2 \, \chi^2 \, \sinh^2 R \tag{37}$$

In flat space, fields that tend to a constant at infinity are on the edge on normalizability. With the help of the delta function, the concept of normalizability can be generalized to continuum-normalizability, and the constant "zero mode" is included in the spectrum of the wave operator, but in hyperbolic space the normalization integral (37) is exponentially divergent for constant χ. The condition for normalizability is that $\chi \to 0$ at least as fast as e^{-R}. The constant mode is therefore non-normalizable.

Normalizable and non-normalizable modes have very different roles in the conventional ADS/CFT correspondence. NM are dynamical excitations with finite energy and can be produced by events internal to the anti de Sitter space. By contrast NNM cannot be excited dynamically. Shifting the value of a NNM is equivalent to changing the boundary conditions from the bulk point of view, or changing the Lagrangian from the boundary perspective. But, as we will see, in the cosmological framework of FSSY, asymptotic warmness blurs this distinction.

5.1 Scalars

Correlation functions of massless (minimally coupled) scalars, χ, depend on time and on the *dimensionless* geodesic distance between points on $\mathscr{H}_3$. In the limit in which the points tend to the holographic boundary Σ at $R \to \infty$, the geodesic distance between points 1 and 2 is given by,

$$l = R_1 + R_2 + \log(1 - \cos\alpha) \quad (38)$$

where α is the angular distance on Ω_2 between 1 and 2. It follows on $O(3,1)$ symmetry grounds that the correlation function $\langle\chi(1)\chi(2)\rangle$ has the form,

$$\begin{aligned} \langle\chi(1)\chi(2)\rangle &= G(T_1,\, T_2,\, l_{1,2}) \\ &= G\{T_1,\, T_2,\, (R_1 + R_2 + \log(1 - \cos\alpha))\}. \end{aligned} \quad (39)$$

Before discussing the data on the Coleman De Luccia background, let us consider the form of correlation functions for scalar fields in anti de Sitter space. We work in units in which the radius of the anti de Sitter space is 1. By symmetry, the correlation function can only depend on l, the proper distance between points. The large-distance behavior of the two-point function has the form

$$\langle\chi(1)\chi(2)\rangle \sim \frac{e^{-(\Delta-1)l}}{\sinh l}. \quad (40)$$

In anti de Sitter space the dimension Δ is related to the mass of χ by

$$\Delta(\Delta - 2) = m^2. \quad (41)$$

We will be interested in the limit in which the two points 1 and 2 approach the boundary at $R \to \infty$. Using (38) gives

$$\langle\chi(1)\chi(2)\rangle \sim e^{-\Delta R_1} e^{-\Delta R_2} (1 - \cos\alpha)^{-\Delta}. \quad (42)$$

It is well known that the "infrared cutoff" R, in anti de Sitter space, is equivalent to an ultraviolet cutoff in the boundary Holographic description [15]. The exponential factors, $\exp(-\Delta R)$ in (42) correspond to cutoff dependent wave function renormalization factors and are normally stripped off when defining boundary correlators. The remaining factor, $(1 - \cos\alpha)^{-\Delta}$ is the conformally covariant correlation function of a boundary field of dimension Δ.

In FSSY it was claimed that in the Coleman De Luccia background, the correlation function contains two terms, one of which was associated with NM and the other with NNM. A third term was found, but ignored on the basis that it was negligible when continued to the boundary. In fact the third term has an interesting significance that we will come back to, but first we will review the terms studied in FSSY.

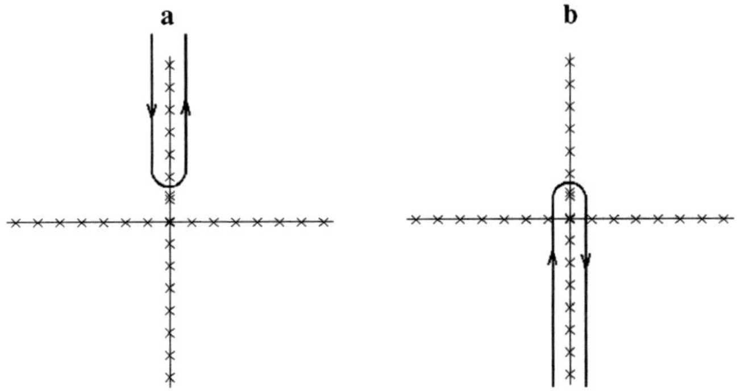

Fig. 6 Contours of integration for the two contributions G_1, G_2

In [27] the correlation function was expressed as a sum of two contour integrals on the k plane – k being an eigenvalue of the Laplacian on $\mathscr{H}_3$. The integral involves a certain reflection coefficient $\mathscr{R}(k)$ for a Schrodinger equation, derived from the wave equation on the Coleman De Luccia instanton. The contour integral is

$$e^{-(T_1+T_2)} \oint_C \frac{dk}{2\imath} \mathscr{R}(k) e^{-ik(T_1+T_2)} \frac{\left(e^{-ikR} - e^{-ikR-2\pi k}\right)}{2 \sinh R \sinh k\pi} \tag{43}$$

The contours of integration are shown in Fig. 6. The integrand has poles at all imaginary values of k, with a double pole at $k = i$. In addition there may be other singularities in the lower half plane. FSSY studied only the terms coming from the upper contour labeled **a** in the figure. It contains two terms related to NM and NNM respectively.

The normalizable contribution, G_1, is an infinite sum, each term having the form (42) with T-dependent coefficients. For late times,

$$\begin{aligned} G_1 &= \sum_{\Delta=2}^{\infty} G_\Delta e^{(\Delta-2)(T_1+T_2)} \frac{e^{-(\Delta-1)l}}{\sinh l} \\ &\to \sum_{\Delta=2}^{\infty} G_\Delta e^{(\Delta-2)(T_1+T_2)} e^{-\Delta R_1} e^{-\Delta R_2} (1-\cos\alpha)^{-\Delta} \end{aligned} \tag{44}$$

where Δ takes on integer values from 2 to ∞, and G_Δ are a series of constants which depend on the detailed CDL solution.

The connection with conformal field theory correlators is obvious; (44) is a sum of correlation functions for fields of definite dimension Δ, but with coefficients which depend on the time T. (It should be emphasized that the dimensions Δ in the present context are not related to bulk four dimensional masses by (41).) Note that the sum in (44) begins at $\Delta = 2$, implying that every term falls at least as fast as $\exp(-2R)$ with respect to either argument. Thus every term is normalizable.

Let us now extrapolate (44) to the surface Σ. Σ can be reached in two ways – the first being to go out along a constant T surface to $R = \infty$. Each term in the correlator has a definite R dependence which identifies its dimension.

Another way to get to Σ is to first pass to light-like infinity, $\mathscr{I}^+$, and then slide down the hat, along a light-like generator, until reaching Σ. For this purpose it is useful to define light-cone coordinates, $T^{\pm} = T \pm R$.

$$G_1 = e^{-(T_1^+ + T_2^+)} \sum_{\Delta} G_{\Delta} e^{(\Delta - 1)(T_1^- + T_2^-)} (1 - \cos\alpha)^{-\Delta} \tag{45}$$

We note that apart from the overall factor $e^{-(T_1^+ + T_2^+)}$, G_1 depends only on T^-, and therefore tends to a finite limit on $\mathscr{I}^+$. If we strip that factor off, then the remaining expression consists of a sum over CFT correlators, each proportional to a fixed power of e^{T^-}. In the limit ($T^- \to -\infty$) in which we pass to Σ, each term of fixed dimension tends to zero as $e^{(\Delta-1)(T_1^- + T_2^-)}$ with the dimension-2 term dominating the others.

The second term in the scalar correlation function discussed by FSSY consists of a single term,

$$G_2 = \frac{e^l}{\sinh l}(T_1 + T_2 + l) \to \left\{T_1^+ + T_2^+ + \log(1 - \cos\alpha)\right\} \tag{46}$$

The contribution (46) does not have the form of a correlator of a conformal field of definite dimension. To understand its significance, consider a canonical massless scalar field in two dimensions. On a two sphere the correlation function is ultraviolet divergent and has the form

$$\log\left\{\kappa^2(1 - \cos\alpha)\right\} \tag{47}$$

where κ is the ultraviolet regulator momentum. If the regulator momentum varies with location on the sphere – for example in the case of a lattice regulator with a variable lattice spacing – formula (47) is replaced by

$$\log\{(1 - \cos\alpha)\} + \log\kappa_1 + \log\kappa_2 \tag{48}$$

Evidently if we identify the UV cutoff κ with T^+,

$$\log\kappa = T^+ \tag{49}$$

the expressions in (46) and (48) are identical. The relation (49) is one of the central themes of this paper, that as we will see, relates RG flow to the observations of the Census Taker.

That the UV cutoff of the 2D boundary theory depends on R is very familiar from the UV/IR connection [15] in anti de Sitter space. In that case the T coordinate is absent and the log of the cutoff momentum in the conformal field theory would just be R. The additional time dependent contribution in (49) will become clear later when we discuss the Liouville field.

The logarithmic ultraviolet divergence in the correlator is a signal that massless 2D scalars are ill defined; the well-defined quantities being derivatives of the field.

When calculating correlators of derivatives, the cutoff dependence disappears. Thus for practical purposes, the only relevant term in (48) is $\log(1-\cos\alpha)$.

The existence of a dimension-zero scalar field on Σ is a surprise. It is obviously associated with bulk field-modes which don't go zero for large R. Such modes are non-normalizable on the hyperbolic plane, and are usually not included among the dynamical variables in anti de Sitter space.

In String Theory the only massless scalars in the hatted vacua would be moduli, which are expected to be "fixed" in the Ancestor. For that reason FSSY considered the effect of adding a four-dimensional mass term, $\mu\chi^2$, in the Ancestor vacuum. The result on the boundary scalar was to shift its dimension from $\Delta = 0$ to $\Delta = \mu$ (for small μ less than the Ancestor Hubble constant the corresponding mode stays non-normalizable). However the correlation function was not similar to those in G_1, each term of which had a dependence on T^-. The dimension μ term depends only on T^+:

$$G_2 \to e^{-\mu T_1^+} e^{-\mu T_2^+} \log(1-\cos\alpha)^{-\mu} \tag{50}$$

The two terms, (45) and (50) depend on different combinations of the coordinates, T^+ and T^-. It seems odd that there is one and only one term that depends solely on T^+ and all the rest depend on T^-. In fact the only reason is that FSSY ignored an entire tower of higher dimension terms, coming from the contour **b** that, like (50), depend only on T^+. From now on we will group all terms independent of T^- into the single expression G_2:

$$G_2 = \sum_{\Delta'} G_{\Delta'} e^{-(\Delta')(T_1^+ + T_2^+)} (1-\cos\alpha)^{-\Delta'} \tag{51}$$

The Δ' include μ, the positive integers and whatever other poles appear for $ik < 1$. In the case $\mu = 0$, the leading term in G_2 is (46).

We will return to the two terms G_1 and G_2 in Sect. 6.5.

5.2 Metric Fluctuations

To prove that there is a local field theory on Σ, the most important test is the existence of an energy-momentum tensor. In the ADS/CFT correspondence, the boundary energy-momentum tensor is intimately related to the bulk metric fluctuations. We assume a similar connection between bulk and boundary fields in the present context. In FSSY, metrical fluctuations were studied in a particular gauge which we will call the *Spatially Transverse-Traceless* (STT) gauge. The coordinates of region I can be divided into FRW time, T, and space x^m where $m = 1, 2, 3$. The STT gauge for metric fluctuations is defined by

$$\begin{aligned} \nabla^m h_{mn} &= 0 \\ h^m_m &= 0 \end{aligned} \tag{52}$$

In the second of equations (52), the index is raised with the aid of the background metric (3). The main benefit of the STT gauge is that metric fluctuations satisfy minimally coupled, massless, scalar equations, and the correlation functions are similar to G_1 and G_2. However the index structure is rather involved. We define the correlator,

$$\begin{aligned} \langle h^{\mu}_{\nu} h^{\sigma}_{\tau} \rangle &= G \left\{ {\mu\sigma \atop \nu\tau} \right\} \\ &= G_1 \left\{ {\mu\sigma \atop \nu\tau} \right\} + G_2 \left\{ {\mu\sigma \atop \nu\tau} \right\}. \end{aligned} \tag{53}$$

The complicated index structure of G was worked out in detail in FSSY. In this paper we quote only the results of interest – in particular those involving elements of $G \left\{ {\mu\sigma \atop \nu\tau} \right\}$ in which all indices lie in the two-sphere Ω_2. Thus we consider the correlation function $G_1 \left\{ {ik \atop jl} \right\}$.

As in the scalar case, G_1 consists of an infinite sum of correlators, each corresponding to a field of dimension $\Delta = 2, 3, 4, \ldots$. The asymptotic T and R dependence of the terms is identical to the scalar case, and the first term has $\Delta = 2$. This is particulary interesting because it is the dimension of the energy-momentum tensor of a two-dimensional boundary conformal field theory. Once again this term is also time-independent.

After isolating the dimension-two term and stripping off the factors $\exp(-2R)$, the resulting correlator is called $G_1 \left\{ {ik \atop jl} \right\} |_{\Delta=2}$. The calculations of FSSY revealed that this term is two-dimensionally traceless, and transverse.

$$\begin{aligned} G_1 \left\{ {ik \atop il} \right\} |_{\Delta=2} = G_1 \left\{ {ik \atop jk} \right\} |_{\Delta=2} &= 0 \\ \nabla_i G_1 \left\{ {ik \atop jl} \right\} |_{\Delta=2} &= 0. \end{aligned} \tag{54}$$

Equation (54) is the clue that, when combined with the dimension-2 behavior of $G_1 \left\{ {ik \atop il} \right\} |_{\Delta=2}$, hints at a local theory on Σ. It insures that it has the precise form of a two-point function for an energy-momentum tensor in a conformal field theory. The only ambiguity is the numerical coefficient connecting $G_1 \left\{ {ik \atop jl} \right\} |_{\Delta=2}$ with $\langle T^i_j T^k_l \rangle$. We will return to this coefficient momentarily.

The existence of a transverse, traceless, dimension-two operator is a necessary condition for the boundary theory on Σ to be local: at the moment it is our main evidence. But there is certainly more that can be learned by computing multipoint functions. For example, from the three-point function $\langle hhh \rangle$ it should be possible verify the operator product expansion and the Virasoro algebra for the energy-momentum tensor.

Dimensional analysis allows us to estimate the missing coefficient connecting the metric fluctuations with T^i_j, and at the same time determine the central charge c. In [27] we found c to be of order the horizon entropy of the Ancestor vacuum. We repeat the argument here:

Assume that the (bulk) metric fluctuation h has canonical normalization, i.e., it has bulk mass dimension 1 and a canonical kinetic term. Either dimensional analysis or explicit calculation of the two point function $\langle hh \rangle$ shows that it is proportional to square of the Ancestor Hubble constant.

$$\langle hh \rangle \sim H^2. \tag{55}$$

Knowing that the three point function $\langle hhh \rangle$ must contain a factor of the gravitational coupling (Planck Length) l_p, it can also be estimated by dimensional analysis.

$$\langle hhh \rangle \sim l_p H^4. \tag{56}$$

Now assume that the 2D energy-momentum tensor is proportional to the boundary dimension-two part of h, i.e., the part that varies like e^{-2R}. Schematically,

$$T = qh \tag{57}$$

with q being a numerical constant. It follows that

$$\begin{aligned} \langle TT \rangle &\sim q^2 H^2 \\ \langle TTT \rangle &\sim q^3 l_p H^4. \end{aligned} \tag{58}$$

Lastly, we use the fact that the ratio of the two and three point functions is parametrically independent of l_p and H because it is controlled by the classical algebra of diffeomorphisms: $[T,T] = T$. Putting these elements together we find,

$$\langle TT \rangle \sim \frac{1}{l_p^2 H^2} \tag{59}$$

Since we already know that the correlation function has the correct form, including the short distance singularity, we can assume that the right hand side of (59) also gives the central charge. It can be written in the rather suggestive form:

$$c \sim Area/G \quad (G = l_p^2) \tag{60}$$

where *Area* refers to the horizon of the Ancestor vacuum. In other words, the central charge of the hypothetical CFT is proportional to the *horizon entropy of the Ancestor* .

5.3 Dimension Zero Term

The term $G_2 \left\{ {ik \atop jl} \right\}$ begins with a term, which like its scalar counterpart, has a non-vanishing limit on Σ. It is expressed in terms of a standard 2D bi-tensor $t \left\{ {ik \atop jl} \right\}$ which is traceless and transverse in the two dimensional sense. If the correlation function were given just by $t\begin{Bmatrix} ik \\ jl \end{Bmatrix}$, it would be a pure gauge artifact. One can see this by considering the linearized expression for the 2D curvature-scalar C,

$$C = \nabla_i \nabla_j h^{ij} - 2\nabla^i \nabla_i \, Tr \, h. \tag{61}$$

The 2D curvature associated with a traceless transverse fluctuation vanishes, and since $t\begin{Bmatrix} ik \\ jl \end{Bmatrix}$ by itself is traceless-transverse with respect to both points, it would be pure gauge if it appeared by itself.

However, the actual correlation function $G_2\begin{Bmatrix} ik \\ jl \end{Bmatrix}$ is given by

$$G_2\begin{Bmatrix} ik \\ jl \end{Bmatrix} = t\begin{Bmatrix} ik \\ jl \end{Bmatrix} \{R_1 + T_1 + R_2 + T_2 + \log(1 - \cos\alpha)\} \tag{62}$$

The linear terms in $R + T$, being proportional to $t\begin{Bmatrix} ik \\ jl \end{Bmatrix}$ are pure gauge, but the finite term

$$t\begin{Bmatrix} ik \\ jl \end{Bmatrix} \log(1 - \cos\alpha) \tag{63}$$

gives rise to a non-trivial 2D curvature-curvature correlation function of the form

$$\langle CC \rangle = (1 - \cos\alpha)^{-2}. \tag{64}$$

One difference between the metric fluctuation h, and the scalar field χ, is that we cannot add a mass term for h in the Ancestor vacuum to shift its dimension.

Finally, as in the scalar case, there is a tower of higher dimension terms in the tensor correlator, $G_2\begin{Bmatrix} ik \\ jl \end{Bmatrix}$ that only depend on T^+.

The existence of a zero dimensional term in $G_2\begin{Bmatrix} ik \\ jl \end{Bmatrix}$, which remains finite in the limit $R \to \infty$ indicates that fluctuations in the boundary geometry – fluctuations which are due to the asymptotic warmness – cannot be ignored. One might expect that in some way these fluctuations are connected with the field U that we encountered in the Holographic version of the Wheeler-DeWitt equation. In the next section we will elaborate on this connection.

That is the data about correlation functions on the boundary sphere Σ that form the basis for our conjecture that there exists a local holographic boundary description of the open FRW universe. There are a number of related puzzles that this data raises: First, how does time emerge from a Euclidean QFT? The bulk coordinate R can be identified with scale size just as in ADS/CFT but the origin of time requires a new mechanism.

The second puzzle concerns the number of degrees of freedom in the boundary theory. The fact that the central charge is the entropy of the Ancestor suggests that there are only enough degrees of freedom to describe the false vacuum and not the much large number needed for the open FRW universe at late time.

6 Liouville Theory

6.1 Breaking Free of the STT Gauge

The existence of a Liouville sector describing metrical fluctuations on Σ seems dictated by both the Holographic Wheeler-DeWitt theory and from the data of the previous section. It is clear that the Liouville field is somehow connected with the non-normalizable metric fluctuations whose correlations are contained in (62), although the connection is somewhat obscured by the choice of gauge in [27]. In the STT gauge the fluctuations h are traceless, but not transverse (in the 2D sense). From the viewpoint of 2D geometry they are not pure gauge as can be seen from the fact that the 2D curvature correlation does not vanish. One might be tempted to identify the Liouville mode with the zero-dimension piece of (62). To do so would of course require a coordinate transformation on Ω_2 in order to bring the fluctuation h_i^j to the "conformal" form $\tilde{h}\delta_i^j$.

This identification may be useful but it is not consistent with the Wheeler-DeWitt philosophy. The Liouville field U that appears in the Wheeler-DeWitt wave function is not tied to any specific spatial gauge. Indeed, the wave function is required to be invariant under gauge transformations,

$$x^{\mu} \to x^{\mu} + f^{\mu}(x) \tag{65}$$

under which the metric transforms:

$$g_{\mu\nu} \to g_{\mu\nu} + \nabla_{\mu} f_{\nu} + \nabla_{\nu} f_{\mu}. \tag{66}$$

Let's consider the effect of such transformations on the boundary limit of h_{ij}. The components of f along the directions in Σ induce 2D coordinate transformation under which h transforms conventionally. Invariance under these transformations merely mean that the action S must be a function of 2D invariants.

Invariance under the shifts f^R and f^T are more interesting. In particular the combination $f^+ = f^R + f^T$ generates non-trivial transformations of the boundary metric h_{ij}. An easy calculation shows that,

$$h_i^j \to h_i^j + f^+(\Omega_2)\delta_i^j. \tag{67}$$

In other words, shift transformations f^+, induce Weyl re-scalings of the boundary metric. This prompts us to modify the definition of the Liouville field from

$$U = T + \tilde{h} \tag{68}$$

to

$$U = T + \tilde{h} + f^+. \tag{69}$$

One might wonder about the meaning of an equation such as (69). The left side of the equation is supposed to be a dynamical field on Σ, but the right side contains an arbitrary function f^+. The point is that in the Wheeler-DeWitt formalism the wave function must be invariant under shifts, but in the original analysis of FSSY a specific gauge was chosen. Thus, in order to render the wave function gauge invariant, one must allow the shift f^+ to be an integration variable, giving it the status of a dynamical field.

A similar example is familiar from ordinary gauge theories. The analog of the Wheeler-DeWitt gauge-free formalism would be the unfixed theory in which one integrates over the time component of the vector potential. The analog of the STT gauge would be the Coulomb gauge. To go from one to the other we would perform the gauge transformation

$$A_0 \rightarrow A_0 + \partial_0 \phi. \tag{70}$$

Integrating over the gauge function ϕ in the path integral would restore the gauge invariance that was given up by fixing Coulomb gauge.

Returning to the Liouville field, since both $\tilde{h}$ and f are linearized fluctuation variables, we see that the classical part of U is still the FRW conformal time.

One important point: because the effect of the shift f^+ is restricted to the trace of h, it does not influence the traceless-transverse (dimension-two) part of the metric fluctuation, and the original identification of the 2D energy-momentum tensor is unaffected.

Finally, invariance under the shift f^- is trivial in this order, at least for the thin wall geometry. The reason is that in the background geometry, the area does not vary along the T^- direction.

Given that the boundary theory is local, and includes a boundary metric, it is constrained by the rules of two-dimensional quantum gravity laid down long ago by Polyakov [32]. Let us review those rules for the case of a conformal "matter" field theory coupled to a Liouville field. Two-dimensional coordinate invariance implies that the central charge of the Liouville sector cancels the central charge of all other fields. We have argued in [27] (and in Sect. 5) that the central charge of the matter sector is of order the horizon area of the Ancestor vacuum, measured in Planck Units. It is obvious from the 4-dimensional bulk viewpoint that the semiclassical analysis that we have relied on, only makes sense when the Hubble radius is much larger than the Planck scale. Thus we take the central charge of matter to satisfy $c \gg 1$. As a consequence, the central charge of the Liouville sector, c_L, must be large and negative. Unsurprisingly, the negative value of c is the origin of the emergence of time.

The formal development of Liouville theory begins by defining two metrics on Ω_2. The first is what I will call the reference metric $\hat{g}_{ij}$. Apart from an appropriate degree smoothness, and the assumption of Euclidean signature, the reference metric is arbitrary but fixed. In particular it is not integrated over in the path integral. Moreover, physical observables must be independent of $\hat{g}_{ij}$.

The other metric is the "real" metric denoted by g_{ij}. The purpose of the reference metric is merely to implement a degree of gauge fixing. Thus one assumes that the real metric has the form,

$$g_{ij} = e^{2U} \hat{g}_{ij}. \tag{71}$$

The real metric – that is to say U – is a dynamical variable to be integrated over.

For positive c_L the Liouville Lagrangian is

$$L_L = \frac{Q^2 \sqrt{\hat{g}}}{8\pi} \left\{ \hat{\nabla} U \hat{\nabla} U + \hat{R} u \right\} \tag{72}$$

where $\hat{R}$, $\hat{\nabla}$, all refer to the sphere Ω_2, with metric $\hat{g}$. The constant Q is related to the central charge c_L by

$$Q^2 = \frac{c_L}{6} \tag{73}$$

The two dimensional cosmological constant has been set to zero for the moment, but it will return to play a surprising role. For future reference we note that the cosmological term, had we included it, would have had the form,

$$L_{cc} = \sqrt{\hat{g}} \lambda e^{2U}. \tag{74}$$

It is useful to define a field $\phi = 4QU$ in order to bring the kinetic term to canonical form. One finds,

$$L_L = \frac{\sqrt{\hat{g}}}{8\pi} \left\{ \hat{\nabla} \phi \hat{\nabla} \phi + Q \hat{R} \phi \right\} \tag{75}$$

and, had we included a cosmological term, it would be

$$L_{cc} = \sqrt{\hat{g}} \lambda \exp \frac{\phi}{2Q}. \tag{76}$$

By comparison with the case of positive c_L, very little is rigorously understood about Liouville theory with negative central charge. In this paper we will make a huge leap of faith that may well come back to haunt us: we assume that the theory can be defined by analytic continuation from positive c_L. To that end we note that the only place that the central charge enters (75) and (76) is through the constants Q and γ, both of which become imaginary when c_L becomes negative. Let us define,

$$Q = i\mathcal{Q}. \tag{77}$$

Equations (75) and (76) become,

$$\begin{aligned} L_L &= \frac{\sqrt{\hat{g}}}{8\pi} \left\{ \hat{\nabla} \phi \hat{\nabla} \phi + i\mathcal{Q} \hat{R} \phi \right\} \\ &= \frac{\sqrt{\hat{g}}}{8\pi} \left\{ \hat{\nabla} \phi \hat{\nabla} \phi + 2i\mathcal{Q} \phi \right\} \end{aligned} \tag{78}$$

(where we have used $\hat{R} = 2$), and

$$L_{cc} = \sqrt{\hat{g}}\lambda \exp\frac{-i\phi}{2\mathcal{Q}}. \tag{79}$$

Let us come now to the role of λ. First of all λ has nothing to do with the four-dimensional cosmological constant, either in the FRW patch or the Ancestor vacuum. Furthermore it is not a constant in the action of the boundary theory. Its proper role is as a *Lagrange multiplier* that serves to specify the time T, or more exactly, the global scale factor. The procedure is motivated by the Wheeler-DeWitt procedure of identifying the scale factor with time. In the present case of the thin-wall limit, we identify $\exp 2U$ with $\exp 2T$. Thus we insert a δ function in the path integral,

$$\delta\left(\int \sqrt{\hat{g}}(e^{2U} - e^{2T})\right) = \int dz \exp iz\left(\int \sqrt{\hat{g}}(e^{2U} - e^{2T})\right) \tag{80}$$

The path integral (which now includes an integration over the imaginary 2D cosmological constant z) involves the action

$$L_L + L_{cc} = \frac{\sqrt{\hat{g}}}{8\pi}\left\{\hat{\nabla}\phi\hat{\nabla}\phi + 2i\mathcal{Q}\phi + 8\pi iz\exp\frac{-i\phi}{2\mathcal{Q}} - 8\pi ize^{2T}\right\} \tag{81}$$

There is a saddle point when the potential

$$V = 2i\mathcal{Q}\phi + 8\pi iz\exp\frac{-i\phi}{2\mathcal{Q}} - 8\pi ize^{2T} \tag{82}$$

is stationary; this occurs at,

$$\begin{aligned} \exp\frac{-i\phi}{2\mathcal{Q}} &= e^{2T} \\ z &= i\frac{\mathcal{Q}^2}{8\pi}e^{-2T} \end{aligned} \tag{83}$$

or in terms of the original variables,

$$\begin{aligned} e^{2U} &= e^{2T} \\ \lambda &= \frac{\mathcal{Q}^2}{8\pi}e^{-2T} \end{aligned} \tag{84}$$

Once λ has been determined by (84), the Liouville theory with that value of λ determines expectation values of the remaining variables as functions of the time. Thus, as we mentioned earlier, the cosmological constant is not a constant of the theory but rather a parameter that we scan in order to vary the cosmic time.

It should be noted that the existence of the saddle point (84) is peculiar to the case of negative c.

6.2 Liouville, Renormalization, and Correlation Functions

6.3 Preliminaries

There are two preliminary discussions that will help us understand the application of Liouville Theory to cosmic holography. The first is about the ADS/CFT connection between the bulk coordinate R, and renormalization-group-running of the boundary field theory. There are three important length scales in every quantum field theory. The first is the "low energy scale;" in the present case the low energy scale is the radius of the sphere which we will call L.

The second important length is the "bare" cutoff scale – where the underlying theory is prescribed. Call it a. The bare input is a collection of degrees of freedom, and an action coupling them. In a lattice gauge theory the degrees of freedom are site and link variables, and the couplings are nearest neighbor to insure locality[5] In a ferromagnet they are spins situated on the sites of a crystal lattice.

The previous two scales have obvious physical meaning but the third scale is arbitrary: a sliding scale called the renormalization or reference scale. We denote it by δ. The reference scale is assumed to be much smaller than L and much larger than a, but otherwise it is arbitrary. It helps to keep a concrete model in mind. Instead of a regular lattice, introduce a "dust" of points with average spacing a. It is not essential that a be uniform on the sphere. Thus the spacing of dust points is a function of position, $a(\Omega_2)$. The degrees of freedom on the dust-points, and their nearest-neighbor couplings, will be left implicit.

Next we introduce a second dust at larger spacing, δ. The δ-dust provides the reference scale. It is well known that for length scales greater than δ, the bare theory on the a-dust can be replaced by a renormalized theory defined on the δ-dust. The renormalized theory will typically be more complicated, containing second, third, and nth neighbor couplings.

Generally, the dimensionless form of the renormalized theory will depend on δ in just such a way that physics at longer scales is exactly the same as it was in the original theory. The dimensionless parameters will flow as the reference scale is changed.

If there is an infrared fixed-point, and if the bare theory is in the basin of attraction of the fixed-point, then as δ becomes much larger than a, the dimensionless parameters will run to their fixed-point values. In that case the continuum limit ($a \to 0$) will be a conformal field theory with $SO(3,1)$ invariance.

Similar things hold in the theory of bulk anti de Sitter space, although in that case the discussion of the bare scale is less relevant – one might as well take the continuum limit $a \to 0$ from the start. In the boundary field theory the infrared scale is provided by the spherical boundary of ADS. From the bulk viewpoint the boundary is at infinite proper distance, at $R = \infty$. However, the time for a signal to reach

[5] Nearest neighbor is common but not absolutely essential. However this subtlety is not important for us.

the boundary and be reflected back to the bulk is finite. In that respect anti de Sitter space behaves like a finite cavity, requiring specific boundary conditions. To be definite, the bulk theory is infrared regulated by replacing Σ with a reference-boundary Σ_0, at finite R. Specifying the boundary conditions on Σ_0 is equivalent to specifying the field theory parameters at scale δ. In parallel with the field theory discussion, the cutoff R, can vary with angular position: $R = R_0(\Omega_2)$. We can now state the UV/IR connection by the simple identification,

$$\delta(\Omega_2) = e^{-R_0(\Omega_2)} \tag{85}$$

A useful slogan is that *"Motion along the R direction is the same as renormalization-group flow."*

Now to the second preliminary – some observations about Lioville Theory. Again it is helpful to have a concrete model. Liouville theory is closely connected with the theory of dense, planar, "fishnet" diagrams [33] such as those which appear in large N gauge theories, and matrix models [34–36]. The fishnet plays the role of the bare lattice in the previous discussion, but now it's dynamical – we sum over all fishnet diagrams, assuming only that the spacing (on the sphere) is everywhere much smaller than the sphere size, L. As before, we call the angular spacing between neighboring points on the sphere, $a(\Omega)$.

Each fishnet defines a metric on the sphere. Let $d\alpha$ be a small angular interval (measured in radians). The fishnet-metric is defined by[6]

$$ds^2 = \frac{d\alpha^2}{a(\Omega)^2} \tag{86}$$

As before we introduce a reference scale δ. It can also be a fishnet, but now it is fixed, its vertices nailed down, not to be integrated over. We continue to assume that δ satisfies the inequalities, $a(\Omega) \ll \delta(\Omega) \ll L$, but otherwise it is arbitrary. The δ-metric is defined by

$$ds_\delta^2 = \frac{d\alpha^2}{\delta(\Omega)^2} \tag{87}$$

We can now define the Liouville field U. All it is is the ratio of the reference and fishnet scales:

$$e^U \equiv \delta / a \tag{88}$$

Using (88) together with $\delta = e^{-R}$, and $ds = \frac{d\alpha}{a}$, we see that U is also given by the relation,

$$ds = d\alpha\, e^{(R_0+U)}. \tag{89}$$

In (89) both R_0 and U are functions of location on Ω_2, but only U is dynamical, i.e., to be integrated over.

---

[6] Strictly speaking there is no need to introduce a discrete fishnet lattice at the scale δ. It is sufficient to just define a continuous function $\delta(\Omega_2)$, and from it define a reference metric by (86).

6.4 Liouville in the Hat

With that in mind, we return to cosmic holography, and consider the metric on the regulated spatial boundary of FRW, Σ_0. In the absence of fluctuations it is

$$ds^2 = e^{2R_0} e^{2T} d^2\Omega_2.$$

In general relativity it is natural to allow both R_0 and T to vary over the sphere, so that

$$ds^2 = e^{2R_0(\Omega_2)} e^{2T(\Omega_2)} d^2\Omega_2 \quad (90)$$

The parallel between (89 and (90) is obvious. Exactly as we might have expected from the Wheeler-DeWitt interpretation, the Liouville field, U, may be identified with time T, at least when both are large.

$$U \approx T \quad (91)$$

To summarize, Let's list a number of correspondences:

$$\begin{aligned} \delta &\leftrightarrow \mu^{-1} \leftrightarrow e^R \\ \lambda &\leftrightarrow e^{-T} \\ a &\leftrightarrow e^{-T^+} = e^{-(T+R)} \end{aligned} \quad (92)$$

One other point about Liouville Theory: the density of vertices of a fishnet is normally varied by changing the weight assigned to vertices. When the fishnet is a Feynman diagram the weight is a coupling constant g. It is well known that the coupling constant and Liouville cosmological constantare alternate descriptions of the same thing. Either can be used to vary the average vertex density – increasing it either by increasing g or decreasing λ. The very dense fishnets correspond to large U and therefore large FRW time, whereas very sparse diagrams dominate the early Planckian era.

6.5 Proactive and Reactive Objects in Quantum Field Theory

There are two kinds of objects in Wilsonian renormalization that correspond quite closely to the terms G_1 and G_2 that we have found in the Sect. 5. I don't know if there is a term for the distinction, but I will call them *"proactive"* and *"reactive"*. Proactive objects are not quantities that we directly measure; they are objects which go into the definition of the theory. The best example is the exact Wilsonian action, defined at a specific reference scale. The form of proactive quantities depends on that reference scale, and so does the value of their matrix elements; indeed their form, varies with δ in such a way as to keep the physics fixed at longer distances.

By contrast, reactive objects are observables whose value does not depend at all on the reference scale. They do depend on the "bare" cutoff scale a through wave function renormalization constants, which typically tend to zero as $a \to 0$. The wave function renormalization constants are usually stripped off when defining a quantum field but we will find it more illuminating to keep them.

The distinction between these two kinds of objects is subtle, and is perhaps best expressed in Polchinski's version of the exact Wilsonian renormalization group [37]. In that scheme, at every scale there is a renormalized description in terms of local defining fields $\phi(x)$, but the proactive action grows increasingly complicated as the reference scale is lowered.

Consider the exact effective action defined at reference scale δ. It is given by an infinite expansion of the form

$$L_W(\delta) = \sum_{\Delta=2}^{\infty} g_\Delta \mathscr{O}_\Delta \tag{93}$$

Where $\mathscr{O}_\Delta$ are a set of operators of dimension Δ, and g_Δ are dimensional coupling constants. The renormalization flow is expressed in terms of the dimensionless coupling constants,

$$\tilde{g}_\Delta = g_\Delta \delta^{(2-\Delta)}. \tag{94}$$

The $\tilde{g}$ satisfy RG equations,

$$\frac{d\tilde{g}}{d\log\delta} = -\beta(\tilde{g}), \tag{95}$$

and at a fixed point they are constant. Thus the dimensional constants g_Δ in the Lagrangian will grow with δ. Normalizing them at the bare scale a, in the fixed-point case we get,

$$g_\Delta = g_a \left\{\frac{\delta}{a}\right\}^{(\Delta-2)} \tag{96}$$

$$L_W(\delta) = \sum_{\Delta=2}^{\infty} \mathscr{O}_\Delta \left\{\frac{\delta}{a}\right\}^{(\Delta-2)}. \tag{97}$$

Now consider the two point function of the effective action, $\langle L_W(\delta) L_W(\delta)\rangle$, evaluated at distance scale $L \gg \delta$

$$\langle L_W(\delta) L_W(\delta)\rangle = \sum_{\Delta=2}^{\infty} \langle \mathscr{O}_\Delta \mathscr{O}_\Delta \rangle \left(\frac{\delta}{a}\right)^{2(\Delta-2)} \tag{98}$$

Suppose the theory is defined on a sphere of radius L and we are interested in the correlator $\langle L_W(\delta) L_W(\delta)\rangle$ between points separated by angle α. The factor $\langle \mathscr{O}_\Delta \mathscr{O}_\Delta \rangle$ is the two-point function of a field of dimension Δ, in a theory on the sphere of size L, with an ultraviolet cutoff at the reference scale δ. Accordingly it has the form

$$\langle \mathcal{O}_\Delta \mathcal{O}_\Delta \rangle = \left(\frac{\delta}{L}\right)^{2\Delta} (1-\cos\alpha)^{-\Delta} \tag{99}$$

where the two factors of $\left(\frac{\delta}{L}\right)^{\Delta}$ are the ultraviolet-sensitive wave function renormalization constants. The final result is

$$\langle L_W(\delta) L_W(\delta) \rangle = \sum_{\Delta=2}^{\infty} C_\Delta \left(\frac{\delta}{a}\right)^{2(\Delta-2)} \left(\frac{\delta}{L}\right)^{2\Delta} (1-\cos\alpha)^{-\Delta} \tag{100}$$

Note the odd dependence of (100) on the arbitrary reference scale δ. That dependence is typical of proactive quantities.

Now consider a reactive quantity such as a fundamental field, a derivative of such a field, or a local product of fields and derivatives. Their matrix elements at distance scale L will be independent of the reference scale (although it will depend of the bare cutoff a) and be of order,

$$\langle \phi \phi \rangle \sim \left(\frac{a}{L}\right)^{2\Delta_\phi} \tag{101}$$

where Δ_ϕ is the operator dimension of ϕ. Thus we see two distinct behaviors for the scaling of correlation functions:

$$\left(\frac{\delta}{a}\right)^{2(\Delta-2)} \left(\frac{\delta}{L}\right)^{2\Delta} \quad \text{proactive} \tag{102}$$

and

$$\left(\frac{a}{L}\right)^{2\Delta_\phi} \quad \text{reactive} \tag{103}$$

The formulas are more complicated away from a fixed point but the principles are the same.

We note that the effective action is not the only proactive object. The energy-momentum tensor, and various currents computed from the effective action, will also be proactive. As we will see these two behaviors – proactive and reactive – exactly correspond to the dependence in (45) and (46).

Now we are finally ready to complete the discussion about the relation between the correlators of Sect. 5 and proactive/reactive operators. Begin by noting that in ADS/CFT, the minimally coupled massless (bulk) scalar is the dilaton, and its associated boundary field is the Lagrangian density. It may seem puzzling that in the present case, an entire infinite tower of operators seems to replace, what in ADS/CFT is a single operator. In the case of the metric fluctuations a similar tower replaces the energy-momentum tensor. The puzzle may be stated another way. The FRW geometry consists of an infinite number of Euclidean ADS time slices. At what time (or what 2D cosmological constant) should we evaluate the boundary limits of the metric fluctuations, in order to define the energy-momentum tensor? As we will see, a parallel ambiguity exists in Liouville theory.

Return now, to the three scales of Liouville Theory: the infrared scale L, the reference scale δ, and the fishnet scale a, with $L \gg \delta \gg a$. It is natural to assume that the basic theory is defined at the bare fishnet scale a by some collection of degrees of freedom at each lattice site, and also specific nearest-neighbor couplings – the latter insuring locality. Now imagine a Wilsonian integration of all degrees of freedom on scales between the fishnet scale and the reference scale, including the fishnet structure itself. The result will be a proactive effective action of the type we described in (93). Moreover the correlation function of L_{eff} will have the form (100). But now, making the identifications (85) and

$$\frac{\delta}{a} = e^{U} = e^{T}, \tag{104}$$

we see that (102) for proactive scaling becomes (for each operator in the product)

$$e^{(\Delta-2)T} e^{-\Delta R} \tag{105}$$

This is in precise agreement with the coefficients in the expansion (44). Similarly the reactive scaling (103) is $e^{-\Delta T^{+}}$, in agreement with the properties of G_2.

It is not obvious to me exactly why the bulk fields should correspond to proactive and reactive boundary fields in the way that they do. I might point out the solutions to the wave equation in the bulk are generally sums of two types of modes,

$$\begin{aligned} \chi_- &\to g_-(\Omega_2)\, F_-(T^-)\, e^{-2T} \\ \chi_+ &\to g_+(\Omega_2)\, F_+(T^+). \end{aligned} \tag{106}$$

Evidently the two types of solutions couple to objects that are reactive and proactive under the RG flow.

What happens to the proactive objects if we approach Σ by sending $T^+ \to \infty$ and $T_- \to -\infty$? In this limit only the dimension-two term survives: exactly what we would expect if the matter action ran toward a fixed point. All of the same things hold true for the tensor fluctuations. Before the limit $T^- \to -\infty$, the energy-momentum tensor consists of an infinite number of higher dimension operators but in the limit, all tend to zero except for the dimension two term.

It should be observed that the higher dimension contributions to $G_1 \left\{ {ik \atop jl} \right\}$ are not transverse in the two-dimensional sense. This is to be expected: before the limit is taken, the Liouville field does not decouple from the matter field, and the matter energy-momentum is not separately conserved. But if the matter theory is at a fixed point, i.e., scale invariant, the Liouville and matter do decouple and the matter energy-momentum should be conserved. Thus, in the limit in which the dimension-two term dominates, it should be (and is) transverse-traceless.

The RG flow is usually thought of in terms of a single independent flow-parameter. In some versions it's the logarithm of the bare cutoff scale, and in other

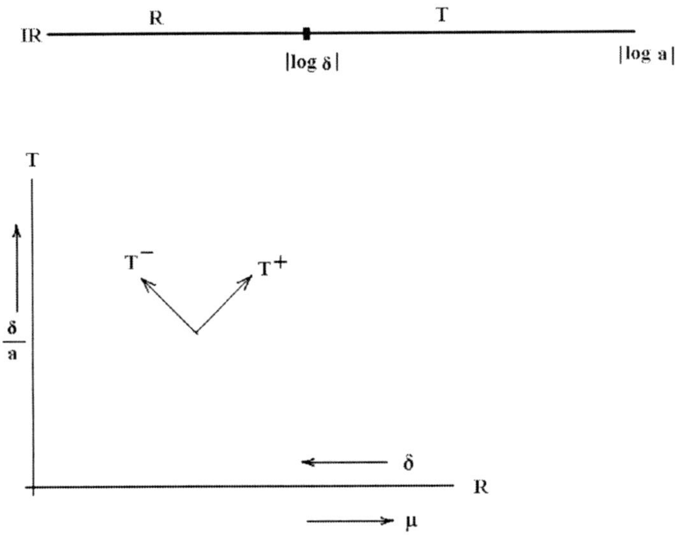

Fig. 7 The Wilson line of scales, and the two dimensional R,T plane

formulations it's the log of the renormalization scale. In the conventional ADS/CFT framework, R can play either role. One can imagine a bare cutoff at some large R_0 or one can push the bare cutoff to infinity and think of R as a running renormalization scale.

However, for our purposes, it is better to keep track of both scales. One can either think of a one-dimensional (logarithmic) axis – we can call it the "Wilson line" – extending from the infrared scale to the fishnet scale a, or a two dimensional R,T plane. In either case the effective action as a function of two independent variables. Figure 7 shows a sketch of the Wilson line and the two dimensional plane representing the two directions R, and T.

The two independent parameters can be chosen to be a and δ, or equivalently R and T. Yet another choice is to work in momentum space. The reference energy-scale is usually called μ.

$$\mu = e^R \tag{107}$$

And in the case of negative central charge, the two-dimensional cosmological constant λ can replace T.

In this light, it is extremely interesting that the distinction between proactive and reactive scaling, corresponds to motion along the two light-like directions T^- and T^+ as depicted in Fig. 8.

It is important to understand that the duality between FRW cosmology and Liouville 2D gravity does not only involve the continuum fixed point theory. As long as T is finite the theory has some memory of the bare theory. It's only in the limit $T \to \infty$ that the theory flows to the fixed point and loses memory of the bare details. We will come back to this point in Sect. 8 when I discuss the Garriga et al. [31] "Persistence of Memory" phenomenon.

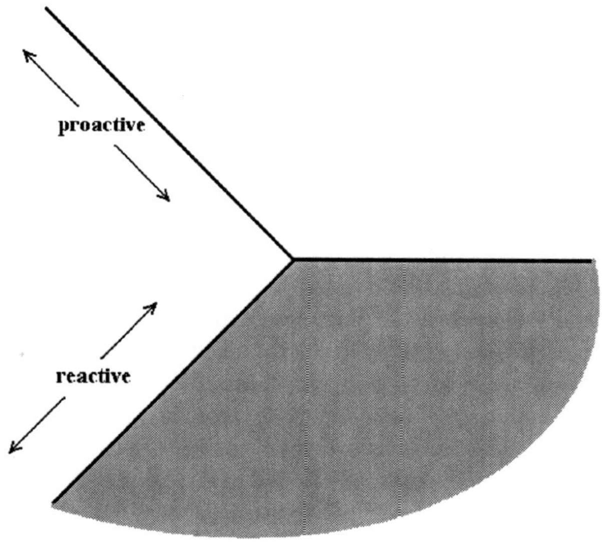

Fig. 8 Proactive and reactive quantities scale with the two light-like directions T^- and T^+

6.6 Boussovian Bounds and The c-Theorem

When a conformal matter theory is coupled to a Liouville field, the two sectors decouple, except for the constraint that the total central charge vanish,

$$c + c_L = 0, \tag{108}$$

where c and c_L refer to matter and Liouville respectively. Moreover the matter central charge is constant since a conformal theory is, by definition, at a fixed-point. From now on when I speak of the central charge I will be referring to the matter sector only.

There is one caveat to this rule that c is constant at a fixed point: it applies straightforwardly as long as the reference scale is much smaller than the infrared scale, i.e., the size of the sphere. However, the finite (coordinate) size of the boundary sphere provides an infrared cutoff that is similar to the confinement scale in a confining gauge theory. As the reference scale becomes comparable to the total sphere, the theory runs out of lower energy degrees of freedom and, with some definition, the c-function will go to zero.

The central charge is a measure of the number of degrees of freedom in an area-cell of size e^{-2T^+}. Naively, the holographic principle would say it is the area in such a cell in Planck units. However, as we have seen in the past, this is not always the case [38, 39].

Recall that motion up and down the T^- axis, at fixed T^+, corresponds to the usual RG flow of the Wilsonian action (keeping the bare scale fixed and letting the reference scale vary). Thus, at a fixed point, the area along a fixed T^+ line should be constant. In the thin wall limit the area is given by

$$A = e^{2T} \sinh^2 R. \tag{109}$$

As long as $R \gg 1$ the area is indeed constant for fixed T^+. However as we approach $R \sim 1$ the area quickly tends to zero, consistent with the remarks above.

More generally, away from fixed points the Zamolodchikov c-theorem requires c to decrease with increasing reference-scale δ. This seems to suggest that the area must monotonically decrease as T^- increases, with T^+ held fixed, which, as we will see raises a paradox; the area is not monotonic beyond the thin-wall approximation.

The study of how area varies along light sheets – "Boussology" – has a long and celebrated history which I will assume you are familiar with [39]. Rather than deal with the equations that determine how area varies I will draw some Bousso diagrams and tell you the conclusions. First the thin-wall case: Figure 9 shows the FRW patch of a thin-wall Coleman De Luccia nucleation. In fact it is the forward light-cone of a point in flat Minkowski space. The red line is a light-sheet of constant T^+. The Bousso wedges indicate the light-like directions along which the area decreases. The entire geometry consists of a single region in which all wedges point toward

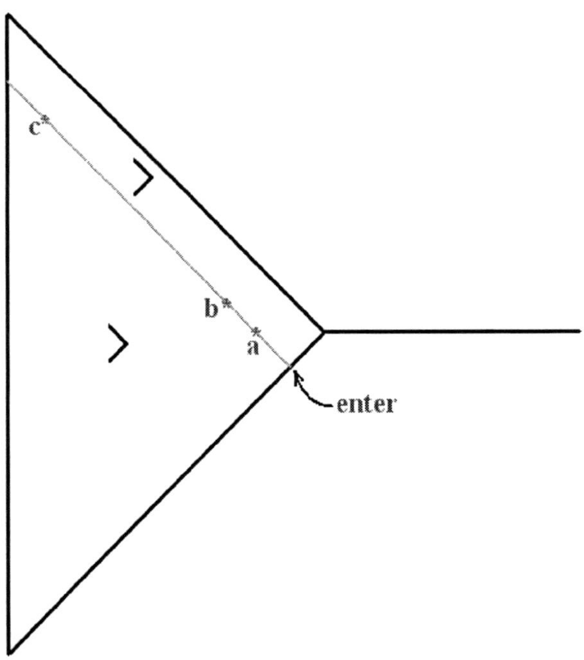

Fig. 9 Bousso diagram for the FRW geometry resulting from a thin-wall CDL bubble. The entire FRW geometry consists of a single region in which the contracting light-sheets all point in the same directions. The thin light-like line is a surface of constant T^+

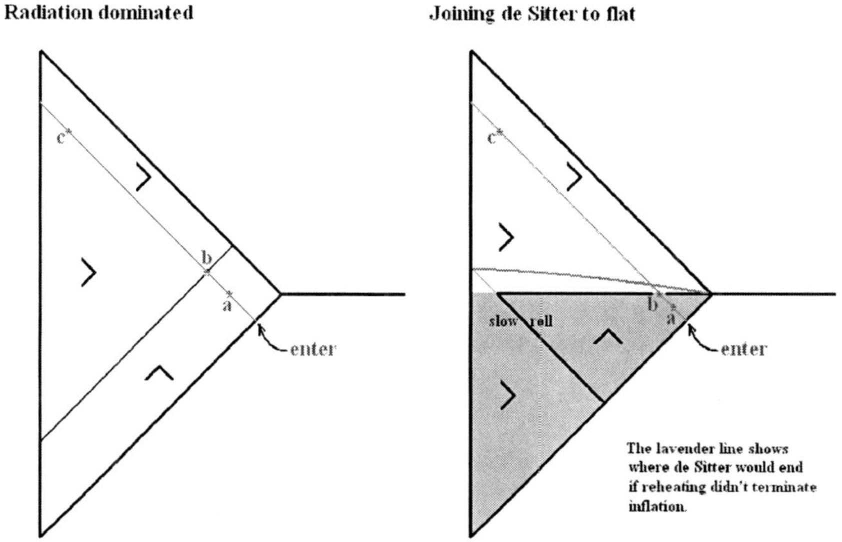

Fig. 10 Bousso diagrams for radiation dominated and slow-roll universes. The slow-roll case has been simplified by attaching a de Sitter region to a patch of flat space–time. The grey area represents the inflating region

the origin – the vertical side of the triangle – at $R = 0$. The area is maximum at "*enter*": it is almost constant from "*enter*" to a and then b, but by c it starts to significantly decrease. All of this is in accord with the c-theorem.

Next, in Fig. 10 some more interesting cosmological examples are shown. On the left is pure radiation dominated FRW. Unlike the thin-wall case it consists of two regions separated by the bold black line. In this case the area *increases* from "*enter*" to a, reaching a maximum at b and then decreasing very imperceptibly to c. At c it starts a quick decent to zero. This looks dangerous.

On the right side of Fig. 10 the blue region is a patch with a finite vacuum energy. It is intended to model an era of slow-roll inflation. For simplicity, at reheating (horizontal edge of the blue region) I have attached it to flat space–time. More realistically one might attach it to the radiation dominated case but the result would be the same.

In this slow-roll case the area starts an exponential increase at "*enter*" and again reaches a maximum at b. Beyond that the behavior is the same as for thin-wall. In fact this behavior of an increase of area, followed by a decrease is very generic. It would seem to violate the c-theorem.

The point, however, is that the identification of c with area is wrong. The number of degrees of freedom of a system is a direct measure of the maximum entropy of that system. The right measure of c, for example at the point a, is not the area of a, but rather the maximum entropy that can pass through the light-sheet from a to the origin at $R = 0$. According to Bousso, that is given by the maximum entropy on the interval between a and b, plus the maximum entropy between b and the origin.

Calling the areas of the various points $A(a),\ A(b),\ A(c)$ the maximum entropy of the light-sheet bounded by a is given by,

$$S_{max}(a) = (A(b) - A(a)) + (A(b)). \tag{110}$$

It is maximized at "*enter*" and monotonically decreases to zero at $R = 0$. Thus, identifying $c = S_{max}$ restores the monotonicity of the central charge.

It's interesting to compare the naive behavior, $c = A$ with the correct formula $c = S_{max}$ in the slow-roll inflationary case. The naive formula would have c exponentially increasing from "*enter*" to b. The correct formula has it exponentially decreasing toward its (inflated) value at b.

7 Scaling and the Census Taker

7.1 Moments

Now we come to the heart of the matter: the intimate connection between the scaling behavior of two-dimensional quantum field theory and the observations of a Census Taker as he moves toward the Census Bureau. In order to better understand the connection between the cutoff-scale and T^+, let's return to the similar connection between cutoff and the coordinate R in anti de Sitter space. We normalize the ADS radius of curvature to be 1; with that normalization, the Planck area is given by $1/c$ where c is the central charge.

Consider the proper distance between points 1 and 2 given by (38). The relation $l = R_1 + R_2 + \log(1 - \cos\alpha)$ is approximate, valid when l and $R_{1,2}$ are all large. When $l \sim 1$ or equivalently, when

$$\alpha^2 \sim e^{-(R_1+R_2)} \tag{111}$$

(38) breaks down. For angles smaller than (111) the distance in anti de Sitter space behaves like

$$l \sim e^R \alpha. \tag{112}$$

Thus a typical correlation function will behave as a power of $(1 - \cos\alpha)$ down to angular distances of order (111) and then fall quickly to zero.

The angular cutoff in anti de Sitter space has a simple meaning. The solid angle corresponding to the cutoff is of order e^{-2R} while the area of the regulated boundary is e^{+2R}. Thus, metrically, the cutoff area is of order unity. This means that in Planck units, the cutoff area is the central charge c of the boundary conformal field theory.

Now consider the cutoff angle implied by (48) and (49). By an argument parallel to the one above, the cutoff angle becomes

$$\alpha^2 \sim e^{-(T_1^+ + T_2^+)} \tag{113}$$

Once again this corresponds to a proper area on Σ_0 (the regulated boundary) which is time-independent, and in Planck units, of order the central charge.

Consider the Census Taker looking back from some late time T_{CT}. For convenience we place the CT at $R = 0$. His backward light-cone is the surface

$$T + R = T_{CT}. \tag{114}$$

The CT can never quite see Σ. Instead he sees the regulated surfaces corresponding to a fixed proper cutoff (Fig. 4). The later the CT observes, the smaller the angular structure that he can resolve on the boundary. This is another example of the UV/IR connection, this time in a cosmological setting.

Let's consider a specific example of a possible observation. The massless scalar field χ of Sect. 5 has an asymptotic limit on Σ that defines the dimension zero field $\chi(\Omega)$. Moments of χ can be defined by integrating it with spherical harmonics,

$$\chi_{lm} = \int \chi(\Omega) Y_{lm}(\Omega) d^2\Omega \tag{115}$$

It is worth recalling that in anti de Sitter space the corresponding moments would all vanish because the normalizable modes of χ all vanish exponentially as $R \to \infty$. The possibility of non-vanishing moments is due entirely to the asymptotic warmness of open FRW.

We can easily calculate[7] the mean square value of χ_{lm} (It is independent of m).

$$\langle \chi_l^2 \rangle = \int \log(1 - \cos\alpha) P_l(\cos\alpha) \sim \frac{1}{l(l+1)} \tag{116}$$

It is evident that at a fixed Census Taker time T_{CT}, the angular resolution is limited by (113). Correspondingly, the largest moment that the CT can resolve corresponds to

$$l_{max} = e^{T_{CT}} \tag{117}$$

Thus we arrive at the following picture: the Census Taker can look back toward Σ but at any given time his angular resolution is limited by (113) and (117). As time goes on more and more moments come into view. Once they are measured they are frozen and cannot change. In other words the moments evolve from being unknown quantum variables, with a gaussian probability distribution, to classical boundary conditions that explicitly break rotation symmetry (and therefore conformal symmetry). One sees from (116) that the symmetry breaking is dominated by the low moments.

But I doubt that this phenomenon ever occurs in an undiluted form. Realistically speaking, we don't expect massless scalars in the non-supersymmetric Ancestor. In Sect. 5 we discussed the effect of a small mass term, in the ancestor vacuum, on the

[7] I am indebted to Ben Freivogel for explaining (116) to me.

correlation functions of χ. The result of such a mass term is a shift of the leading dimension[8] from 0 to μ. This has an effect on the moments. The correlation function becomes

$$e^{-\mu T_1^+} e^{-\mu T_2^+} (1-\cos\alpha)^{-\mu}. \tag{118}$$

and the moments take the form

$$\langle \chi_l^2 \rangle = e^{-2\mu T_{CT}} \int (1-\cos\alpha)^{\mu} \, P_l(\cos\alpha) \tag{119}$$

The functional form of the l dependence changes a bit, favoring higher l, but more importantly, the observable effects decrease like $e^{-2\mu T_{CT}}$. Thus as T_{CT} advances, the asymmetry on the sky decreases exponentially with conformal time. Equivalently it decreases as a power of proper time along the CT's world-line.

7.2 Homogeneity Breakdown

Homogeneity in an infinite FRW universe is generally taken for granted, but before questioning homogeneity we should know exactly what it means. Consider some three-dimensional scalar quantity such as energy density, temperature, or the scalar field χ. Obviously the universe is not uniform on small scales, so in order to define homogeneity in a useful way we need to average χ over some suitable volume. Thus at each point X of space, we integrate χ over a sphere of radius r and then divide by the volume of the sphere. For a mathematically exact notion of homogeneity the size of the sphere must tend to infinity. The definition of the average of χ at the point X is

$$\overline{\chi(X)} = \lim_{r\to\infty} \frac{\int \chi d^3x}{V_r} \tag{120}$$

Now pick a second point Y and construct $\overline{\chi(Y)}$. The difference $\overline{\chi(X)} - \overline{\chi(Y)}$ should go to zero as $r \to \infty$ if space is homogeneous. But as the spheres grow larger than the distance between X and Y, they eventually almost completely overlap. In Fig. 11 we see that the difference between $\overline{\chi(X)}$ and $\overline{\chi(y)}$ is due to the two thin crescent-shaped regions, 1 and 3. It seems evident that the overwhelming bulk of the contributions to $\overline{\chi(X)}$, $\overline{\chi(Y)}$ come from the central region 3, which occupies almost the whole figure. The conclusion seems to be that the averages, if they exist at all, must be independent of position. Homogeneity while true, is a triviality.

This is correct in flat space, but surprisingly it can break down in hyperbolic space.[9] The reason is quite simple: despite appearances the volume of regions 1

---

[8] In the de Sitter/CFT correspondence [40], the dimension of a massive scalar becomes complex when the mass exceeds the Hubble scale. In our case the dimension remains real for all μ. I thank Yasuhiro Sekino for this observation.

[9] L.S. is grateful to Larry Guth for explaining this phenomenon, and to Alan Guth for emphasizing its importance in cosmology.

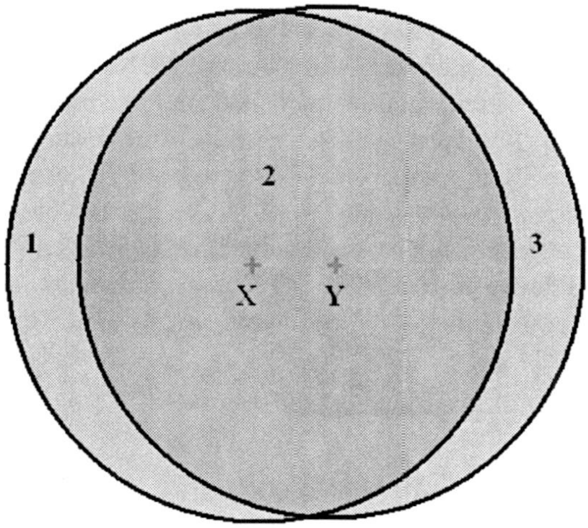

Fig. 11 Two large spheres centered at X and Y

and 3 grow just as rapidly as the volume of 2. The ratio of the volumes is of order

$$\frac{V_1}{V_2} = \frac{V_3}{V_2} \sim \frac{l}{R_{curvature}} \tag{121}$$

and remains finite as $r \longrightarrow \infty$.

To be more precise we observe that

$$\begin{aligned} \overline{\chi(X)} &= \frac{\int_1 \chi + \int_2 \chi}{V_1 + V_2} \\ \overline{\chi(Y)} &= \frac{\int_3 \chi + \int_2 \chi}{V_3 + V_2} \end{aligned} \tag{122}$$

and that the difference $\overline{\chi(X)} - \overline{\chi(Y)}$ is given by

$$\overline{\chi(X)} - \overline{\chi(Y)} = \frac{\int_1 \chi}{V_1 + V_2} - \frac{\int_1 \chi}{V_1 + V_2} \tag{123}$$

which, in the limit $r \longrightarrow \infty$ is easily seen to be proportional to the dipole-moment of the boundary theory,

$$\overline{\chi(X)} - \overline{\chi(Y)} = l \int \chi(\Omega) \cos\theta d^2\Omega = l\, \chi_{1,0}, \tag{124}$$

where l is the distance between X and Y.

Since, as we have already seen for the case $\mu = 0$, the mean square fluctuation in the moments does not go to zero with distance, it is also true that average value of $|\overline{\chi(X)} - \overline{\chi(Y)}|^2$ will be nonzero. In fact it grows with separation.

However there is no reason to believe that a dimension zero scalar exists. Moduli, for example, are expected to be massive in the Ancestor, and this shifts the dimension of the corresponding boundary field. In the case in which the field χ has dimension μ, the effect (non-zero rms average of moments) persists in a somewhat diluted form. If a renormalized field is defined by stripping off the wave function normalization constants, $\exp(-\mu T^+)$, the squared moments still have finite expectation values and break the symmetry. However, from an observational point of view there does not seem to be any reason to remove these factors. Thus it seems that as the Census Taker time tends to infinity, the observable asymmetry will decrease like $\exp(-2\mu T_{CT})$.

8 Bubble Collisions and Other Matters

The Census Taker idea originated with attempts to provide a measure on the Landscape. By looking back toward Σ, the Census Taker can see into bubbles of other vacua – bubbles that in the past collided with his hatted vacuum. By counting the bubbles of each type on the sky, he can try to define a measure on the Landscape. Whether or not this can be done, it is important to our program to understand the representation of bubble collisions in the language of the boundary holographic field theory.

Long ago, Guth and Weinberg [41] recognized that a single isolated bubble is infinitely unlikely, and that a typical "pocket universe" will consist of a cluster of an unbounded number of colliding bubbles, although if the nucleation rate is small the collisions will in some sense be rare. To see why such bubble clusters form it is sufficient to recognize why a single bubble is infinitely improbable. In Fig. 12 the main point is illustrated by drawing a time-like trajectory that approaches Σ from within the Ancestor vacuum. The trajectory has infinite proper length, and assuming that there is a uniform nucleation rate, a second bubble will eventually swallow the trajectory and collide with the original bubble. Repeating this process will produce an infinite bubble cluster.

More recently Garriga, Guth, and Vilenkin [31] have argued that the multiple bubble collisions must spontaneously break the $SO(3,1)$ symmetry of a single bubble, and in the process render the (pocket) universe inhomogeneous and anisotropic. The breaking of symmetry in [31] was described, not as spontaneous breaking, but as explicit breaking due to initial conditions. However, spontaneous symmetry breaking is nothing but the memory of a temporary explicit symmetry breaking, if the memory does not fade with time. For example, a small magnetic field in the very remote past will determine the direction of an infinite ferromagnet for all future time. Spontaneous symmetry breaking *is* "The Persistence of Memory."

The actual observability of bubble collisions depends on the amount of slow-roll inflation that took place after tunneling. Much more than 60 e-foldings would

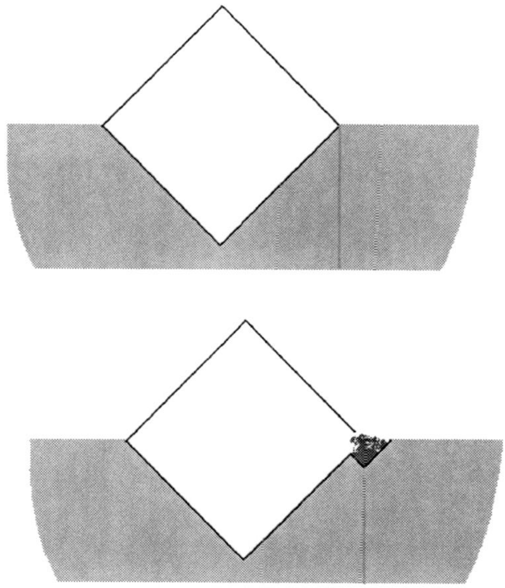

Fig. 12 The *top figure* represents a single nucleated bubble. The vertical trajectory is a time-like curve of infinite length approaching Σ. Because there is a constant nucleation rate along the curve, it is inevitable that a second bubble will nucleate as in the *lower figure*. The two bubbles will collide

probably wipe out any signal, but our interest in this paper is conceptual. We will take the viewpoint that anything within the past light-cone of the Census Taker is in principle observable.

In the last section we saw that perturbative infrared effects are capable of breaking the $SO(3,1)$ symmetry, and it is an interesting question what the relation between these two mechanisms is. The production of a new bubble would seem to be a non-perturbative effect that adds to the perturbative symmetry breaking effects of the previous section. Whether it adds distinctly new effects that are absent in perturbation theory is not obvious and may depend on the specific nature of the collision. Let us classify the possibilities.

8.1 Collisions with Identical Vacua

The simplest situation is if the true-vacuum bubble collides with another identical bubble, the two bubbles coalescing to form a single bubble, as in the top of Fig. 13.

The surface Σ is defined by starting at the tip of the hat and tracking back along light-like trajectories until they end – in this case at a false vacuum labeled **F.** The collision is parameterized by the space-like separation between nucleation points. Particles produced at the collision of the bubbles just add to the particles that were produced by ordinary FRW evolution. The main effect of such a collision is to create

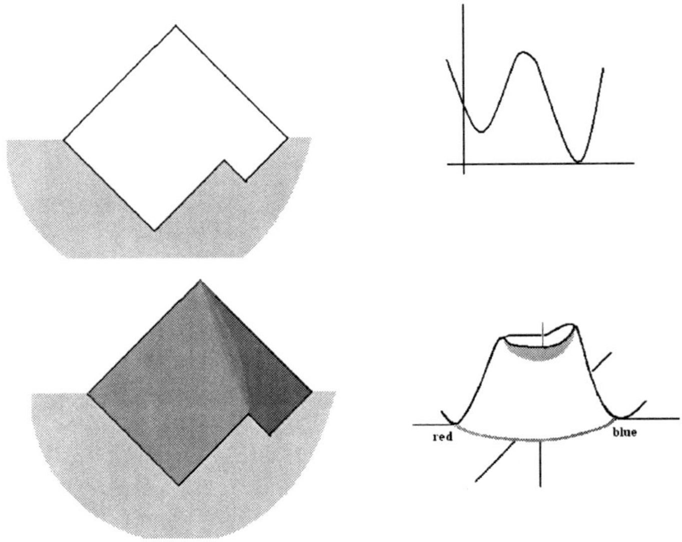

Fig. 13 In the *top figure* two identical bubbles collide. This would be the only type of collision in a simple landscape with two discrete minima – one of positive energy and one of zero energy. In the *lower figure* a more complicated situation is depicted. In this case the false vacuum F can decay to two different true vacua, "light" and "dark," each with vanishing energy. The two true vacua are connected by a flat direction, but CDL instantons only lead to the light and dark points

a very distorted boundary geometry, if the nucleation points are far apart. When they are close the double nucleation blends in smoothly with the single bubble. These kind of collisions seem to be no different than the perturbative disturbances caused by the non-normalizable mode of the metric fluctuation. Garriga, Guth, and Vilenkin, compute that the typical observer will see multipole moments on the sky, but as we've seen, similar multipole moments can also occur perturbatively.

In the bottom half of Fig. 13 we see another type of collision in which the colliding bubbles correspond to two different true vacua: red (r) and blue (b). But in this case red and blue are on the same moduli-space, so that they are connected by a flat direction.[10] Both vacua are included within the hat. In the bulk space–time they bleed into each other, so that as one traverses a space-like surface, blue gradually blends into purple and then red.

On the other hand the surface Σ is sharply divided into blue and red regions, as if by a one dimensional domain wall. This seems to be a new phenomenon that does not occur in perturbation theory about either vacuum.

As an example, consider a case in which a red vacuum-nucleation occurs first, and then much later a blue vacuum bubble nucleates. In that case the blue patch on the boundary will be very small and The Census Taker will see it occupying

[10] I assume that there is no symmetry along the flat direction, and that there are only two tunneling paths from the false vacuum, one to red, and one to blue.

tiny angle on the sky. How does the boundary field theorist interpret it? The best description is probably as a small blue instanton in a red vacuum. In both the bulk and boundary theory this is an exponentially suppressed, non-perturbative effect.

However, in a conformal field theory the size of an instanton is a modulus that must be integrated over. As the instanton grows the blue region engulfs more and more of the boundary. Eventually the configuration evolves to a blue 2D vacuum, with a tiny red instanton. One can also think of the two configurations as the observations of two different Census Takers at a large separation from one another. Which one of them is at the center, is obviously ambiguous.

The same ambiguous separation into dominant vacuum, and small instanton, can be seen another way. The nucleation sites of the two bubbles are separated by a space-like interval. There is no invariant meaning to say that one occurs before the other. A element of the de Sitter symmetry group can interchange which bubble nucleates early and which nucleates later.

Nevertheless, a given Census Taker will see a definite pattern on the sky. One can always define the CT to be at the center of things, and integrate over the relative size of the blue and red regions. Or one can keep the size of the regions fixed – equal for example – and integrate over the location of the CT.

From both the boundary field theory, and the bubble nucleation viewpoints, the probability for any finite number of red-blue patches is zero. Small red instantons will be sprinkled on every blue patch and vice versa, until the boundary becomes a fractal. The fractal dimensions are closely connected to operator dimensions in the boundary theory. Moreover, exactly the same pattern is expected from multiple bubble collisions.

But the Census taker has a finite angular resolution. He cannot see angular features smaller than $\delta\alpha \sim \exp(-T_{CT})$. Thus he will see a finite sprinkling of red and blue dust on the sky. At T_{CT} increases, the UV cutoff scale tends to zero and the CT sees a homogeneous"purple" fixed-point theory.

The red and blue patches are reminiscent of the Ising spin system (coupled to a Liouville field). As in that case, it makes sense to average over small patches and define a continuous "color field" ranging from intense blue to intense red. It is interesting to ask whether Σ would look isotropic, or whether there will be finite multipole moments of the renormalized color color field (as in the case of the χ field). The calculations of Garriga, Guth, and Vilenkin suggest that multipole moments would be seen. But unless for some reason there is a field of exactly zero dimension, the observational signal should fade with Census Taker time.

There are other types of collisions that seem to be fundamentally different from the previous. Let us consider a model landscape with three vacua – two false, B and W (Black and White); and one true vacuum T. Let the vacuum energy of B be bigger than that of W, and also assume that the decays $B \to W$, $B \to T$, and $W \to T$ are all possible. Let us also start in the Black vacuum and consider a transition to the True vacuum. The result will be a hat bounded by Σ.

However, if a bubble of W forms, it may collide with the T bubble as in Fig. 14. The W bubble does not end in a hat but rather, on a space-like surface. By contrast, the true vacuum bubble does end in a hat. The surface Σ is defined as always, by

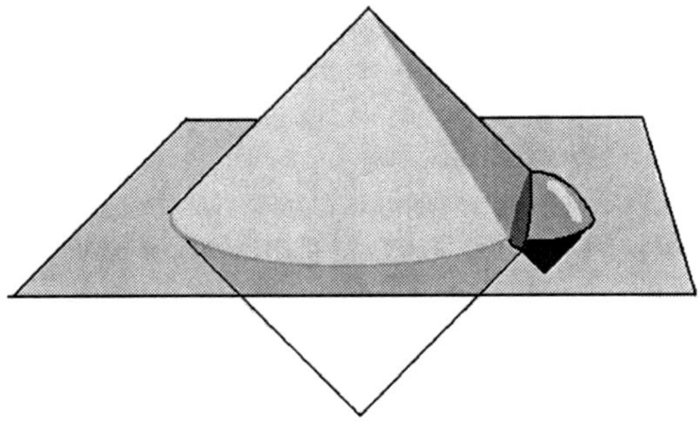

Fig. 14 A bubble of True vacuum forms in the Black false vacuum and then collides with a bubble of White vacuum. The true vacuum is bounded by a hat but the White vacuum terminates in a space-like surface. Some generators of the hat intersect the Black vacuum and some intersect the White. Thus Σ, shown as the red curve, is composed of two regions

following the light-like generators of the hat backward until they enter the bulk – either Black or White – as in Fig. 14.

In this case a portion of the boundary Σ butts up against B, while another portion abuts W. In some ways this situation is similar to the previous case where the boundary was separated into red and blue regions, but there is no analogue of the gradual bleeding of vacua in the bulk. In the previous case the Census Taker could smoothly pass from red to blue. But in the current example, the CT would have to crash through a domain wall in order to pass from T to W. Typically this happens extremely fast, long before the CT could do any observation. In fact if we define Census Takers by the condition that they eventually reach the Census Bureau, then they simply never enter W.

From the field theory point of view this example leads to a paradox. Naively, it seems that once a W patch forms on Σ, a B region cannot form inside it. A constraint of this type on field configurations would obviously violate the rules of quantum field theory; topologically (on a sphere) there is no difference between a small W patch in a B background, and a small B patch in a W background. Thus field configurations must exist in which a W region has smaller Black spots inside it. There is no way consistent with locality and unitarity to forbid bits of B in regions of W.

Fortunately the same conclusion is reached from the bulk point of view. The rules of tunneling transitions require that if the transition $B \to W$ is possible, so must be the transition $W \to B$, although the probability for the latter would be smaller (by a large density of states ratio). Thus one must expect B to invade regions of W.

As the Census Taker time advances he will see smaller and smaller spots of each type. If one assumes that there are no operators of dimension zero, then the pattern should fade into a homogeneous average grey, although under the conditions I described it will be almost White.

The natural interpretation is that the boundary field theory has two phases of different free energy, the B free energy being larger than that of W. The dominant configuration would be the ones of lower free energy with occasional fluctuations to higher free energy.

8.2 The Persistence of Memory

The "Persistence of Memory" reported in [31] had nothing to do with whether or not the Census Taker's sees a fading signal: Garriga, Guth, and Vilenkin were not speaking about Census Taker time at all. They were referring to the fact that no matter how long after the start of eternal inflation a bubble nucleates, it will remember the symmetry breaking imposed by the initial conditions; not whether the signal fades with T_{CT}. Returning to Fig. 12, one might ask why no bubble formed along the red trajectory in the infinitely remote past. The authors of [31] argue that eternal inflation does not make sense without an initial condition specifying a past surface on which no bubbles had yet formed. That surface invariably breaks the O(3,1) symmetry and distinguishes a "preferred Census Taker" who is at rest in the frame of the initial surface. He alone sees an isotropic sky whereas all the other Census Takers see non-zero anisotropy. What's more the effect persists no matter how late the nucleation takes place.

As before, when the Census Taker's time advances, the asymmetry should become diluted if there are no dimension zero operators, but the existence of a preferred Census Taker at finite time makes this symmetry breaking seem different than what we have discussed up to now.

Let us consider how this phenomenon fits together with the RG flow discussed earlier.[11] Begin by considering the behavior for finite δ in the limit of small a. It is reasonable to suppose that in integrating out the many scales between a and δ, the theory would run to a fixed point. Now recall that this is the limit of very large T. If in fact the theory has run to a fixed point it will be conformally invariant. Thus we expect that the symmetry $O(3,1)$ will be unbroken at very late time.

On the other hand consider the situation of δ/a near 1. The reference and bare scales are very close and very few degrees of freedom have been integrated out. There is no reason why the effective action should be near a fixed point. The implication is that at very early time (recall, $\delta/a = e^T$) the physics on a fixed time slice will not be conformally invariant. Near the beginning of an RG flow the effective action is strongly dependent on the bare theory. The implication of a breakdown of conformal symmetry is that there is no symmetry between Census Takers at different locations in space. In such situations the center of the (deformed) anti de Sitter space is, indeed, special.

Shenker and I suggest that the GGV boundary condition at the onset of eternal inflation is the same thing as the initial condition on the RG flow. In other words, varying the GGV boundary condition is no different from varying the bare fishnet theory.

Is it possible to tune the bare action so that the theory starts out at the fixed point? If this were so, it would be an initial condition that allowed exact conformal invariance for all time. Of course it would involve an infinite amount of fine tuning and is probably not reasonable. But there may be reasons to doubt that it is possible altogether, even though in a conventional lattice theory it is possible.

---

[11] These observations are based on work with Steve Shenker.

The difficulty is that the bare and renormalized theories are fundamentally different. The bare theory is defined on a variable fishnet whose connectivity is part of the dynamical degrees of freedom. The renormalized theory is defined on the fixed reference lattice. The average properties of the underlying dynamical fishnet are replaced by conventional fields on the reference lattice. Under these circumstances it is hard to imagine what it would mean to tune the bare theory to an exact fixed point.

The example of the previous subsection involving two false vacua, B and W, raises some interesting questions. First imagine starting with GGV boundary conditions such that, on some past space-like surface, the vacuum is pure Black and that a bubble of true vacuum nucleates in that environment. Naively the boundary is mostly black. That means that *in the boundary theory* the free energy of Black must be lower than that of White.

But we argued earlier that white instantons will eventually fill Σ with an almost white, very light grey color, exactly as if the initial GGV condition were White. That means that White must have the lower potential energy. What then is the meaning of the early dominance of B from the 2D field theory viewpoint?

The point is that it is possible for two rather different bare actions to be in the same broad basin of attraction and flow to the same fixed point. The case of Black GGV conditions corresponds to a bare starting point (in the space of couplings) where the potential of B is lower than W. During the course of the flow to the fixed point the potential changes so that at the fixed-point W has the lower energy.

On the other hand, White GGV initial conditions corresponds to starting the flow at a different bare point – perhaps closer to the fixed point – where the potential of W is lower.

This picture suggests a powerful principle. Start with the space of two-dimensional actions, which is broad enough to contain a very large Landscape of 2D theories. With enough fields and couplings the space could probably contain everything. As Wilson explained [42], the space divides itself into basins of attraction. Each initial state of the universe is described either as a GGV initial condition, or as a bare starting point for an RG flow. The endpoints of these flows correspond to the possible final states-the hats – that the Census Taker can end up in.

We have not exhausted all the kinds of collisions that can occur – in particular collisions with singular, negative cosmological constant vacua. A particularly thorny situation results if there is a BPS domain wall between the negative and zero CC bubbles, then as shown by Freivogel et al. [43] the entire hat may disappear in a catastrophic crunch. A possible interpretation is that the catastrophe is due to the existence of a relevant operator which destabilizes the fixed point. These and other issues will be taken up in a paper with Shenker.

8.3 A Remark About Supersymmetry

Most likely, the only 4D vacua with exactly vanishing cosmological constant are supersymmetric. Does that mean that the boundary theory on Σ is also supersymmetric? The answer is no: correlators on Σ are largely determined by the properties of the non-supersymmetric Ancestor vacuum. For example, the gravitino will be massive in the Ancestor, and the methods of [27] would give different dimensions, Δ, for the graviton and gravitino fluctuations.

In fact Σ is contiguous with the de Sitter Ancestor and has every reason to strongly feel the supersymmetry breaking. It is the region close to the tip of the hat where the physics should be dominated by the properties of the supersymmetric terminal vacuum. If one looks at the expansion (45), it is clear that the tip of the hat is dominated by the asymptotically high dimensional terms. Thus we expect supersymmetry to manifests itself asymptotically, in the spectrum and operator products of high dimensional operators.

8.4 Flattened Hats and Other Tragedies

In a broad sense this paper is about phenomenology: the Census Taker could be us. If we lived in an ideal thin-wall hat we would see, spread across the sky, correlation functions of a holographic quantum field theory. We could measure the dimensions of operators both by the time dependence of the received signals, and their angular dependence. Bubble collisions would appear as patches resembling instantons.

Unfortunately (or perhaps fortunately) we are insulated from these effects by two forms of inflation – the slow-roll inflation that took place shortly after bubble nucleation – and the current accelerated expansion of the universe. The latter means that we don't live in a true hatted geometry. Rather we live in a flattened hat, at least if we ignore the final decay to a terminal vacuum.

The Penrose diagram in Fig. 15 shows an Ancestor, with large vacuum energy, decaying to a vacuum with a very small cosmological constant. The important new feature is that the hat is replaced by a space-like future infinity. Consider the Census Taker's final observations as he arrives at the flattened hat. It is obvious from Fig. 15

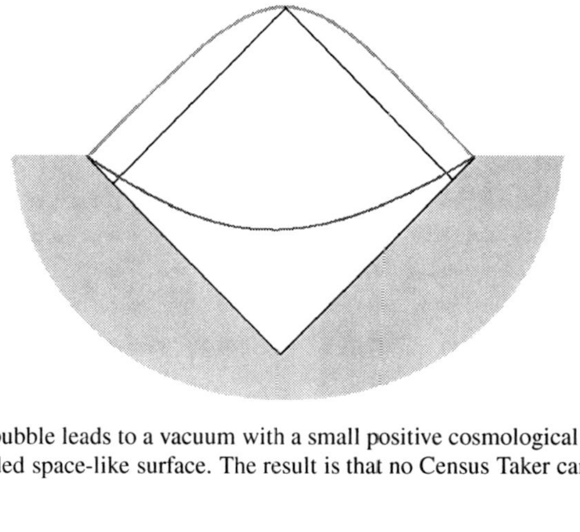

Fig. 15 If a CDL bubble leads to a vacuum with a small positive cosmological constant, the hat is replaced by a rounded space-like surface. The result is that no Census Taker can look back to Σ

that he cannot look back to Σ. His past light cone is at a finite value of T^+. Thus for each time-slice T, there is a maximum radial variable $R = R_0(T)$ within his ken, no matter how long he waits. In other words there is an unavoidable ultraviolet cutoff. It is completely evident that a final de Sitter bubble must be described by a theory with no continuum limit; in other words not only a non-local theory, but one with no ultraviolet completion.

This suggests that de Sitter Space may have an intrinsic imprecision. Indeed, as Seiberg has emphasized, the idea of a metastable vacuum is imprecise, even in condensed matter physics where they are common.[12]

The more tragic fact is that all of the memory of a past bubble may, for observational purposes, be erased by the slow-roll inflation that took place shortly after the Coleman De Luccia tunneling event – unless it lasted for the minimum permitted number of e-foldings [23]. In principle the effects are imprinted on the sky, but in an exponentially diluted form.

8.5 Note About W

What I have described is only half the story: the half involving S, the real part of $\log \Psi$. The other half involves W, the phase of the wave function. Knowing S is enough to compute the expectation values of the fields y at a given value of time, or, strictly speaking, at a value of the scale factor. Scanning over scale factors can be done by varying the Liouville cosmological constant.

However, quantum mechanics cannot be complete without the phase of the wave function. In particular, the values of conjugate momenta requires knowledge of W. The same is true for products of fields at different times.

9 Some Conclusions

I've given some circumstantial evidence that there is a duality between cosmology on the Landscape, and two-dimensional conformal field theory, with a Liouville field. The data that supports the theory are the computations done in [27], but more importantly, a compelling physical picture accompanies the data. The most pertinent observation is that the sky is a two-sphere. Moreover, it is covered with interesting observable correlations; in principle we can look back through the surface of last scattering and observe these correlations at any past time. The Liouville field, or alternately, the Liouville cosmological constant (not to be confused with the four-dimensional cosmological constant) is the dual of that time, along the observers backward light-cone.

[12] I am grateful to Nathan Seiberg for discussions on this point.

As we view the deep sky from increasingly late times, we can in principle see greater angular detail on Σ_0, i.e., spatial "almost-infinity." The increasing angular resolution defines a renormalization group flow that begins with some bare action, and ends at an infrared fixed point. The starting point of the flow is equivalent to the boundary condition, whose memory, Garriga, Guth, and Vilenkin have argued, is persistent. The phenomena of symmetry breaking – the picking out of a special Census Taker at the center of things – by the GGV's initial condition, is equivalent to the breaking of conformal symmetry, at the start of a RG flow. As in anti de Sitter space, breaking conformal symmetry makes the center of ADS (at $R = 0$) special.

Our considerations are based on the Holographic Principle but with a new twist. The asymptotic warmness of space requires a field in order to represent geometric fluctuations at infinity. By now we are used to one or more spatial directions emerging holographically, but this new Liouville degree of freedom creates a new emergence – of time.

Examples of cosmic phenomena that can be simply interpreted as two-dimensional field theory phenomena are the bubble collisions of Guth and Weinberg [41], which appear as instantons on the two-dimensional sky; the fading of the initial conditions with Census Taker time is connected with the spectrum of conformal dimensions – in particular the lack of dimension zero scalars; and ordinary slow-roll inflation corresponds to an exponential decrease in the Zamolodchikov c-function [44]. It will be interesting to try to interpret CMB fluctuations in this language but this has not been studied – at least that I know of.

10 Warning to All Census Takers

Memo from the Director

Keep in mind that when you look back toward Σ, what you see will be influenced not only by conditions on the regulated boundary, but also by the gravitational field between you and Σ_0. Angular separations detected by you, must be corrected for nearby gravitational distortions such as lensing and gravitational waves. Please make all corrections before reporting your data.

Acknowledgments I am enormously grateful to my collaborators, Ben Freivogel, Yasuhiro Sekino, Chen-Pin Yeh, Steve Shenker, and Alex Maloney. This lecture represents an ongoing discussion with them. Any (positive) reference to it should be accompanied by [1] and [27].

Discussions with Larry Guth and Alan Guth were important to my understanding of the breaking of symmetries in Hyperbolic space.

I would also like to acknowledge helpful conversations with Raphael Bousso, another believer in the causal patch; Simeon Hellerman; and Matt Kleban for various insights; and finally, Nathan Seiberg for explaining the intrinsic imprecision of metastable states.

Any conceptual or computational errors in the paper are of course my own.

References

1. A. Maloney, S. Shenker, L. Susskind, unpublished.
2. N. Kaloper, M. Kleban, L. Sorbo, Observational implications of cosmological event horizons, Phys. Lett. B 600 (2004) 7, [arXiv:astro-ph/0406099].
3. M. Kleban, Inflation unloaded, [arXiv:hep-th/0412055].
4. R. Bousso, J. Polchinski, Quantization of four-form fluxes and dynamical neutralization of the cosmological constant, JHEP 0006 (2000) 006, hep-th/0004134.
5. S. Kachru, R. Kallosh, A. Linde, S. P. Trivedi: De Sitter vacua in string theory. Phys. Rev. D 68 (2003) 046005, hep-th/0301240.
6. L. Susskind, The anthropic landscape of string theory, [arXiv:hep-th/0302219].
7. M. R. Douglas, The statistics of string/M theory vacua, JHEP 0305 (2003) 046, hep-th/0303194.
8. L. Susskind, L. Thorlacius, J. Uglum, The stretched horizon and black hole complementarity, Phys. Rev. D48 (1993) 3743–3761, arXiv:hep-th/9306069.
9. Y. Kiem, H. L. Verlinde, E. P. Verlinde, Quantum horizons and complementarity, in *High Energy Physics and Cosmology*, ed. by E. Gava, A. Masiero, K. S. Narain, S. Randjbar-Daemi, Q. Shafi (ICTP Series in Theoretical Physics, ICTP, Trieste, Italy, June 13–July 29, 1994), Vol. 11, pp 586–606.
10. T. Banks, W. Fischler, M theory observables for cosmological space–times, hep-th/0102077.
11. G. 't Hooft, Dimensional reduction in quantum gravity, gr-qc/9310026.
12. L. Susskind, The world as a hologram, J. Math. Phys. 36 (1995) 6377–6396, hep-th/9409089
13. J. M. Maldacena, The large N limit of superconformal field theories and supergravity, Int. J. Theor. Phys. 38 (1999) 1113–1133, hep-th/9711200.
14. E. Witten, Anti-de Sitter space and holography. Adv. Theor. Math. Phys. 2 (1998) 253–291, hep-th/9802150.
15. L. Susskind, E. Witten, The holographic bound in Anti-de Sitter space, (1998), hep-th/9805114.
16. J. R. Gott, Nature 295 (1982) 304.
17. A. H. Guth, E. Weinberg, Nucl. Phys. B 212 (1983) 321.
18. P. J. Steinhardt, The very early universe, ed. by G. W. Gibbons, S. W. Hawking, S. T. C. Siklos (Cambridge University Press, Cambridge, 1983).
19. J. R. Gott, T. S. Statler, Phys. Lett. 136B (1984) 157.
20. M. Bucher, A. S. Goldhaber, A. D. Linde, Eternal chaotic inflation, Mod. Phys. Lett. A 1 (1986) 81–85.
21. A. D. Linde, Eternally existing selfreproducing chaotic inflationary universe, Phys. Lett. B (1986).
22. S. R. Coleman, F. De Luccia, Gravitational effects on and of vacuum decay, Phys. Rev. D 21 (1980) 3305.
23. B. Freivogel, M. Kleban, M. R. Martinez, L. Susskind, Observational consequences of a landscape, JHEP 0603 (2006) 039, arXiv:hep-th/0505232.
24. S. W. Hawking, Comm. Math. Phys. 43 (1975) 199.
25. S. W. Hawking, Phys. Rev. D14 (1976) 2460.
26. R. Bousso, B. Freivogel, I-S. Yang, Eternal inflation: the inside story, Phys. Rev. D 74 (2006) 103516, arXiv:hep-th/0606114.
27. B. Freivogel, Y. Sekino, L. Susskind, C-P. Yeh, A holographic framework for eternal inflation, Phys. Rev. D 74 (2006) hep-th/0606204.
28. T. Banks, W. Fischler, L. Susskind, Quantum cosmology in (2+1)-dimensions and (3+1)-dimensions. Nucl. Phys. B 262 (1985) 159.
29. R. Arnowitt, S. Deser, C. W. Misner, Gravitation: an introduction to current research, ed. by L. Witten (Wiley, New York, 1962), Chapter 7, pp 227–265, [arXiv:gr-qc/0405109].
30. T. Banks, TCP, quantum gravity, the cosmological constant and all that... Nucl. Phys. B 249 (1985) 332.
31. J. Garriga, A. H. Guth, A. Vilenkin, Eternal inflation, bubble collisions, and the persistence of memory, arXiv:hep-th/0612242.

32. A. M. Polyakov, Quantum geometry of bosonic strings. Phys. Lett. B 103 (1981) 207–210.
33. A. Kraemmer, H. B. Nielsen, L. Susskind, Nucl. Phys. B 28 (1971) 34.
34. G. 't Hooft, Nucl. Phys. B 72 (1974) 461.
35. M. R. Douglas, S. H. Shenker, Nucl. Phys. B 335 (1990) 635.
36. D. J. Gross, A. A. Migdal, Phys. Rev. Lett. 64 (1990) 127.
37. J. Polchinski, Nucl. Phys. B 231 (1984) 269–295.
38. W. Fischler, L. Susskind, Holography and cosmology, (1998), hep-th/9806039.
39. R. Bousso, The holographic principle, Rev. Mod. Phys. 74 (2002) 825, [arXiv:hep-th/0203101].
40. A. Strominger, The dS/CFT correspondence, Journal-ref: JHEP 0110 (2001) 034, arXiv:hep-th/0106113.
41. A. H. Guth, E. J. Weinberg, Could the universe have recovered from a slow first order phase transition? Nucl. Phys. B 212 (1983) 321.
42. K. G. Wilson, Rev. Mod. Phys. 47 (1975) 773.
43. B. Freivogel, G. T. Horowitz, S. Shenker, Colliding with a crunching bubble, arXiv:hep-th/0703146.
44. A. B. Zamolodchikov, Irreversibility' of the flux of the renormalization group in A 2-D field theory, JETP Lett. 43 (1986) 730–732.

Static Wormholes in Vacuum and Gravity in Diverse Dimensions

Ricardo Troncoso

Abstract Static wormhole solutions in vacuum for gravity in diverse dimensions are discussed. In dimensions greater than four the theory corresponds to a particular case of the Lovelock action, so that it admits a unique AdS vacuum. One of the wormhole solutions connects two asymptotically locally AdS spacetimes so that both asymptotic regions are connected by light signals in a finite time. The Euclidean continuation of the wormhole can be seen as instanton with vanishing action, and the mass can also be obtained from a surface integral which is shown to vanish. Its stability against free scalar field perturbations is guaranteed provided the squared mass is bounded from below by a negative quantity which could be more stringent than the Breitenlohner–Freedman bound. An exact expression for the spectrum is found analytically. For nonminimal coupling, stability can also be achieved for scalar fields with slow fall-off, and three different quantizations can be carried on, being characterized by the fall-off of the scalar field, which can be fast or slow with respect to each asymptotic region. In four dimensions a static spherically symmetric wormhole solution for conformal gravity in vacuum is found, whose neck connects two static homogeneous universes of constant spatial curvature. Time runs at different rates on each side of the neck. The extension with radial electric or magnetic fields is also discussed and it turns out to have "charge without charge." The solutions can be further generalized to the case of necks with genus greater than one. It is shown that the wormholes in vacuum generically correspond to the matching of different Einstein spacetimes at infinity by means of improper conformal transformations.

R. Troncoso
Centro de Estudios Científicos (CECS), Casilla 1469, and Centro de Ingeniería de la Innovación del CECS (CIN), Valdivia, Chile
e-mail: troncoso@cecs.cl

M. Henneaux, J. Zanelli (eds.), *Quantum Mechanics of Fundamental Systems: The Quest for Beauty and Simplicity*, DOI 10.1007/978-0-387-87499-9_17,

1 Introduction

This article is dedicated to Claudio Bunster on the occasion of his 60th birthday. It is a great honor to take this opportunity to express my gratitude to him, who in my opinion has been the greatest national physicist ever, for his wise guidance and intrepid support through the years. As a Chilean, I can further tell that Claudio's contributions have been well far beyond theoretical physics, helping our country to be ready to face future challenges through science.

Gravity in diverse dimensions is a subject in which Claudio has done major contributions, encouraging in many ways the following work, that is being made along different fronts in collaboration with my colleagues Diego Correa, Gustavo Dotti, Julio Oliva and David Tempo.

The pursuit for wormhole solutions, which are handles in the spacetime topology, it is as old as General Relativity and it has appeared in theoretical physics within different subjects, ranging from the attempt of describing physics as pure geometry, as in the Einstein–Rosen bridge model of a particle [1], to the concept of "charge without charge" [2], as well as in different issues concerning the Euclidean approach to quantum gravity (see, e.g., [3]). More recently, the systematic study of this kind of objects was pushed forward by the works of Morris, Thorne and Yurtsever [4,5]. However, one of the obstacles to circumvent, for practical affairs, is the need of exotic forms of matter, since it is known that the required stress-energy tensor does not satisfy the standard energy conditions (see, e.g., [6]). This makes the stability as well as the existence of wormholes to be somehow controversial.

The need of exotic matter is also required to construct static wormholes for General Relativity in higher dimensions. Nonetheless, in higher-dimensional spacetimes, if one follows the same basic principles of General Relativity to describe gravity, the Einstein theory is not the only possibility. Indeed, the most general theory of gravity in higher dimensions leading to second order field equations for the metric is described by the Lovelock action which possesses nonlinear terms in the curvature [7]. Within this framework, it is worth pointing out that in five dimensions it has been found that the so-called Einstein–Gauss–Bonnet theory, being quadratic in the curvature, admits static wormhole solutions in vacuum [8]. This solution was found allowing a cosmological (volume) term in the Einstein–Gauss–Bonnet action, and choosing the coupling constant of the quadratic term such that the theory admits a single anti-de Sitter (AdS) vacuum. The wormhole connects two asymptotically locally AdS spacetimes each with a geometry at the boundary that is not spherically symmetric. It is worth to remark that in this case, no energy conditions can be violated since the whole spacetime is devoid of any kind of stress-energy tensor. Generically, the mass of the wormhole appears to be positive for observers located at one side of the neck, and negative for the ones at the other side, such that the total mass always vanishes. This provides a concrete example of "mass without mass."

These solutions extend to higher dimensions for special cases of the Lovelock class of theories, also selected by demanding the existence of a unique AdS vacuum [8–10].

It is then natural to wonder whether static wormholes in vacuum could also exist in four dimensions, which of course should be in some framework beyond General Relativity.

The next section is devoted to review the five-dimensional case including some interesting remarks about the behavior of geodesics, as well as the proof of the finiteness of their Euclidean action. The mass of these solutions is also obtained from surface integrals. Section 3 is devoted to explain how these results can be extended to higher dimensions, and includes a summary of the stability proof of static wormholes in vacuum against scalar field perturbations. The possibility of having wormholes in vacuum in four dimensions within conformal gravity is explored through a concrete example in Sect. 4.

2 Static Wormholes in Vacuum: The Five Dimensional Case

The action for the Einstein–Gauss–Bonnet theory with a volume term in five dimensions reads

$$I = \kappa \int d^5x\sqrt{-g}\left(R - 2\Lambda + \alpha\left(R^2 - 4R_{\mu\nu}R^{\mu\nu} + R_{\alpha\beta\gamma\delta}R^{\alpha\beta\gamma\delta}\right)\right), \tag{1}$$

where κ is related to the Newton constant, Λ to the cosmological term, and α is the Gauss–Bonnet coupling. As it was shown in [9], for static geometries whose the spacelike section is a warped product of the real line with a nontrivial manifold, this theory admits additional freedom to fix the metric at the boundary for the special case when the Gauss–Bonnet coupling is properly tuned in terms of the cosmological and Newton constants. In this case it is useful to express the action (1) as

$$I_5 = \kappa \int \varepsilon_{abcde}\left(R^{ab}R^{cd} + \frac{2}{3l^2}R^{ab}e^c e^d + \frac{1}{5l^4}e^a e^b e^c e^d\right)e^e, \tag{2}$$

where $R^{ab} = d\omega^{ab} + \omega^a{}_f\omega^{fb}$ is the curvature two-form for the spin connection ω^{ab}, and e^a is the vielbein. The coupling of the Gauss–Bonnet term turns out to be such that the theory possesses a unique AdS vacuum of radius l. In the absence of torsion, the field equations can be simply written as

$$\mathscr{E}_a := \varepsilon_{abcde}\bar{R}^{bc}\bar{R}^{de} = 0, \tag{3}$$

where $\bar{R}^{bc} := R^{bc} + \frac{1}{l^2}e^b e^c$.

If the boundary metric is chosen as a compact three-dimensional manifold of negative Ricci scalar, then two different kinds of wormhole solutions in vacuum are obtained.

The first one is described by the following metric

$$ds_5^2 = l^2\left[-\cosh^2(\rho - \rho_0)\,dt^2 + d\rho^2 + \cosh^2(\rho)\,d\Sigma_3^2\right], \tag{4}$$

where ρ_0 is an integration constant and $d\Sigma_3^2$ stands for the metric of the base manifold Σ_3 which can be chosen to be locally of the form $\Sigma_3 = S^1 \times H_2/\Gamma$. Here Γ is a freely acting discrete subgroup of $O(2,1)$ and the radius of the hyperbolic manifold H_2 turns out to be $3^{-1/2}$, so that the Ricci scalar of Σ_3 has the value of -6, as required by the field equations. The metric (4) describes a static wormhole with a neck of radius l, located at the minimum of the warp factor of the base manifold, at $\rho = 0$. Since $-\infty < \rho < \infty$, the wormhole connects two asymptotically locally AdS spacetimes so that the geometry at the boundary is locally given by $\mathbb{R} \times S^1 \times H_2$. Actually, it is simple to check that the field equations are solved provided the base manifold Σ_3 has a negative constant Ricci scalar satisfying

$$\tilde{R} = -6. \tag{5}$$

It worth pointing out that Σ_3 is not an Einstein manifold, and that any nontrivial solution of the corresponding Yamabe problem (see, e.g., [11]) provides a suitable choice for Σ_3.

The causal structure of the wormhole is depicted in Fig. 1, where the dotted vertical line shows the position of the neck, and the solid bold lines correspond to the asymptotic regions located at $\rho = \pm\infty$, each of them resembling an AdS spacetime but with a different base manifold since the usual sphere S^3 must be replaced by Σ_3. The line at the center stands for $\rho = \rho_0$.

It is apparent from the diagram that null and timelike curves can go forth and back from the neck. Furthermore, note that radial null geodesics are able to connect both asymptotic regions in finite time. Indeed, one can see from (4) that the coordinate time that a photon takes to travel radially from one asymptotic region, $\rho = -\infty$, to the other at $\rho = +\infty$ is given by

$$\Delta t = \int_{-\infty}^{+\infty} \frac{d\rho}{\cosh(\rho - \rho_0)} = \left[2\arctan\left(e^{\rho-\rho_0}\right)\right]_{-\infty}^{+\infty} = \pi,$$

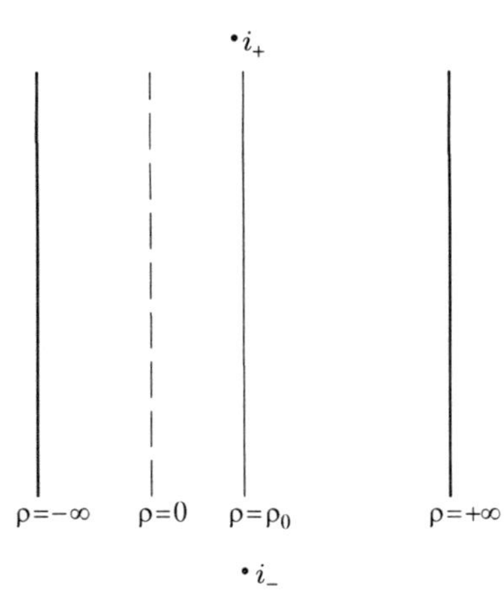

Fig. 1 Penrose diagram for the wormhole (4)

which does not depend on ρ_0. Thus, a static observer located at $\rho = \rho_0$ says that this occurred in a proper time given by πl. Note also that this observer actually lives on a static timelike geodesic, and it is easy to see that a small perturbation along ρ makes him to oscillate around $\rho = \rho_0$. This means that gravity is pulling towards the fixed hypersurface defined by $\rho = \rho_0$ which is parallel to the neck. Hence, the constant ρ_0 corresponds to a modulus parametrizing the proper distance between this hypersurface and the neck. Actually, one can explicitly check that radial timelike geodesics are always confined (see below).

The second wormhole in vacuum is described by the metric

$$ds^2 = l^2 \left[-e^{2\rho} dt^2 + d\rho^2 + \cosh^2(\rho) d\Sigma_3^2 \right], \tag{6}$$

possessing also a throat located at $\rho = 0$. Its Riemann tensor is given by

$$\begin{aligned} R^{t\rho}{}_{t\rho} &= -\frac{1}{l^2},\ R^{\rho i}{}_{\rho j} = -\frac{1}{l^2}\delta^i_j,\ R^{ti}{}_{tj} = -\frac{1}{l^2}\tanh(\rho)\,\delta^i_j, \\ R^{ij}{}_{kl} &= \frac{1}{l^2}\frac{\tilde{R}^{ij}{}_{kl}}{\cosh^2(\rho)} - \frac{1}{l^2}\tanh^2(\rho)\left(\delta^i_k\delta^j_l - \delta^i_l\delta^j_k\right), \end{aligned} \tag{7}$$

where latin indices run along the base manifold. At the asymptotic regions $\rho \to \pm\infty$ the curvature components approach

$$\begin{aligned} R^{t\rho}{}_{t\rho} &= -\frac{1}{l^2},\ R^{\rho i}{}_{\rho j} = -\frac{1}{l^2}\delta^i_j,\ R^{ti}{}_{tj} \simeq \mp\frac{1}{l^2}\delta^i_j, \\ R^{ij}{}_{kl} &\simeq -\frac{1}{l^2}\left(\delta^i_k\delta^j_l - \delta^i_l\delta^j_k\right), \end{aligned} \tag{8}$$

This makes clear that the wormhole (6) connects an asymptotically locally AdS spacetime (at $\rho \to \infty$) with another nontrivial smooth spacetime at the other asymptotic region ($\rho \to -\infty$). Note that although the metric looks singular at $\rho \to -\infty$, the geometry is well behaved at this asymptotic region. This is seen by noting that the basic scalar invariants can be written in terms of contractions of the Riemann tensor with the index position as in (7), whose components have well defined limits (given in (8)). Thus, the invariants cannot diverge. As an example, the limits of some invariants are

$$\lim_{\rho\to-\infty} R^{\alpha\beta}{}_{\alpha\beta} = -\frac{8}{l^2},\ \lim_{\rho\to-\infty} R^{\alpha\beta}{}_{\gamma\delta}R^{\gamma\delta}{}_{\alpha\beta} = \frac{40}{l^4},\ \lim_{\rho\to-\infty} C^{\alpha\beta}{}_{\gamma\delta}C^{\gamma\delta}{}_{\alpha\beta} = \frac{8}{l^4} \tag{9}$$

where $C^{\alpha\beta}{}_{\gamma\delta}$ is the Weyl tensor. Some differential invariants have also been computed and they become all well behaved as $\rho \to -\infty$.

2.1 *Geodesics Around Wormholes in Vacuum*

The class of metrics that describe the wormhole solutions presented here is of the form

$$ds^2 = -A^2(\rho)\,dt^2 + l^2\left[d\rho^2 + \cosh^2(\rho)\,d\Sigma^2\right], \tag{10}$$

where the lapse function $A(\rho)$ can be obtained from (4) and (6).

2.1.1 Radial Geodesics

The radial geodesics are described by the following equations

$$\dot{t} - \frac{E}{A^2} = 0\,, \tag{11}$$

$$l^2\dot{\rho}^2 - \frac{E^2}{A^2} + b = 0, \tag{12}$$

where dot stands for derivatives with respect to the proper time, the velocity is normalized as $u_\mu u^\mu = -b$, and the integration constant E corresponds to the energy. As one expects, (12) tells that gravity is pulling towards the fixed hypersurface defined by $\rho = \rho_0$, where ρ_0 is a minimum of $A^2(\rho)$.

As it can be seen from (4) we have $A^2(\rho) = l^2\cosh^2(\rho - \rho_0)$, then the equations for radial geodesics (11) and (12) reduce to

$$\dot{\rho}^2 - \frac{E^2}{l^4\cosh^2(\rho - \rho_0)} = -\frac{b}{l^2}, \tag{13}$$

$$\dot{t} - \frac{E}{l^2\cosh^2(\rho - \rho_0)} = 0. \tag{14}$$

These equation immediately tell us that the ρ coordinate of a radial geodesic behaves as a classical particle in a Pöschl–Teller potential, so that timelike geodesics are confined and they oscillate around the hypersurface $\rho = \rho_0$. Consequently, an observer sitting at $\rho = \rho_0$ lives in a timelike geodesic (here $d\tau/dt = l$, τ the proper time of this static observer), and radial null geodesics connect both asymptotic regions (i.e., $\rho = -\infty$ with $\rho = +\infty$) in a finite coordinate time which does not depend on ρ_0 (the static observer at $\rho = \rho_0$ says that this occurred in a proper time $\Delta\tau = \pi l$). These observations give a meaning of ρ_0: gravity is pulling towards the fixed hypersurface defined by $\rho = \rho_0$, which is parallel to the neck at $\rho = 0$, and therefore ρ_0 is a modulus parameterizing the proper distance from this hypersurface to the neck.

The geodesic structure of the second wormhole (6) is quite different from the previous one. In this case, the radial geodesic equations (11) and (12) read

$$\dot{\rho}^2 - \frac{e^{-2\rho}E^2}{l^4} = -\frac{b}{l^2}, \tag{15}$$

$$l^2\,\dot{t} - e^{-2\rho}E = 0. \tag{16}$$

As expected, the behavior of the geodesics at $\rho \to +\infty$ is like in an AdS spacetime. Moreover, since gravity pulls towards the asymptotic region $\rho \to -\infty$, radial timelike geodesics always have a turning point and they are doomed to approach to $\rho \to -\infty$ in the future. Note that the proper time that a timelike geodesic takes to reach the asymptotic region at $\rho = -\infty$, starting from $\rho = \rho_f$ is finite and given by

$$\Delta\tau = \int_{\rho=-\infty}^{\rho=\rho_f} \frac{l^2 d\rho}{\sqrt{E^2 e^{-2\rho} - l^2}} = \frac{\pi l}{2} - l\tan^{-1}\left(\sqrt{\frac{E^2}{l^2}e^{-2\rho_f} - 1}\right) < \infty. \tag{17}$$

It is easy to check that null radial geodesics can also reach the asymptotic region at $\rho = -\infty$ in a finite affine parameter. This, together with the fact that spacetime is regular at this boundary, seems to suggest that it could be analytically continued through this surface. However, since the warp factor of the base manifold blows up at $\rho = -\infty$, this null hypersurface should be regarded as a spacetime boundary.

2.1.2 Gravitational Vs. Centrifugal Forces

An interesting effect occurs for geodesics with nonzero angular momentum. One can see that for the class of spacetimes (10), there is a region where the gravitational and centrifugal effective forces point in the same direction. These are expulsive regions that have a single turning point for any value of the conserved energy, and within which bounded geodesics cannot exist.

Let us then consider metrics of the form (10) with the further restriction that the base manifold Σ_3 have a Killing vector ξ. Choosing adapted coordinates $y = (x^1, x^2, \phi)$ such that $\xi = \partial/\partial\phi$, the base manifold metric is $d\Sigma_3^2 = \tilde{g}_{ij}(x)dy^i dy^j$ and the spacetime geodesics with x fixed are described by the following equations

$$\dot{t} = \frac{E}{A^2};\ \dot{\phi} = \frac{L}{C^2}$$
$$l^2\dot{\rho}^2 = -b + \frac{E^2}{A^2} - \frac{L^2}{C^2}. \tag{18}$$

Here we have used the fact that, if u^a is the geodesic tangent vector, then $\xi^a u_a = \mathscr{L}$ is conserved, and $\dot{\phi} = \mathscr{L}/(C^2\tilde{g}_{\phi\phi}(x)) =: L/C^2$, where $C := l\cosh(\rho)$. If ξ is a $U(1)$ Killing vector then $\mathscr{L}$ is a conserved angular momentum. Examples are not hard to construct. For wormholes, we need a nonflat three-manifold with $\tilde{R} = -6$ and a $U(1)$ isometry, an example being $S^1 \times H_2/\Gamma$, where Γ is a freely acting discrete subgroup of $O(2,1)$, and the metric is locally given by

$$d\Sigma_3^2 = \frac{1}{3}\left(dx_1{}^2 + \sinh^2(x_1)\,dx_2{}^2\right) + d\phi^2. \tag{19}$$

The motion along the radial coordinate in proper time is like that of a classical particle in an effective potential given by the r.h.s. of (18). This effective potential, has a minimum at $\rho = \bar{\rho}$ only if the following condition is fulfilled

$$\frac{A'(\bar{\rho})}{A(\bar{\rho})^3}E^2 = \frac{L^2 \tanh\bar{\rho}}{\cosh^2\bar{\rho}}. \tag{20}$$

This expresses the fact that the gravitational effective force is canceled by the centrifugal force if the orbit sits at $\rho = \bar{\rho}$. The class of spacetimes under consideration have regions $\mathscr{U}$ where the sign of $A^{-3}A'$ is opposite to that of $\tanh\bar{\rho}$, i.e., the effective gravitational and centrifugal forces point in the same direction. Within these regions, there is at most a single turning point, and consequently bounded orbits cannot exist.

In the case of the wormhole (4) the centrifugal force reverses its sign at the neck, located at $\rho = 0$, and the gravitational force does it at $\rho = \rho_0$, so that both forces point towards the same direction for ρ between zero and ρ_0. Thus, the expulsive region $\mathscr{U}$ is nontrivial whenever $\rho_0 \neq 0$. This situation is depicted in Fig. 2a.

In the case of the wormhole solution (6) the region $\mathscr{U}$ is defined $\rho \leq 0$ (see Fig. 2b).

2.2 *Regularized Euclidean Action*

Here it is shown that the wormhole geometries described above have vanishing Euclidean action. It has been shown that the action (2) can be regularized by adding a suitable boundary term in a background independent way, which depends only on

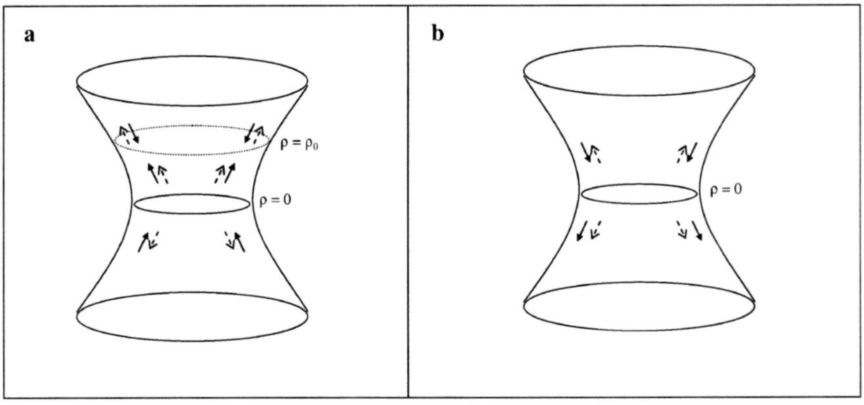

Fig. 2 Gravitational vs. centrifugal forces for wormholes in vacuum. In this diagram, *black* and *dashed arrows* represent the effective gravitational and centrifugal forces, respectively. Figures (**a**) and (**b**), correspond to the wormholes (4) and (6), respectively

the extrinsic curvature and the geometry at the boundary [12–14]. The total action then reads

$$I_T = I_5 - B_4, \tag{21}$$

where the boundary term is given by

$$B_4 = \kappa \int_{\partial M} \varepsilon_{abcde} \theta^{ab} e^c \left(R^{de} - \frac{1}{2} \theta^d{}_f \theta^{fe} + \frac{1}{6l^2} e^d e^e \right), \tag{22}$$

and θ^{ab} is the second fundamental form. The total action (21) attains an extremum for solutions of the field equations provided

$$\delta I_T = \kappa \int_{\partial M} \varepsilon_{abcde} \left(\delta\theta^{ab} e^c - \theta^{ab} \delta e^c \right) \left(\bar{R}^{de} - \frac{1}{2} \theta^d{}_f \theta^{fe} - \frac{1}{2l^2} e^d e^e \right) = 0, \tag{23}$$

where $\bar{R}^{ab} := R^{ab} + \frac{1}{l^2} e^a e^b$. Therefore, following Regge and Teitelboim [15], the value of the regularized Euclidean action makes sense for solutions which are bona fide extrema, i.e., for solutions such that condition (23) is fulfilled.

The Euclidean continuation of the class of wormholes is described by metrics of the form

$$ds^2 = A^2(\rho)\, d\tau^2 + l^2 d\rho^2 + C^2(\rho)\, d\Sigma_3^2, \tag{24}$$

where $0 \leq \tau \leq \beta$ is the Euclidean time, and the functions A and C correspond to the ones appearing in (4) and (6).

As shown in [9], it is simple to verify that the wormholes solutions under discussion are truly extrema of the total action (21).

2.2.1 Wormholes in Vacuum as Extrema of the Regularized Action

For the class of solutions under consideration, the curvature two-form satisfies

$$\bar{R}^{01} = \bar{R}^{1m} = 0, \tag{25}$$

and the condition (23) reduces to

$$\delta I_T = \kappa\beta \left[F\, \mathscr{I}_3 + 6\, G\, \mathscr{V}_3 \right]_{\partial\Sigma}, \tag{26}$$

where β is the Euclidean time period, $\mathscr{V}_3$ is the volume of the base manifold, and $\partial\Sigma$ is the boundary of the spatial section. In (26) $\mathscr{I}_3$ is defined by

$$\mathscr{I}_3 := \int_{\Sigma_3} \sqrt{\tilde{g}} \tilde{R}\, d^3x, \tag{27}$$

and the functions F and G in (26) are given by

$$F := \frac{2}{l}\left[A'\delta C - A\delta C' + C'\delta A - C\delta A'\right], \tag{28}$$

$$\begin{aligned} G :=& \left[A'\left(C^2 - C'^2\right) + 2C'\left(CA - C'A'\right)\right]\frac{\delta C}{l^3} \\ &- \left[A\left(C^2 - C'^2\right) + 2C\left(CA - C'A'\right)\right]\frac{\delta C'}{l^3} \\ &+ C'\left(C^2 - C'^2\right)\frac{\delta A}{l^3} - C\left(C^2 - C'^2\right)\frac{\delta A'}{l^3}. \end{aligned} \tag{29}$$

Here we work in the minisuperspace approach, where the variation of the functions A and C correspond to the variation of the integration constants, and prime ($'$) denotes derivative with respect to ρ.

Now it is simple to evaluate the variation of the action (26) explicitly.

The Euclidean continuation of both wormhole solutions in (4) and (6) can be written as

$$ds^2 = l^2\left[(\cosh\rho + a\sinh\rho)^2 d\tau^2 + d\rho^2 + \cosh^2\rho d\Sigma_3^2\right], \tag{30}$$

where the metrics (4) and (6) are recovered for $a^2 < 1$ and $a^2 = 1$, respectively, and β is arbitrary. In this sense, the wormhole (6) can be regarded as a sort of extremal case of the wormhole (4). In this case, since the boundary is of the form $\partial\Sigma = \Sigma_3^+ \cup \Sigma_3^-$ it is useful to introduce the regulators $\rho_\pm$, such that $\rho_- \le \rho \le \rho_+$. Using the fact that the base manifold has a negative constant Ricci scalar given by $\tilde{R} = -6$, the variation of the action (26) reduces to

$$\delta I_T = 6\kappa\beta l\ \delta a\ \left[\mathscr{V}_3\right]_{\rho_-}^{\rho_+} = 0. \tag{31}$$

Note that, as in the case for the black hole, the boundary term vanishes regardless the position of the regulators ρ_- and ρ_+.

Having shown that the wormholes are truly extrema of the action, it makes sense to evaluate the regularized action on these solutions.

2.2.2 Euclidean Action for Wormholes in Vacuum

For the class of solutions of the form (24), which satisfy (25), the bulk and boundary contributions to the regularized action $I_T = I_5 - B_4$, given by (2) and (22) respectively, reduce to

$$I_5 = \kappa\beta\left[H\ \mathscr{I}_3 + 6\ J\ \mathscr{V}_3\right], \tag{32}$$

$$B_4 = \kappa\beta\left[h\ \mathscr{I}_3 + 6\ j\ \mathscr{V}_3\right]_{\partial\Sigma}. \tag{33}$$

The functions H and J in the bulk term are defined by

$$H := -\frac{8}{l} \int AC \, d\rho \,, \tag{34}$$

$$J := \frac{4}{l^3} \int \left[\left(C^2\right)' (AC)' - \frac{4}{3} AC^3 \right] \, d\rho, \tag{35}$$

where the integrals are taken along the whole range of ρ. For the boundary term (33), the functions h and j are respectively defined by

$$h = -\frac{2}{l} (AC)' , \tag{36}$$

$$j = -\frac{1}{l^3} \left[(AC)' \left(\frac{C^2}{3} - C'^2 \right) + \left(C^2\right)' \left(\frac{AC}{3} - A'C' \right) \right] . \tag{37}$$

Now it is straightforward to evaluate the regularized Euclidean action for the class of solutions under consideration.

The Euclidean continuation of the wormhole metrics (4) and (6) are smooth independently of the Euclidean time period β. The Euclidean action $I_T = I_5 - B_4$, is evaluated introducing regulators such that $\rho_- \leq \rho \leq \rho_+$.

In the case of the Euclidean wormhole (4) the regularized Euclidean action vanishes regardless the position of the regulators, since

$$I_5 = B_4 = 2\kappa l \beta \mathcal{V}_3 \left[3 \sinh(\rho_0) + 8 \cosh^3(\rho) \sinh(\rho - \rho_0) \right]_{\rho_-}^{\rho_+} . \tag{38}$$

Consequently, the total mass of this spacetime also vanishes, since $M = -\frac{\partial I_T}{\partial \beta} = 0$.

For the wormhole (6) the Euclidean action reads

$$I_T = 6\kappa \beta \mathcal{V}_3 \left[(J - j) - (H - h) \right] , \tag{39}$$

with

$$H = -2l \left(e^{2\rho} + 2\rho \right) \Big|_{\rho_-}^{\rho_+} \,, \tag{40}$$

$$J = -\frac{1}{3} l \left(-e^{4\rho} + 3e^{2\rho} + 12\rho - e^{-2\rho} \right) \Big|_{\rho_-}^{\rho_+} \,, \tag{41}$$

$$h = -2l \, e^{2\rho} \Big|_{\rho_-}^{\rho_+} \,,$$

$$j = -\frac{1}{3} l \left(-e^{-4\rho} + 3e^{2\rho} - e^{-2\rho} \right) \Big|_{\rho_-}^{\rho_+} . \tag{42}$$

The regularized action vanishes again independently of $\rho_\pm$, and so does it mass.

It is worth pointing out that both wormholes can be regarded as instantons with vanishing Euclidean action.

The mass for the spacetime metrics discussed here can also be obtained from a suitable surface integral coming from a direct application of Noether's theorem to the regularized action functional.

2.2.3 Mass from a Surface Integral

As it was shown that the wormhole solutions are truly extrema of the regularized action, one is able to compute the mass from the following surface integral

$$Q(\xi) = \frac{\kappa}{l}\int_{\partial\Sigma} \varepsilon_{abcde}\left(I_\xi \theta^{ab} e^c + \theta^{ab} I_\xi e^c\right)\left(\tilde{R}^{de} + \frac{1}{2}\theta^d{}_f \theta^{fe} + \frac{1}{2l^2} e^d e^e\right), \tag{43}$$

obtained by the straightforward application of Noether's theorem.[1] Here $\xi = \partial_t$ is the timelike Killing vector.

For a metric of the form (24), satisfying (25), (43) gives

$$M = 2\frac{\kappa}{l}\left[\left(A'C - C'A\right)\left(\mathscr{I}_3 + \frac{3}{l^2}\left(C^2 - C'^2\right)\mathscr{V}_3\right)\right]_{\partial\Sigma}, \tag{44}$$

which can be explicitly evaluated for the wormhole solutions.

As explained in [8], for the wormhole solution in (4), one obtains that the contribution to the total mass coming from each boundary reads

$$M_\pm = Q_\pm(\partial_t) = \pm 6\kappa \mathscr{V}_3 \sinh(\rho_0), \tag{45}$$

where $Q_\pm(\partial_t)$ is the value of (43) at $\partial\Sigma_\pm$, which again does not depend on ρ_+ and ρ_-. The opposite signs of $M_\pm$, are due to the fact that the boundaries of the spatial section have opposite orientation. The integration constant ρ_0 can be regarded as a parameter for the apparent mass at each side of the wormhole, which vanishes only when the solution acquires reflection symmetry, i.e., for $\rho_0 = 0$. This means that for a positive value of ρ_0, the mass of the wormhole appears to be positive for observers located at ρ_+, and negative for the ones at ρ_-, with a vanishing total mass $M = M_+ + M_- = 0$.

For the wormhole (6) the total mass also vanishes since the contribution to the surface integral (43) coming from each boundary reads

$$M_\pm = \mp 6\kappa \mathscr{V}_3, \tag{46}$$

so that $M = M_+ + M_- = 0$.

Note that $M_\pm$ are concrete examples of Wheeler's conception of *"mass without mass"*.

[1] The action of the contraction operator I_ξ over a p-form $\alpha_p = \frac{1}{p!}\alpha_{\mu_1\cdots\mu_p} dx^{\mu_1}\cdots dx^{\mu_p}$ is given by $I_\xi \alpha_p = \frac{1}{(p-1)!}\xi^\nu \alpha_{\nu\mu_1\cdots\mu_{p-1}} dx^{\mu_1}\cdots dx^{\mu_{p-1}}$, and $\partial\Sigma$ stands for the boundary of the spacelike section.

As explained in the next section, these results found here are not peculiarities of five-dimensional gravity, and similar structures can be found in higher dimensional spacetimes [9, 10].

3 Higher-Dimensional Wormholes in Vacuum

As shown in [8], the five-dimensional static wormhole solution in vacuum, given by (4), can be extended as an exact solution for a very special class of gravity theories among the Lovelock family in higher odd dimensions $d = 2n+1$. In analogy with the procedure in five dimensions, the theory can be constructed so that the relative couplings between each Lovelock term are chosen so that the action has the highest possible power in the curvature and possesses a unique AdS vacuum of radius l. The field equations then read

$$\mathscr{E}_A := \varepsilon_{ab_1\cdots b_{2n}} \bar{R}^{b_1 b_2} \cdots \bar{R}^{b_{2n-1} b_{2n}} = 0, \tag{47}$$

which are solved by the straightforward extension of (4) to higher dimensions

$$ds^2 = l^2 \left[-\cosh^2(\rho - \rho_0)\, dt^2 + d\rho^2 + \cosh^2(\rho)\, d\Sigma_{2n-1}^2 \right],$$

where ρ_0 is an integration constant, and $d\Sigma_{2n-1}^2$ stands for the metric of the base manifold. In the generic case, the base manifold must solve the following equation[2]

$$\varepsilon_{m_1\cdots m_{2n-1}} \bar{R}^{m_1 m_2} \cdots \bar{R}^{m_{2n-3} m_{2n-2}} \tilde{e}^{m_{2n-1}} = 0, \tag{48}$$

where $\tilde{e}^m$ is the vielbein of Σ_{2n-1}. Note that this is a single scalar equation which admits a wide class of solutions. It is simple to verify that $\Sigma_{2n-1} = H_{2n-1}/\Gamma$ and $\Sigma_{2n-1} = S^1 \times H_{2n-2}/\Gamma$ give a solution provided the radii of the hyperbolic spaces H_{2n-1} and H_{2n-2} are given by $r_{H_{2n-1}} = 1$ and $r_{H_{2n-2}} = (2n-1)^{-1/2}$, respectively.[3] In order to describe a wormhole the hyperbolic spaces must be quotiented by a freely acting discrete subgroup Γ of $O(2n-2,1)$, otherwise the spacetime would possess only one conformal boundary (see, e.g., [16]). Further examples in five and seven dimensions for base manifolds corresponding to all the possible products of constant curvature spaces in five and seven dimensions have been constructed in [17]. Note that generically, Σ_{2n-1} is not an Einstein manifold.

The metric in higher odd dimensions then describes a static wormhole with a neck of radius l connecting two asymptotic regions which are locally AdS spacetimes, so that the geometry at the boundary can be given by $\mathbb{R} \times S^1 \times H_{2n-2}/\Gamma$. The

[2] Equation (48) corresponds to the trace of the Euclidean field equations for the same theory in $2n-1$ dimensions with a unit AdS radius.

[3] As explained in [8], the field equations acquire certain class of degeneracy around the solution with $\Sigma_{2n-1} = H_{2n-1}$.

wormhole in higher dimensions shares the features described in the five-dimensional case, including the meaning of the parameter ρ_0, and its causal structure is depicted in Fig. 1.

As in the five-dimensional case, the Euclidean continuation of the wormhole metric is smooth and it has an arbitrary Euclidean time period. The Euclidean action can be regularized in higher odd dimensions in a background independent way as in [12–14], by the addition of a suitable boundary term which is the analogue of (22), and can also be written in terms of the extrinsic curvature and the geometry at the boundary. The nonvanishing components of the second fundamental form θ^{ab} acquire the same form as in the five-dimensional case, so that it is easy to check that the regularized action has an extremum for the wormhole solution. As in the five-dimensional case, the Euclidean continuation of the wormhole can be seen as an instanton with a regularized action that vanishes independently of the position of the boundaries, so that its mass is also found to vanish. This means that AdS spacetime has a greater action than the wormhole, but a lower "vacuum energy."

The wormhole mass for the Lorenzian solution can also be shown to vanish making use of a surface integral which is the extension of (43) to higher odd dimensions [12–14]. The contribution to the total mass coming from each boundary does not depend on the location of the boundaries and is given by

$$Q_{\pm}(\partial_t) = \pm \alpha_n \mathcal{V}_{d-2}\, \kappa \sinh(\rho_0)\,,$$

so that for a nonvanishing integration constant ρ_0, the wormhole appears to have "mass without mass." Here $\alpha_n := \left[(1-2n)^{n-1} - 2^n(1-n)^{n-1}\right](2n-1)!$

It is simple to show that for different base manifolds, the Euclidean action also vanishes, and the surface integrals for the mass possess a similar behavior.

Wormholes in vacuum have also been shown to exist for a wider class of theories for all dimensions greater than four [10], which are selected as in [18] by the requirement of having a unique maximally symmetric vacuum solution.

3.1 Stability of Asymptotically Locally AdS Wormholes in Vacuum Against Scalar Field Perturbations

It is natural to wonder whether the wormholes described here can be regarded as a stable solutions providing a suitable ground state in order to define a field theory on it. This has been explored in [17] for the case in which the apparent mass at each side of the wormhole (4) vanishes ($\rho_0 = 0$), so that the solution acquires reflection symmetry. In this case the metric reads

$$ds_d^2 = l^2\left[-\cosh^2\rho\, dt^2 + d\rho^2 + \cosh^2\rho\, d\Sigma_{d-2}^2\right]. \tag{49}$$

As a first step in this direction, it has been proved that the wormholes described by the metric (49) are stable against scalar field perturbations, provided its squared mass satisfies a lower bound. The bound is generically more restrictive than that discovered by Breitenlohner and Freedman for AdS spacetime [19], given by $m^2 > m_{BF}^2$ with

$$m_{BF}^2 := -\frac{1}{l^2}\left(\frac{d-1}{2}\right)^2 . \tag{50}$$

The strategy followed to prove their stability is similar to the one used by Breitenlohner and Freedman for AdS spacetime [19–21]. Thus, one can see that for a free massive, minimally coupled scalar field, stability of the metric (49) is guaranteed provided the squared mass is bounded from below by a negative quantity which depends on the lowest eigenvalue of the Laplace operator on the base manifold. Remarkably, the Klein–Gordon equation for the scalar field can be solved analytically on the background metric (49), so that an exact expression for the spectrum can be found requiring the energy flux to vanish at each boundary. These boundary conditions single out scalar fields with fast fall-off. It can also be shown that in the presence of nonminimal coupling with the scalar curvature, stability can also be achieved for scalar fields with slow fall-off provided the squared mass also satisfies certain negative upper bound. It is worth to remark that, unlike the case of AdS spacetime, the Ricci scalar of the wormhole is not constant, so that the nonminimal coupling contributes to the field equation with more than just a shift in the mass. Nevertheless, in this case an exact expression for the spectrum can also be found, and three different quantizations for the scalar field can be carried on, being characterized by the fall-off of the scalar field, which can be fast or slow in each one of the asymptotic regions.

As it occurs for asymptotically AdS spacetimes [22–26], it is natural to expect that these results can be extrapolated to scalar fields with a selfinteraction that can be unbounded from below. In this case, it would be interesting to explore the subtleties due to the presence of a nontrivial potential, since the asymptotic form of the scalar field obtained through the linear approximation could no longer be reliable. Indeed, for certain critical values of the mass, the nonlinear terms in the potential could become relevant in the asymptotic region, such that the scalar field would be forced to develop additional logarithmic branches [22]. These effects should also be sensitive to the spacetime dimension, and for certain critical values of the mass, they would be particularly relevant in the sense of the dual conformal field theory. Nonetheless, note that the existence of asymptotically AdS wormholes raises some puzzles concerning the AdS/CFT correspondence [16, 27, 28].

It is also natural to wonder about the stability of the wormhole against gravitational perturbations. However, this is not a simple task, since for the class of theories under consideration, the degrees of freedom of the graviton could depend on the background geometry (see, e.g., [29, 30]), so that the dynamics of the perturbations has to be analyzed from scratch. Nevertheless, it is simple to check that the wormhole solves the field equations of the corresponding locally supersymmetric extension in five [31] and higher odd dimensions [32, 33]. If the wormhole had some unbroken supersymmetries, its stability would be guaranteed nonperturbatively. However, a quick analysis shows that the wormhole in vacuum breaks all the supersymmetries. Nonetheless, one cannot discard that supersymmetry could be restored by switching on the torsion as in [34], or some nontrivial gauge fields without backreaction [35, 36].

It would also be interesting to explore whether stability holds for the whole class of wormholes discussed here.

For pure Gauss–Bonnet gravity, it has also been shown that wormhole solutions with a jump in the extrinsic curvature along a "thin shell of nothingness" exist [37], and this has also been extended to the full Einstein–Gauss–Bonnet theory in five dimensions [38]. For this theory, it is possible to have wormholes made of thin shells of matter fulfilling the standard energy conditions [39, 40]. For smooth matter distributions, wormholes that do not violate the weak energy condition also exist, provided the Gauss–Bonnet coupling constant is negative and bounded according to the shape of the solution [41, 42].

4 Static Wormholes in Vacuum in Four Dimensions

It is very simple to obtain a wormhole solution in four dimensions from the compactification of the five-dimensional ones in vacuum discussed here, since it has been shown that they always admits a base manifold with a S^1 factor. The dimensionally reduced metrics in four dimensions then possess similar geometrical and causal behavior as their five-dimensional seeds. Nonetheless, the dimensionally reduced solution is supported by a nontrivial dilaton field with a clumsy nonvanishing stress-energy tensor.

If one wonders about the possibility of realizing this class of wormholes in vacuum in four dimensions, one knows that it should be performed within some scenario beyond general Relativity. Our previous experience has shown that one of the key features allowing this possibility is the relaxation of the asymptotic conditions for the metric. An interesting theory possessing this last feature is conformal gravity (see, e.g., [43, 44]), whose Lagrangian can be written as the square of the Weyl Tensor, so that the action reads

$$I = \alpha \int d^4x \sqrt{-g} C^{\alpha\beta\mu\nu} C_{\alpha\beta\mu\nu}.$$

Apart from being invariant under diffeomorphisms, this action is also invariant under local rescalings of the metric

$$g_{\mu\nu} \to \Omega^2(x) g_{\mu\nu},$$

and the field equations are given by the vanishing of the Bach tensor, i.e.,

$$\left(\nabla^\mu \nabla^\nu + \frac{1}{2} R^{\mu\nu}\right) C^\alpha{}_{\mu\beta\nu} = 0.$$

It is simple to verify that any Einstein space with cosmological constant is a solution of the theory in vacuum.

Conformal gravity was intensively studied in the past and it has been show to be renormalizable [45, 46] – note that the coupling α is dimensionless. Nevertheless, the theory possesses fourth-order field equations for the metric, so that it has ghosts as it is generically expected. In the context we are interested in, regardless the theory is suitable or not as the ultimate one to describe gravity, it certainly deserves to be studied.

It is also worth pointing out that it has been recently shown that for smooth matter distributions, wormholes that do not violate the weak energy condition near the throat can exist in conformal gravity [47].

For our purposes, it is worth pointing out that the most general spherically symmetric solution of conformal gravity (see, e.g., [48]) possesses a relaxed asymptotic behavior as compared with general relativity. Thus, for the reasons explained above, the door is open to look for wormholes in vacuum within this theory.

As shown in [49], conformal gravity admits a static spherically symmetric wormhole solution in vacuum, whose metric reads

$$ds^2 = -\left(1+a^2\tanh(\rho)\right)dt^2 + \frac{1}{\left(1+a^2\tanh(\rho)\right)}d\rho^2 + l_0^2\cosh(\rho)^2 d\Omega^2, \quad (51)$$

where $d\Omega^2$ stands for the line element of S^2, and the range of the radial coordinate is given by $-\infty < \rho < \infty$. The wormhole possesses a single integration constant a that parametrizes the radius of the neck given by

$$l_0^2 = \frac{1}{\sqrt{3a^4+1}},$$

being located at $\rho = 0$. The wormhole connects two static homogeneous universes of constant spatial curvature with different radii, as it can be seen from the asymptotic behavior of the curvature

$$\lim_{\rho\to\infty\pm} R^{mn}{}_{kl} = -\left(1\pm a^2\right)\delta^{mn}{}_{kl},$$

and time runs at different rates at each side of the neck, since

$$\lim_{\rho\to\infty\pm} g_{tt} = -\left(1\pm a^2\right).$$

The case of $a^2 \to -a^2$ just amounts to reflection symmetry on the radial coordinate ρ.

For $a^2 < 1$ The wormhole interpolates between static universes with spatial geometries given by hyperbolic spaces of radii $\left(1\pm a^2\right)^{-1/2}$. In the case of $a = 0$ the metric acquires a simple form, which reads

$$ds^2 = -dt^2 + d\rho^2 + \cosh(\rho)^2 d\Omega^2,$$

and it can be regarded as "the groundstate" with $l_0 = 1$ and $C^{\alpha\beta}_{\mu\nu} \neq 0$. For $a = 1$ the wormhole interpolates between flat space and a static universe with spatial geometry given by a hyperbolic space of radius $2^{-1/2}$. In the case of $a^2 > 1$ the wormhole develops a cosmological horizon at one side of the neck, located at $\rho = \rho_+$, with

$$\rho_+ = -\tanh^{-1}\left(\frac{1}{a^2}\right),$$

and interpolates between an Einstein Universe ($R \times S^3$) for $\rho = -\infty$, and $R \times H_3$.

The causal structure of the wormhole coincides with the one of Minkowski spacetime in two dimensions for $a^2 \leq 1$, and for $a^2 > 1$ reduces to the one of two-dimensional Rindler spacetime.

The extension of the wormhole (51) with radial electric or magnetic fields can also be found, and it turns out to have electric or magnetic "charge without charge." The solutions can be further generalized to the case of necks with genus greater than one. It can also be shown that the wormholes in vacuum generically correspond to the matching of different Einstein spacetimes at infinity by means of improper conformal transformations.

As an ending remark, it is worth pointing out that the definition of mass in conformal gravity is a very subtle issue, which is not free of controversy [50–53]. Thus, in order to have a suitable analysis of the solutions the construction of finite conserved charges written as suitable surface integrals for conformal gravity should be addressed.

Acknowledgments This work was partially funded by FONDECYT grants 1051056, 1061291, 1071125, 1085322. The Centro de Estudios Científicos (CECS) is funded by the Chilean Government through the Millennium Science Initiative and the Centers of Excellence Base Financing Program of Conicyt. CECS is also supported by a group of private companies which at present includes Antofagasta Minerals, Arauco, Empresas CMPC, Indura, Naviera Ultragas and Telefónica del Sur. CIN is funded by Conicyt and the Gobierno Regional de Los Ríos.

References

1. A. Einstein and N. Rosen, Phys. Rev. **48**, 73 (1935).
2. C. W. Misner and J. A. Wheeler, Ann. Phys. **2**, 525 (1957).
3. G. W. Gibbons and S. W. Hawking (eds.) "*Euclidean quantum gravity*", World Scientific, Singapore, (1993).
4. M. S. Morris, K. S. Thorne and U. Yurtsever, Phys. Rev. Lett. **61**, 1446 (1988).
5. M. S. Morris and K. S. Thorne, Am. J. Phys. **56**, 395 (1988).
6. M. Visser, "*Lorentzian wormholes: from Einstein to Hawking*", AIP, Woodbury, NY, (1995).
7. D. Lovelock, J. Math. Phys. **12**, 498 (1971).
8. G. Dotti, J. Oliva and R. Troncoso, Phys. Rev. D **75**, 024002 (2007).
9. G. Dotti, J. Oliva and R. Troncoso, Phys. Rev. D **76**, 064038 (2007).
10. G. Dotti, J. Oliva and R. Troncoso, work in progress.
11. J. Lee and T. Parker, Bull. Am. Math. Soc. **17**, 37 (1987).
12. P. Mora, R. Olea, R. Troncoso and J. Zanelli, JHEP **0406**, 036 (2004) [arXiv:hep-th/0405267].

13. P. Mora, R. Olea, R. Troncoso and J. Zanelli, "Vacuum energy in odd-dimensional AdS gravity" arXiv:hep-th/0412046.
14. P. Mora, R. Olea, R. Troncoso and J. Zanelli, JHEP **0602**, 067 (2006) [arXiv:hep-th/0601081].
15. T. Regge and C. Teitelboim, Ann. Phys. (NY) **88**, 286 (1974).
16. J. M. Maldacena and L. Maoz, JHEP **0402**, 053 (2004).
17. D. H. Correa, J. Oliva and R. Troncoso, JHEP **08**, 081 (2008) arXiv:0805.1513 [hep-th].
18. J. Crisóstomo, R. Troncoso and J. Zanelli, Phys. Rev. D **62**, 084013 (2000).
19. P. Breitenlohner and D. Z. Freedman, Phys. Lett. B **115**, 197 (1982).
20. P. Breitenlohner and D. Z. Freedman, Ann. Phys. **144**, 249 (1982).
21. L. Mezincescu and P. K. Townsend, Ann. Phys. **160**, 406 (1985).
22. M. Henneaux, C. Martinez, R. Troncoso and J. Zanelli, Ann. Phys. **322**, 824 (2007).
23. M. Henneaux, C. Martinez, R. Troncoso and J. Zanelli, Phys. Rev. D **65**, 104007 (2002).
24. M. Henneaux, C. Martinez, R. Troncoso and J. Zanelli, Phys. Rev. D **70**, 044034 (2004).
25. T. Hertog and K. Maeda, JHEP **0407**, 051 (2004).
26. A. J. Amsel and D. Marolf, Phys. Rev. D **74**, 064006 (2006) [Erratum-ibid. D **75**, 029901 (2007)].
27. E. Witten and S. T. Yau, Adv. Theor. Math. Phys. **3**, 1635 (1999).
28. N. Arkani-Hamed, J. Orgera and J. Polchinski, JHEP **0712**, 018 (2007).
29. J. Saavedra, R. Troncoso and J. Zanelli, J. Math. Phys. **42**, 4383 (2001).
30. O. Miskovic, R. Troncoso and J. Zanelli, Phys. Lett. B **615**, 277 (2005).
31. A. H. Chamseddine, Nucl. Phys. B **346**, 213 (1990).
32. R. Troncoso and J. Zanelli, Phys. Rev. D **58**, 101703 (1998).
33. R. Troncoso and J. Zanelli, Int. J. Theor. Phys. **38**, 1181 (1999).
34. F. Canfora, A. Giacomini and R. Troncoso, Phys. Rev. D **77**, 024002 (2008) [arXiv:0707.1056 [hep-th]].
35. O. Miskovic, R. Troncoso and J. Zanelli, Phys. Lett. B **637**, 317 (2006).
36. F. Canfora, A. Giacomini, J. Oliva and R. Troncoso, work in progress.
37. E. Gravanis and S. Willison, Phys. Rev. D **75**, 084025 (2007).
38. C. Garraffo, G. Giribet, E. Gravanis and S. Willison, J. Math. Phys. **49**, 042502 (2008).
39. M. Thibeault, C. Simeone and E. F. Eiroa, Gen. Rel. Grav. **38**, 1593 (2006).
40. M. G. Richarte and C. Simeone, Phys. Rev. D **76**, 087502 (2007).
41. B. Bhawal and S. Kar, Phys. Rev. D **46**, 2464 (1992).
42. H. Maeda and M. Nozawa, Phys. Rev. D **78**, 024005 (2008).
43. E. S. Fradkin and A. A. Tseytlin, Phys. Rept. **119**, 233 (1985).
44. N. Berkovits and E. Witten, JHEP **0408**, 009 (2004) [arXiv:hep-th/0406051].
45. K. S. Stelle, Phys. Rev. **D16**, 953 (1977).
46. K. S. Stelle, Gen. Rel. Grav. **9**, 353 (1978).
47. F. S. N. Lobo, Class. Quant. Grav. **25**, 175006 (2008).
48. R. J. Riegert. Phys. Rev. Lett. **53**, 315 (1984).
49. J. D. Tempo, J. Oliva, and R. Troncoso, "Static wormholes in vacuum for conformal gravity," Proceedings of "7th Alexander Friedmann International Seminar on Gravitation and Cosmology," Joao Pessoa, Brazil, July 2008. To be published in Int. J. Mod. Phys. A.
50. D. G. Boulware, G. T. Horowitz, and A. Strominger, Phys. Rev. Lett. **50**, 1726 (1983).
51. S. Deser and B. Tekin, Phys. Rev. Lett. **89**, 101101 (2002).
52. S. Deser and B. Tekin Phys. Rev. D **67**, 084009 (2003).
53. S. Deser and B. Tekin, Phys. Rev. D **75**, 084032 (2007) [arXiv:gr-qc/0701140].

Part 3
CLOSING

Claudio Bunster: A Personal Recollection

Jorge Zanelli

When I met Claudio for the first time, by the end of the winter 1978, I felt I had known him for a long time. Indeed, I was quite familiar with Claudio's remarkable work on classical electrodynamics, and he was a living legend among students and faculty at the University of Chile. At the time Claudio graduated, undergraduates were required to write a short thesis summarizing some topic and possibly including some experimental results. In his undergraduate thesis, Claudio solved an old open problem in a field where many old pros had tried their artillery before. This greatly exceeded the requirements for graduation and was in stark contrast with the scientific background of the country, where the great majority of university professors never published a single scientific paper in their entire life.

So, this is how I got to know Claudio, that legendary undergraduate student who managed to publish the solution of an old outstanding problem [1] (Fig. 1).

When I started working on my PhD, one of the hot references among the graduate students in theoretical physics was the little book Claudio had recently published with Andrew Hanson and Tullio Regge on constrained Hamiltonian systems [2] which I had studied thoroughly and contained many enlightening examples that were so necessary to understand Dirac's famous little book [3]. I had no image in my mind of the real Claudio, but I expected him to be like one of those pedantic young professors who haven't completely recovered from their own fame (Fig. 2). Instead, what I met looked more like a cross between a poet, a snake charmer and a car mechanic. In the early 1980s Chile was under extreme political tension and the possibilities for developing a scientific career were as remote as pursuing a career in Greek literature in Afghanistan today. At the time, there was a minute Chilean scientific community, mostly working abroad, and the very few resident scientists were found at the two main universities in Santiago. The natural thing to do would have been to set up the institute in one of these universities, which meant accepting the rules of the central government of the time, or to start an operation under the

J. Zanelli
Centro de Estudios Científicos, Valdivia, Chile
e-mail: z-at-cecs-dot-cl

M. Henneaux, J. Zanelli (eds.), *Quantum Mechanics of Fundamental Systems: The Quest for Beauty and Simplicity*, DOI 10.1007/978-0-387-87499-9_18,

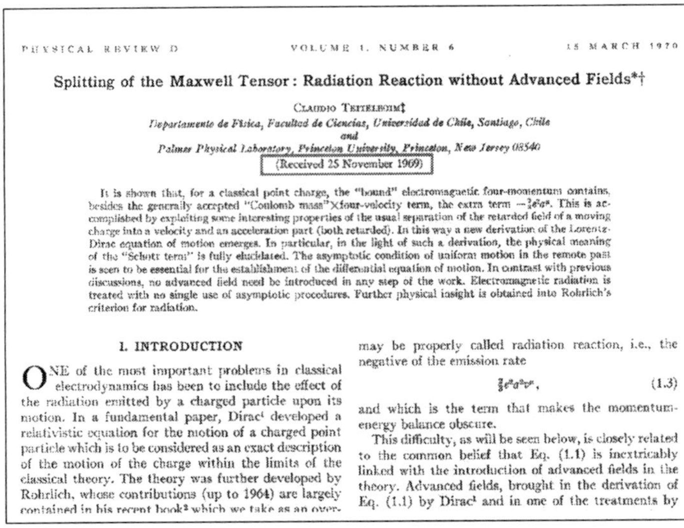

PHYSICAL REVIEW D VOLUME 1, NUMBER 6 15 MARCH 1970

Splitting of the Maxwell Tensor: Radiation Reaction without Advanced Fields*†

CLAUDIO TEITELBOIM‡
Departamento de Física, Facultad de Ciencias, Universidad de Chile, Santiago, Chile
and
Palmer Physical Laboratory, Princeton University, Princeton, New Jersey 08540

(Received 25 November 1969)

It is shown that, for a classical point charge, the "bound" electromagnetic four-momentum contains, besides the generally accepted "Coulomb mass"×four-velocity term, the extra term $-\frac{2}{3}e^2a^\mu$. This is accomplished by exploiting some interesting properties of the usual separation of the retarded field of a moving charge into a velocity and an acceleration part (both retarded). In this way a new derivation of the Lorentz-Dirac equation of motion emerges. In particular, in the light of such a derivation, the physical meaning of the "Schott term" is fully elucidated. The asymptotic condition of uniform motion in the remote past is seen to be essential for the establishment of the differential equation of motion. In contrast with previous discussions, no advanced field need be introduced in any step of the work. Electromagnetic radiation is treated with no single use of asymptotic procedures. Further physical insight is obtained into Rohrlich's criterion for radiation.

I. INTRODUCTION

ONE of the most important problems in classical electrodynamics has been to include the effect of the radiation emitted by a charged particle upon its motion. In a fundamental paper, Dirac[1] developed a relativistic equation for the motion of a charged point particle which is to be considered as an exact description of the motion of the charge within the limits of the classical theory. The theory was further developed by Rohrlich, whose contributions (up to 1964) are largely contained in his recent book[2] which we take as an over-

may be properly called radiation reaction, i.e., the negative of the emission rate

$$\tfrac{2}{3}e^2a^2v^\mu, \qquad (1.3)$$

and which is the term that makes the momentum-energy balance obscure.

This difficulty, as will be seen below, is closely related to the common belief that Eq. (1.1) is inextricably linked with the introduction of advanced fields in the theory. Advanced fields, brought in the derivation of Eq. (1.1) by Dirac[1] and in one of the treatments by

Fig. 1 The first paper published by Claudio Bunster when he arrived in Princeton to work with J. A. Wheeler

Fig. 2 Claudio discussing with Jeanette Nelson and Romualdo Tabensky in Princeton, ca. 1978

umbrella of some of the ONG's run by the political opposition. Claudio and his two partners, the biologist Ramon Latorre and the geophysicist Armando Cisternas, chose to invent a third possibility, and against all expectations, they managed to raise some funds to start a new, independent, world class, research institute. This testifies to the other outstanding feature of our celebrated friend: he doesn't follow standard rules. The Centro de Estudios Científicos de Santiago, also known by its tongue-in-cheek acronym, CECS, has been attracting top scientists and scholars from all

Fig. 3 The rented two-story building that housed CECS in Santiago from its foundation until it moved 850 km south

Fig. 4 Claudio introducing the next speaker at a conference in 1991, as the speaker (*left*) looks on

over the world to this remote place on earth, where most of them would have only considered visiting for holidays. After running successfully for 15 years, CECS was a landmark for science in Latin America (Figs. 3 and 4). It had acquired a presence that was admired and respected in Chile and by our neighbors. A standard person would feel satisfied and would expect to run the institute by automatic pilot, thanks to the considerable momentum it had acquired. Not Claudio. When things looked like standard scenarios, he found it unbearable and had to move on. So at the end of 1999, Claudio made a strong bet that the center could be relocated in Valdivia, get funding and be even more successful than in Santiago. Today, this looks like a great decision, but it was far from obvious 8 years ago. It was an act of poetic madness. Within a few years, CECS had renovated several buildings and built new ones for the researchers, administration, wet labs and lecture rooms. The institute in Valdivia was again on firm ground and continued to be one of the leaders in the fields of Biology and Theoretical Physics. Again, that looks like an achievement to feel proud and satisfied, now it seemed that one could start to quietly enjoy the fruits of success. This would be the normal path for most people, but not Claudio. He wanted new challenges and is always looking for new adventures.

In 2002, Claudio managed to get funds to start a new scientific area at CECS in the field of Glaciology an Climate Change. This was not a news item as it is today, but he had the foresight 8 years ago when he invited a group of glaciologists whom he hardly knew to organize a conference at the newly established institute in Valdivia. At the same time, he mobilized resources to start a transgenic facility at CECS that was to become a pioneer in Latin America as well, giving the biologists at CECS a unique edge in the field, as well as harnessing the opportunity for new developments in areas of molecular biology and genomics with many potential applications (Fig. 5).

Nowadays, CECS has grown to be a leading institute that has helped considerably to put Chile in the scientific map of the world. It has a unique mixture of fields that came together by the potential of their researchers and thanks to the vision and leadership of Claudio Bunster. His quest for beauty and simplicity, together with his constant pursuit of an ultimate frontier have taken him and his team to explore the confines of the universe, the mysteries of life on Earth, and the challenges this life is facing at present. Nobody would doubt the beauty and mathematical simplicity of

Fig. 5 The remodeling of the Schuster Hotel in 2000, which houses CECS in Valdivia today

Fig. 6 The breathtaking beauty of Antarctica seen from the expedition camp in Patriot Hills, 2004

black holes and magnetic poles; we all marvel at the beauty of living matter; no one could remain indifferent by the magnificence of a glacier.

The achievements of Claudio Bunster in these 60 years seem more than enough for most of us, but I am certain they are not enough for him. I wonder where he will take us next (Fig. 6).

Acknowledgments The Centro de Estudios Científicos (CECS) is funded by the Chilean Government through the Millennium Science Initiative and the Centers of Excellence Base Financing Program of Conicyt. CECS is also supported by a group of private companies which at present includes Antofagasta Minerals, Arauco, Empresas CMPC, Indura, Naviera Ultragas and Telefónica del Sur. CIN is funded by Conicyt and the Gobierno Regional de Los Ríos. This work was partially funded by FONDECYT grants 1061291, 7070117.

References

1. C. Teitelboim, *Splitting of the Maxwell tensor: radiation reaction without advanced fields*, Phys. Rev. **D1**, 1572 (1970).
2. A. J. Hanson, T. Regge and C. Teitelboim, *Constrained Hamiltonian Systems*, Academia dei Lincei, Roma, 1976.
3. P. A. M. Dirac, *Lectures on Quantum Mechanics*, Dover, New York, 2001.

Profile of Claudio Bunster

In life and work, Claudio Bunster prefers extreme challenges. Bunster, a physicist who contemplates brain-warping theories of space and time, returned to his native Chile from the United States precisely when most intellectuals would have stayed clear – during the middle of the Pinochet dictatorship. Shut out of the universities by the military government, he broke the conventional mold by founding a research institute that he then moved from Santiago, the capital, to Valdivia in southern Chile, against the flow of minds and money. He led the presidential science advisory committee during the administration of Eduardo Frei, and served on the Dialogue Board of Human Rights to reconcile Chilean military and civil society.

For his achievements, Bunster was elected to the National Academy of Sciences in 2005. In his Inaugural Article, which appeared in the July 24, 2007 issue of PNAS [1], he showed that after a black hole swallows a magnetic monopole, the space–time singularity starts rotating.

Learning Physics Despite the Hurdles

During high school in Santiago, Bunster taught himself physics. He had to. "My teachers were extremely boring," he says. At first, he did not know what physics was, just that he liked the name. There was something magic in the way "física" rolled off the tongue. Then he became intrigued more and more by the nature of time. When he was 15, he found a book on relativity theory in a bookstore. "I remember being astonished to think," he says, "that when I saw a leaf in a tree moving with the wind, I was observing the past, since it had taken some time for the light to travel from the leaf to my eye."

This is a Profile of a recently elected member of the National Academy of Sciences to accompany the member's Inaugural Article on page 12243 in issue 3D of volume 104.

M. Henneaux, J. Zanelli (eds.), *Quantum Mechanics of Fundamental Systems: The Quest for Beauty and Simplicity*, DOI 10.1007/978-0-387-87499-9_19,

Fig. 1 Claudio Bunster

When he entered the University of Chile in Santiago in 1965, he enrolled in the new and experimental "Institute of Sciences," an oddity in the system of Chilean universities that, even now, he considers "little Soviet Unions": stolid, inflexible bureaucracies that never take risks.

Nevertheless, he pursued on his own what he calls *the* problem of classical electrodynamics: the electron, a charged particle, interacts with its own field. The self-energy stored in this interaction is infinite if calculated by standard methods. "It seemed to me that one should eliminate the concept of field altogether," Bunster says.

When someone told Bunster that John Archibald Wheeler and Richard Feynman had done just that and published their results in a 1945 paper in *Reviews of Modern Physics* [2], he rode buses all over Santiago trying desperately to find back issues of the journal. He finally found what he was looking for in the library of the Institute of Cosmic Rays. "I read the paper with emotion," he says. "There it was, not only done, but done with a style I had not seen before. Deep, elegant physics that read as literature. Equations written with words instead of symbols. Generosity in giving credit to the work of others."

Then the book *Geometrodynamics*, a collection of articles by Wheeler and colleagues [3], fell into his hands. "I was mesmerized," Bunster says. "Here was a contemporary theoretical physicist doing general relativity as frontier science, mixing gravity with the rest of physics – daring, magic! I thought I should find a way to study under that man." Wheeler, one of the most prominent physicists of the century, was a professor at Princeton University, the global epicenter of physics, a continent away. It seemed impossible that Bunster would ever be able to approach him.

Exciting Times

Just when Bunster thought he would never get to Princeton, he "had an unbelievable piece of luck," he explains. In a footnote to one of Wheeler's papers, the theorist acknowledged a French mathematician named André Avez. Bunster knew that Avez happened to be in Santiago at the time. Through his mathematics teacher, he arranged an introduction. Avez and Bunster had lunch and spoke for several hours. At the end of their conversation, Bunster recalls Avez saying: "Well, I will write on your behalf to Princeton. The food there will not be as good as in Chile but you will learn the best mathematics and physics in the world." Several weeks later, Bunster received a letter from Wheeler himself indicating that he had requested the Graduate School Office to mail Bunster an application package. Bunster still remembers Wheeler writing that he knew of "the many things [Bunster] had already learned and the many books [he] had already read." But Wheeler warned that competition to get into the Princeton Ph.D. program was fierce.

"I applied, and waited anxiously for the postman every day," Bunster says. "I even went to the corner to intercept him. And then, one day, the letter arrived." He had been accepted.

Times were exciting at Princeton. Wheeler's group consisted of eight or ten students and postdocs, with others "orbiting around." The eminent professor had just coined the name "black hole" for the space–time singularity that forms when a star collapses on itself and crams so much mass into so small a space that the intense gravity prevents light from escaping.

Although astronomers have now found many promising candidates, when Bunster was in Wheeler's lab, black holes were pure theory. The equations predicted them, but no experimental evidence had been found. "The black hole has come from conjecture to maybe being responsible for the structure of the universe," Bunster says. Being such extreme objects, black holes naturally attracted his attention when he joined Wheeler's group. Another topic that fascinated him was quantum gravity, the branch of physics that seeks to unify quantum mechanics with general relativity. "I was interested in looking at space–time as a derived – as opposed to fundamental – concept, he says, pointing out that today many contemporary physicists look to string theory to explain the basis for space and time.

"Space–time is like a layer cake made of pancakes," he says. "Each pancake is space at a given time. And therefore, in quantum mechanics, Wheeler said there could be no space–time, because there is no precise history of space. He invented something called 'superspace,' which was a big space in which each point was a three-dimensional space." In his doctoral dissertation, Bunster showed how what we experience as space and time would emerge in the classical limit of quantum superspace.

Bunster spent 1 year after completing his doctorate as a postdoc in Wheeler's group before being promoted to assistant professor. There were some very bright minds around. Bunster often discussed ideas with the legendary Freeman Dyson. "I treasure those conversations with him," he says. But his most productive collaboration was with a young Italian named Tullio Regge. "He had very original ideas, and so at a young age he became a professor" at the Institute for Advanced Study (IAS) – a center for theoretical research in Princeton, NJ. At this time, and until 2005 (when he discovered the identity of his biological father), Claudio Bunster was known as Claudio Teitelboim. Many terms in the field of quantum gravity bear the Regge–Teitelboim moniker: the Regge–Teitelboim model, Regge–Teitelboim equations, and so on.

Torn Between Two Worlds

Wheeler left Princeton for the University of Texas at Austin in 1977. Soon after, Bunster joined Regge, Dyson, and others at the IAS. It was a top-notch forum, he says, "where one was welcome to pursue crazy lines of thought." Also, the IAS was very flexible with duties – there were no classes to teach – and permitted Bunster to travel home regularly so he could lay the groundwork for a research institute that he planned to establish there. In Chile the political situation had imploded.

On September 11, 1973, the military overthrew the government of Salvador Allende. At first a junta governed Chile, and then Augusto Pinochet ruled alone with a ruthless hand. Chilean expatriates were aware of the turmoil in their country and Bunster was torn between the crisis in his homeland and the intellectual haven of Princeton. "It was the end of an illusion," he says of the coup, "and with hindsight it's easy to see that [the Allende plan] would not have worked, but there was an illusion of change, and then all that collapsed and there was all this brutality. The sense of impotence, being away and not being able to do anything about it, was very strong."

During this time, Wheeler had not forgotten about his former student. Indeed, he tried very hard to persuade Bunster to join his group at Austin. A magazine called *The Texas Monthly* began to appear regularly in Bunster's mailbox on Wheatsheaf Lane in Princeton. The chance to rejoin Wheeler was an offer Bunster couldn't refuse, and he moved to Austin in 1980. He was productive in Texas for a while, but Chile began to loom ever larger in his thoughts. So he negotiated a deal in which University of Texas, Austin, allowed him to spend half the year in Texas and the other half in Chile. With seed money from the Tinker Foundation, he founded an independent research institute in Santiago called the Centro de Estudios Cientificos de Santiago (CECS).

Unfortunately, the University of Texas, Austin, arrangement did not work out. "Half a man is less than half a man," Bunster quips. He returned to the IAS at Princeton, which he knew was

> "I had shown independence of thought and a willingness to go into battle."

generous with its flexibility. For several years, he shuttled between Princeton and Santiago. But as the CECS became a more concrete entity, he increasingly felt the gravitational pull of Chile, where, even though banned from teaching at the universities, he eventually settled for good.

"It cannot be a coincidence that I did so during the hard times of the Pinochet dictatorship," he says. "Not because I was in favor of it. Quite the opposite. I felt that precisely at that time I could be more useful, and this could be best accomplished by having an independent place where colleagues from all over the world could visit and bring some light into the darkness."

The Pinochet regime would have preferred darkness. "Sometimes one could feel the presence of a government agent lurking around," Bunster says. But the CECS was tolerated, perhaps because of its lack of a conventional political agenda. "There were phone threats, but for those times that was peanuts," says Bunster.

Building a Ship of Small Draft

Bunster used the IAS – an example of a center of excellence where international visitors would be welcome – as inspiration for building the CECS. But beyond the ideals of excellence and openness, the CECS, which was set up in a rented house in the Santiago suburbs, was not a heavily planned venture. Its motto is a quote from Captain James Cook: "If you plan to make a voyage of discovery, choose a ship of small draught."

"Big ships cannot go into shallow waters or narrow passages," Bunster explains. "They are not easy to maneuver. The universities are not ships of small draft in Chile."

The CECS has three main areas of concentration: theoretical physics, molecular physiology, and glaciology and climate change. "We say we are interested in the cosmos, the planet and life," says Bunster. "You know, it sounds very much like a policy but it was just a string of coincidences. Instead of choosing research fields, the focus was put on finding first-rate people willing to take risks and open new paths. The people were primary, not the fields. The rest was the magic of life." These days the CECS is home to 15 leading researchers, plus postdocs and graduate students for a total staff of 80. Apart from Bunster's status as director, the institute's governing structure is horizontal – all scientists bear the title of "researcher" and have equal standing. Leadership is exerted by achievement and stature, and not by bureaucratic rank, Bunster says.

In 2000, Bunster organized the institution's move from Santiago to the southern city of Valdivia. "This country is very centralized, and the migration of a group of first-rate people from the capital to the south goes against the usual stream," he says. "It was a revolutionary move, and it worked very well." Drawing on his contacts from his time at Princeton and University of Texas, Austin, Bunster has built the CECS into an internationally recognized retreat. There is no tenure and no permanent source of money, but the researchers often feel exhilarated by their freedom. A sense of solidarity prevails.

"Sometimes there is an Antarctic exploration and one has to choose between a new transgenic facility or going to the south pole," explains Bunster. "So the biologists say, well, go to the south pole, we'll do our lab next year." Money does remain an issue, though. "The Center is fully equipped," Bunster says, "but we had a budget for six years that burned in three because we did the work planned for six and more. We think that life is too short. So I am at the moment trying to keep the boat moving."

Public Service and Science

Once Bunster returned to Chile, he could not help but become enmeshed in public service. After a plebiscite denied Pinochet a further role in the Chilean government in 1988, Patricio Aylwin was elected the new president and served until 1994 in a period during which Chile made the transition from military to civilian rule. Then, when Eduardo Frei came to power in 1994, he named Bunster as his science advisor and head of the presidential advisory committee for scientific matters. Bunster believes Frei chose him because "I had shown independence of thought and a willingness to go into battle." This was an opportunity to revamp national science support, and Bunster took the initiative, developing the Presidential Chairs on Science: prestigious, well funded research positions. A committee of distinguished foreign scientists (to bypass established local power groups) chose the honorees, who received their awards from the Chilean president himself in a ceremony at the presidential palace. "It was also novel," Bunster says, "that there were very few strings attached. The money was not minor by Chilean standards" – $100,000 US a year for 3 years.

Bunster was able to further expand this support. During one of his regular stays at the IAS, he had a conversation with Phillip Griffiths, then IAS director, about taking the Presidential Chairs to a higher level by giving support to groups rather than individuals. Griffiths, who was interested in fostering science in developing countries, had the support of James Wolfensohn, then president of the World Bank. With a World Bank loan, Bunster was able to kick-start the Millennium Science Initiative (MSI), which makes grants to centers of excellence in Chile, currently on the order of $2 million US per year for 10 years. The CECS has received MSI funding since 2000. Following the Chilean prototype, the MSI has been extended to several other countries.

At the end of his administration, Frei established the Dialogue Board on Human Rights to break the barrier between the military and civil society. "Democracy was weak," Bunster says, "and I strongly believed that one should talk to the military, rather than keeping them isolated." Bunster was nominated to the Board and thought hard about how science might help reconcile the military to the people. One avenue that seemed promising was glaciology. The military possessed the logistics and transport to enable research in harsh territory.

"When the Dialogue Board was established," he remembers, "I was one of the few people who had not been on the side of the Pinochet government, and yet had a relationship of mutual trust and respect with the military that had grown from working in the field, shoulder to shoulder with them, taking risks together and getting to know each other through long conversations inside a tent with the wind and the snow howling outside." At one meeting Bunster recalls that the Board debated for days behind closed doors, "wondering whether to dig into the brutalities of the past, or, without ignoring those, to focus on showing a road to the future." Pressing for the committee to reach a decision, Bunster told them that "in quantum mechanics, a photon has no reality or meaning until it is emitted by an atom." Thereupon a general stood up in agreement. His branch of the armed forces adhered to the quantum theory, he said. "That was very important, because he was 'on the other side,'" Bunster says. The Dialogue Board was thereby able to send a message to the Chilean people that the barrier had been broken between the military and society. Bunster has since overseen collaborations between the Chilean military and CECS researchers conducting research on climate change in Antarctica and on the Patagonian ice fields.

The Confluence of Extremes

Through all his forays into public service and his work to create and administer the CECS, Bunster has continued his research. In his Inaugural Article [1], he and colleague Marc Henneaux considered what would happen when two extreme objects are brought together: a black hole and a magnetic monopole. Magnetic monopoles (particles at which magnetic field lines would originate, or end, as electric field lines do at electrons and protons) have not yet been observed in experiments, but they keep turning up in calculations. "Black holes are known because they hide things," Bunster says. "So it's quite natural to think that maybe the magnetic poles are hiding inside black holes."

One reason why Bunster might think so is that research has shown that black holes, such as the candidate Sagittarius A* at the center of the Milky Way, rotate. Indeed, the radiation emitted from matter falling into black hole candidates shows that they are rotating. However, "it's not clear where the rotation comes from," says Bunster. He thinks it might be caused by monopoles. "And so we played this game and we let the monopole fall radially into a non-rotating black hole," he says. One might think that the black hole would just swallow the monopole. But the result of Bunster's and Henneaux's calculations was that the black hole began to rotate.

Exactly how to show that black holes really do contain magnetic monopoles is a different challenge, but one that work such as Bunster's may eventually set up for astronomers to conquer. Black holes, quantum superspace, Chilean geography, the fight against oppressive dictatorship: all his life, Claudio Bunster has sought out the extreme, taken its measure, and solved it with innovation and enterprise.

Postscript. On May 12, 2008, Bunster spoke at John Archibald Wheeler's memorial service in Princeton, NJ. Wheeler, Bunster said, often quoted Teddy Roosevelt's exhortation "Do what you can, where you are, with what you have."

"I followed his advice," said Bunster. "At latitude 40S there is a science institute where one can find the footprints of Wheeler in every corner."

Kaspar Mossman, *Science Writer*

References

1. Bunster C, Henneaux M (2007) A monopole near a black hole. *Proc Natl Acad Sci U.S.A.* 104:12243–12249.
2. Wheeler JA, Feynman RP (1945) Interaction with the absorber as the mechanism of radiation. *Rev Mod Phys* 17:157–181.
3. Wheeler JA (1962) *Geometrodynamics* (Academic Press, New York).

Index

A

ADM with dynamical lapse and shift, 77–79
  hyperbolicity of, 79–81
  process for extending formulation for, 81–85
    hyperbolicity of, 85–88
AdS/CFT duality, 34
Affleck, I., 19
Ananth, S., 66, 67
Ancestor vacuum, 219
Anti-de Sitter (AdS) spacetimes, 102
Anti-de Sitter (AdS) vacuum, 268
Antiholomorphic diffeomorphisms, 93
Approximation
  linear, 281
  mini-superspace, 223
  to Wheeler-DeWitt equation, 224
Auxiliary degrees of freedom, 34

B

Bach tensor, 282
Bagger, J., 119
Bañados, M., 94
Bañados–Teitelboim–Zanelli (BTZ) black hole, 94
  Carter–Penrose diagram for, 95
  horizons and constraints for, 97–100
Bazhanov, V.V., 19
Bekenstein–Hawking entropy, 92–94, 96, 103
Bekenstein, J.D., 91, 95, 172
Belinskii, Khalatnikov and Lifschitz (BKL) analysis, 156–157
Belinsky, V.A., 155–163
Bergmann–Komar formulation, of Dirac brackets, 99
Bianchi identities, 32, 56
Black holes
  analogy, 22–24
  conformal field theory derivation of, 94–96
  degrees of freedom of, 96, 217
  entropy, 91
    "universality" of, 100–101
  negative mass, 120
  quantum, 92
  solutions, 171–172
    case $\Lambda = \alpha = 0$, 172
    case of Λ and α not simultaneously null, 173–174
  statistical mechanics of, 101
  as thermodynamic objects, 91
Bocharova, N., 172
Bohr–Sommerfeld quantization, 121
Boltzmann factors, for divergent Casimir energies, 22
Bosonic field, 38
Bosonic Landau model, 183
Bosonic spinor, 111
Bosonic string
  fermions and supersymmetry from
    fermionic content of, 139
    fermionic D-Branes, 134
    fermionic subspaces of, 126–128
    tadpole-free and anomaly-free fermionic open strings, 137–139
  generalized Kac–Moody approach
    D_{24}^{+++} and Bosonic String, 142–144
    $E_{11} \equiv E_8^{+++}$ and 11-dimensional supergravity, 139–142
Boundary conditions, 32, 149

Bousso diagram
for FRW geometry, 248
for radiation dominated and slow-roll universes, 249
Boussovian Bounds, 247–250
Bronnikov, K., 172
Brown, J.D., 95
BSSN system, 72
Bubble collisions
flattened hats and other tragedies, 261–262
with identical vacua, 255–258
and other matters, 254–255
and persistence of memory, 259–260
and supersymmetry, 260–261
Bubble nucleation, 215, 219

C
Calabi–Yau moduli space, 21
Cardy formula, 93
Cardy, J.A., 93
Cartan subalgebra, motion in, 160–162
Casimir energies, Boltzmann factors for divergent, 22
Casimir operators, 190
Cauchy–Kowalewski theorem, 76
Census Bureau, 215–217, 250
CFT automorphisms, 24
Chan–Paton group, 126, 137
Chan–Paton multiplicity, 21
Chapligyn formulation, 31
Charge conjugation, 134–135
Chern–Simons (CS) action
as Brane Couplings, 114–115
non-Abelian connections coupling with 2p-Barnes, 118–119
Chern–Simons (CS) supergravity
BPS spectrum of, 109
for M-Algebra, 109–111
Chern–Simons (CS) theories, 101, 107
degrees of freedom, 115–116
difficulties with standard M-Brane construction in, 112–114
(2n+1)-dimensional Abelian, 116–117
and 2p Branes, 116–117
Chirality, 135–136
Christensen, S.M., 94
Christoffel symbol *Γ(g)*, 31, 72, 84
Cimate change, problems of, 4
Clebsch–Gordan coefficient, 38
Coleman De Luccia bubble, 218
Coleman De Luccia instanton, 227
Minkowski continuation of, 228
Coleman De Luccia nucleation, 248
Condensed-matter systems, 24
Configuration space torus geometry, 130–133
Conformal connection functions, 72
Conformal defects, in sigma model, 19
Conformal field theory, 93–94
Conformal scalar coupling, 169–170
Conformal transformations, 18
Cremmer, E., 108
Cremmer–Julia–Scherk (CJS) supergravity, 108
c-Theorem, 247–250
Zamolodchikov (*see* Zamolodchikov c-Theorem)

D
Dark matter, 31
Degree of freedom, 35, 45, 102, 226
in black holes, 217
of boundary theory, 220
fermionic, 162
Dialynas, T.E., 172
Diffeomorphism invariance, 29
Dimension zero term, 234–235
Dirac–Born–Infeld equations, 18, 21
Dirac, P.A.M., 34
Dirac quantization, 200
Dirac square root of gravity, 7
Dynkin diagram, 140

E
Eddington theory, 28
Ehlers–Geroch transformations, of general relativity, 20
Einstein–Gauss–Bonnet theory, 268
Einstein–Hilbert action, 121
Einstein–Hilbert–Eddington–Born–Infeld action, 31
Einstein–Rosen bridge model, of particle, 268
Einstein tensor, 32
Electro-magnetic duality transformations, 54
Emparan, R., 101, 103
Energy density, level of, 32
Energy flux, 281
Energy-momentum tensor, 169, 232
Equations of motion, 56, 82
Euclidean anti de Sitter space, 220
"Euclidean" black hole spacetime, 98

F
Fadeev–Popov-like determinant, 102
Fermionic Dp-brane partition functions, 136

Fermionic Landau Model, 183–185
Fermions, 7, 17
Field theories, light-frame formulation of, 34–38
Fierz, M., 200
Fourier transform, 37
Freivogel, B., 260
Frohlich, J., 21
Fulling, S.A., 94

G
Gamma-driver shift, 88
Garriga, J., 227, 254
Gauge theory, 17
Gauge transformations, 203
Gauss–Bonnet coupling, 269
Gauss–Bonnet gravity, 282
General relativity
  Ehlers–Geroch transformations of, 20
  equations of, 222
  Hamiltonian formulation of, 71
  'unbroken' ground state of, 28
Geodesics
  around wormholes in vacuum
    gravitational *vs.* centrifugal forces, 273–274
    radial geodesics, 272–273
  motion, 162
Glaciology, 4
Goldstone mechanism, 102
Gravitational *vs.* centrifugal forces, 273–274
Gravitinos, 54, 62
Ground state, 189, 280
Gupta–Bleuer method, 188
Guth, A.H., 254

H
Hamiltonian density, 222
Hamiltonian systems, 4, 17, 18, 21, 35, 289
  Dirac's reasoning for constrained, 79
  in quadratic form, 43–44
  strong hyperbolicity for quasilinear, 73–77
Hamilton Jacobi equation, 150
Harvey, J.A., 24
Hawking radiation, 93, 97
  flux, 94
Hawking, S.W., 91, 95
Heavy quarks of QCD, 17
Henneaux, M., 6, 7, 27, 95, 301
Hermitian conjugates, 187
Hilbert space metric operator, 190
Holographic boundary, 229
Holographic field theory, 217
Holographic principle, 214
Holographic Wheeler-DeWitt equation, 221–223, 236
  holographic WDW, 225–226
  wave function, 226–227
  Wheeler-DeWitt and emergence of time, 223–225
Holomorphic diffeomorphisms, 93
Hubble scale, of Ancestor, 218
Hubble volume, 151

I
Isometry transformations, 188

J
Julia, B., 108

K
Kac–Moody algebra, 19, 126, 139
  emergence of Coxeter groups and, 157–160
  Weyl group of, 162
Kac–Moody currents, 18
Kaloper, N., 102
Kaluza–Klein gauge group, 97
Khalatnikov, I.M., 155–163
Killing spinors, 111, 120
Klein–Gordon equation, for scalar field, 281
Klein–Gordon operator, 94
Klimcík, C., 172
Kondo problem, 19

L
Lagrange multiplier, 225, 239
Lagrangian density, 77
Lambert, N., 119
Landau Models, 181
Lie-algebra, 114
Lie derivative, 84
Lifshitz, E.M., 155–163
Light-cone decomposition, 35
Light-cone formulations (LC_2 formulations), 34
Light-cone supersymmetry algebra, 39
Light-frame formulation
  of field theories, 34–38
  of higher dimensional theories, 44–45
    eleven-dimensional supergravity, 50
    generalized derivatives, 48–50

superpoincaré algebra in 11 dimensions, 50–52
superpoincaré algebra in ten dimensions, 46–48
ten-dimensional superYang–Mill theory, 45–46
of supersymmetric field theories, 38–42
maximally supersymmetric Yang–Mills theory, 42–43
maximal supergravity, 43
Linear theory, for spin two
gravitating magnetic pole, 209–210
spatial rotations and quantization condition, 210–212
Liouville field, 226
Liouville Lagrangian, 238
Liouville, renormalization, and correlation functions, 240
Liouville theory, and breaking free of STT gauge, 236–239
Loop operators, in 2d CFT, 17–19
Loop quantum gravity, 101, 103
Lorentz force law, 198
Lorentz generators, 40
Lorentzian Kac–Moody algebras, 155
Cartan matrix of, 158
Weyl group of, 159

M
Magnetic flux, 23
Magnetic monopoles, in electromagnetism, 205–207
Magnetic poles, 198
in classical electromagnetism, 198–199
Dirac string and rotations, 202–203
electromagnetic potential, 200–202
field angular momentum, 199–200
test electric charge, 200, 203–205
gravitating, 209–210
Majorana condition, 39
Majorana supercharge, 109
M-Algebra, CS supergravity for, 108
Mateos, D., 101, 103
Maxwell's equations, 198
Melnikov, V., 172
Millennium Science Initiative (MSI), 300
Minkowski metric, 37
Minkowski space, 215
Möbius amplitudes, 138
Monodromy transformations, 21
Moore, G.W., 24
Morris, M.S., 268
M-Theory, 33

N
Natural numbers, logarithms of, 24
Newtonian constant, in 26 dimensions, 136
Non-linear sigma model, 17
Non-normalizable modes (NNM), 228
Non-zero anti-commutators, 187
Normalizable modes (NM), 221, 228, 251

O
Olive, D.I., 113

P
Partial differential equations (PDEs), 72
quasilinear system of, 73
Penrose diagram, for wormhole, 270
Petkova, V.B., 19
Planck density, 151
Planck length, 234
Poincare algebra, 35
Poisson bracket algebra, 93
Polyakov, A.M., 237
Pöschl–Teller potential, 272
Preons, in CS M-Theory, 111–112

Q
Quantum cosmology, 221
Quantum field theory, 24
proactive and reactive objects in, 242–247
Quantum gravity, 96
Quantum Mechanics of Fundamental Systems, 11
Quantum-monodromy matrix, 19
Quantum theory, 205
of gravity, 221

R
Rarita–Schwinger spinor, 50
Regge, T., 275
Renormalization-group (RG) flow, 217
Ricci tensor, 28, 171
Riemannian manifold, 31
Riemann tensor, 120
Rindler spacetime, 284

S
Scaling and Census Taker
homogeneity breakdown, 252–254
moments, 250–252
Scherk, J., 108
Schrödinger equation, 195, 206, 225

Shenker, S., 217
SLAC citation density, 7
Space–time index, 38
Spatially Transverse-Traceless (STT) gauge, 232
Spectrum-generating symmetries, interfaces as, 19–22
Spin-statistic theorem, 130
Spontaneous magnetization, 30
Static wormhole solutions. *See also* Wormholes in vacuum
  regularized euclidean action, 274–275
  in vacuum for gravity, 267, 269–272
Stoke's theorem, 18, 117, 201
String Theory, 221
Strominger, A., 92
SUGRA, validation of, 7
Super-Flag, 186–191
Supergravity, 33
  exceptional symmetries in maximal, 53–54
    covariant, 54–57
    $E_{7(7)}$ invariance on light-cone, 57–59
    $E_{7(7)}$ light-cone superspace, 60–63
    superspace action, 63–67
    vector and scalar LC_2 Hamiltonians, 59–60
Superplane Landau Model, 185–186
SuperPoincaré algebra, 64
Super-Yang Mills theory, 17
  ten dimensional, 45–46

T
T-duality transformation, 23
Teitelboim, C., 94, 275
Tensions, 136–137
Terning, J., 102
Thorne, K.S., 268
Topological loops, eigenvalues of, 18

U
Universality, problem of, 91–93

V
Vafa, C., 92
Verlinde algebra, 22
Vilenkin, A., 254
Virasoro algebra, 20, 93
Virasoro generators, 18

W
Wave function, (anti-)periodicity of, 208
Weinberg, E.J., 254
Wess–Zumino–Witten model, 101
Weyl tensor, 271
Weyl transformations, 93
Wheeler DeWitt equation, 149
Wheeler, J.
  and Hugh Everett III, 195–196
  and recertification of GR as true physics, 195
  and teaching, 194
Wilson, K.G., 17
Wilson loops, 17
Wilzcek, F., 207
Witten, E., 113
Wormholes in vacuum, 275–276
  Euclidean action for, 276–278
  in four dimensions, 282–284
  higher-dimensional, 279–280
  stability of asymptotically locally AdS, 280–282

Y
Yang–Mills Theory, 33
  maximally supersymmetric, 42–43
  ten-dimensional super, 45–46
Yurtsever, U., 268

Z
Zamolodchikov c-Theorem, 248
Zanelli, J., 94
Zuber, J.B., 19

Printed in the United States
148373LV00001B/88/P

9 780387 874982